PRAISE FOR THE *MANGA GUIDE* SERIES

"Highly recommended."
—CHOICE MAGAZINE ON *THE MANGA GUIDE TO DATABASES*

"Stimulus for the next generation of scientists."
—SCIENTIFIC COMPUTING ON *THE MANGA GUIDE TO MOLECULAR BIOLOGY*

"A great fit of form and subject. Recommended."
—OTAKU USA MAGAZINE ON *THE MANGA GUIDE TO PHYSICS*

"The art is charming and the humor engaging. A fun and fairly painless lesson on what many consider to be a less-than-thrilling subject."
—SCHOOL LIBRARY JOURNAL ON *THE MANGA GUIDE TO STATISTICS*

"This is really what a good math text should be like. Unlike the majority of books on subjects like statistics, it doesn't just present the material as a dry series of pointless-seeming formulas. It presents statistics as something *fun*, and something enlightening."
—GOOD MATH, BAD MATH ON *THE MANGA GUIDE TO STATISTICS*

"I found the cartoon approach of this book so compelling and its story so endearing that I recommend that every teacher of introductory physics, in both high school and college, consider using it."
—AMERICAN JOURNAL OF PHYSICS ON *THE MANGA GUIDE TO PHYSICS*

WOW!

"A single tortured cry will escape the lips of every thirty-something biochem major who sees *The Manga Guide to Molecular Biology*: 'Why, oh why couldn't this have been written when I was in college?'"
—THE SAN FRANCISCO EXAMINER

"A lot of fun to read. The interactions between the characters are lighthearted, and the whole setting has a sort of quirkiness about it that makes you keep reading just for the joy of it."
—HACK A DAY ON *THE MANGA GUIDE TO ELECTRICITY*

"*The Manga Guide to Databases* was the most enjoyable tech book I've ever read."
—RIKKI KITE, LINUX PRO MAGAZINE

"For parents trying to give their kids an edge or just for kids with a curiosity about their electronics, *The Manga Guide to Electricity* should definitely be on their bookshelves."
—SACRAMENTO BOOK REVIEW

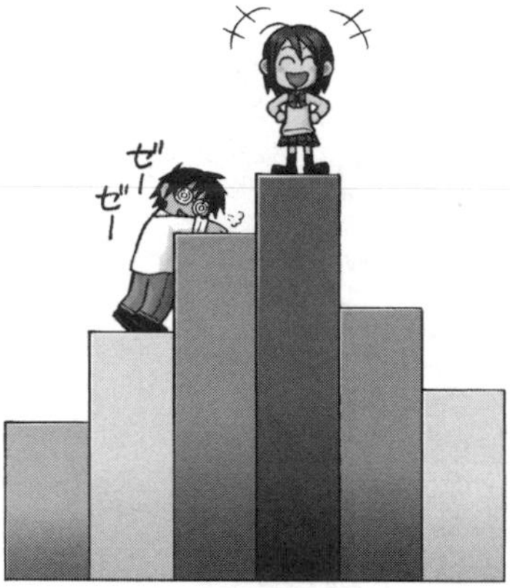
ゼーゼー

THE MANGA GUIDE™ TO
STATISTICS

SHIN TAKAHASHI
TREND-PRO CO., LTD.

no starch
press

The Manga Guide to Statistics is a translation of the Japanese original, *Manga de Wakaru Tokeigaku*, published by Ohmsha, Ltd. of Tokyo, Japan, © 2004 by Shin Takahashi and TREND-PRO Co., Ltd.

This English edition is co-published by No Starch Press and Ohmsha, Ltd.

11 3 4 5 6 7 8 9

ISBN-10: 1-59327-189-1
ISBN-13: 978-1-59327-189-3

SUSTAINABLE FORESTRY INITIATIVE Certified Fiber Sourcing
Label applies to the text stock www.sfiprogram.org

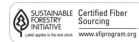

Publisher: William Pollock
Author: Shin Takahashi
Illustrator: Iroha Inoue
Scenario Writer: re_akino
Producer: TREND-PRO Co., Ltd.
Production Editor: Megan Dunchak
Developmental Editor: Tyler Ortman
Technical Reviewers: Tom Bowler and Whitney Ortman
Compositor: Riley Hoffman
Proofreader: Christine Lias
Indexer: Karin Arrigioni

For information on book distributors or translations, please contact No Starch Press, Inc. directly:
No Starch Press, Inc.
555 De Haro Street, Suite 250, San Francisco, CA 94107
phone: 415.863.9900; fax: 415.863.9950; info@nostarch.com; http://www.nostarch.com/

Library of Congress Cataloging-in-Publication Data

Takahashi, Shin.
The manga guide to statistics / Shin Takahashi and Trend-pro Co. -- 1st ed.
p. cm.
Includes index.
ISBN-13: 978-1-59327-189-3
ISBN-10: 1-59327-189-1
1. Mathematical statistics--Comic books, strips, etc. 2. Mathematical statistics--Caricatures and cartoons.
I. Trend-pro Co. II. Title.
QA276.T228 2009
519.5--dc22

2008042157

TABLE OF CONTENTS

PREFACE vii

OUR PROLOGUE:
STATISTICS WITH ♥ HEART-POUNDING EXCITEMENT ♥ 1

1
DETERMINING DATA TYPES 13

1. Categorical Data and Numerical Data 14
2. An Example of Tricky Categorical Data 20
3. How Multiple-Choice Answers Are Handled in Practice 28
Exercise and Answer 29
Summary 29

2
GETTING THE BIG PICTURE: UNDERSTANDING NUMERICAL DATA 31

1. Frequency Distribution Tables and Histograms 32
2. Mean (Average) 40
3. Median 44
4. Standard Deviation 48
5. The Range of Class of a Frequency Table 54
6. Estimation Theory and Descriptive Statistics 57
Exercise and Answer 57
Summary 58

3
GETTING THE BIG PICTURE: UNDERSTANDING CATEGORICAL DATA 59

1. Cross Tabulations 60
Exercise and Answer 64
Summary 64

4
STANDARD SCORE AND DEVIATION SCORE 65

1. Normalization and Standard Score 66
2. Characteristics of Standard Score 73
3. Deviation Score 74
4. Interpretation of Deviation Score 76
Exercise and Answer 78
Summary 80

5
LET'S OBTAIN THE PROBABILITY 81

1. Probability Density Function 82
2. Normal Distribution 86

3. Standard Normal Distribution 89
    Example I 95
    Example II 97
4. Chi-Square Distribution 99
5. t Distribution. 106
6. F Distribution 106
7. Distributions and Excel. 107
Exercise and Answer. 108
Summary 109

6
LET'S LOOK AT THE RELATIONSHIP BETWEEN TWO VARIABLES 111

1. Correlation Coefficient 116
2. Correlation Ratio. 121
3. Cramer's Coefficient 127
Exercise and Answer. 138
Summary 142

7
LET'S EXPLORE THE HYPOTHESIS TESTS 143

1. Hypothesis Tests. 144
2. The Chi-Square Test of Independence 151
    Explanation. 152
    Exercise. 157
    Thinking It Over 158
    Answer 160
3. Null Hypotheses and Alternative Hypotheses 170
4. P-value and Procedure for Hypothesis Tests 175
5. Tests of Independence and Tests of Homogeneity 184
    Example 184
    Procedure 185
6. Hypothesis Test Conclusions. 187
Exercise and Answer. 188
Summary 189

APPENDIX
LET'S CALCULATE USING EXCEL 191

1. Making a Frequency Table 192
2. Calculating Arithmetic Mean, Median, and Standard Deviation 195
3. Making a Cross Tabulation 197
4. Calculating the Standard Score and the Deviation Score. 199
5. Calculating the Probability of the Standard Normal Distribution. 204
6. Calculating the Point on the Horizontal Axis of the Chi-Square Distribution. 205
7. Calculating the Correlation Coefficient 206
8. Performing Tests of Independence. 208

INDEX 213

PREFACE

This is an introductory book on statistics. The intended readers are:

- Those who need to conduct data analysis for research or business
- Those who do not necessarily need to conduct data analysis now but are interested in getting an idea of what the world of statistics is like
- Those who have already acquired general knowledge of statistics and want to learn more

Statistics is one of the areas of mathematics most closely related to everyday life and business. Familiarizing yourself with statistics may come in handy in situations like:

- Estimating how many servings of fried noodles you can sell at a food stand you are planning to set up during a school festival
- Estimating whether you will be able to pass a certification exam
- Comparing the probability that a sick person will get better between a case in which medicine X is used and a case in which it is not used

This book consists of seven chapters. Basically, each chapter is organized in the following sections:

- Cartoon
- Text explanation to supplement the cartoon
- Exercise and answer
- Summary

You can learn a lot by just reading the cartoon section, but deeper understanding and knowledge will be acquired if you read the other sections as well.

I would be very pleased if you start feeling that statistics is fun and useful after reading this book.

I would like to thank the staff in the development department of Ohmsha, Ltd., who offered me the opportunity to write this book. I would also like to thank TREND-PRO, Co., Ltd. for making my manuscript into a cartoon, the scenario writer, re_akino, and the illustrator, Iroha Inoue. Last but not least, I would like to thank Dr. Sakaori Fumitake of the College of Social Relations at Rikkyo University. He provided me with invaluable advice while I was preparing the manuscript for this book.

SHIN TAKAHASHI

OUR PROLOGUE: STATISTICS WITH ♥ HEART-POUNDING EXCITEMENT ♥

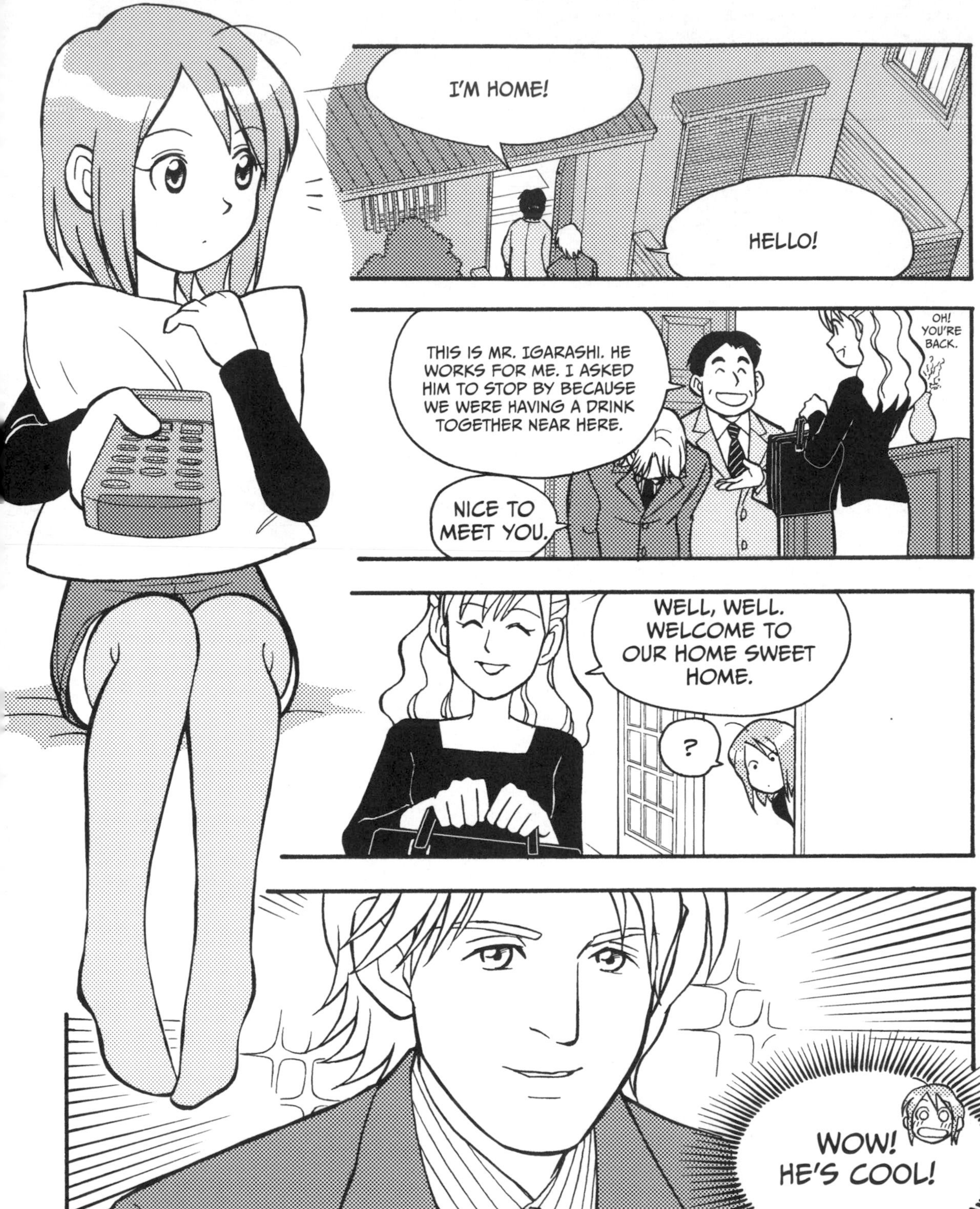
I'M HOME!
HELLO!
THIS IS MR. IGARASHI. HE WORKS FOR ME. I ASKED HIM TO STOP BY BECAUSE WE WERE HAVING A DRINK TOGETHER NEAR HERE.
OH! YOU'RE BACK.
NICE TO MEET YOU.
WELL, WELL. WELCOME TO OUR HOME SWEET HOME.
?
WOW! HE'S COOL!

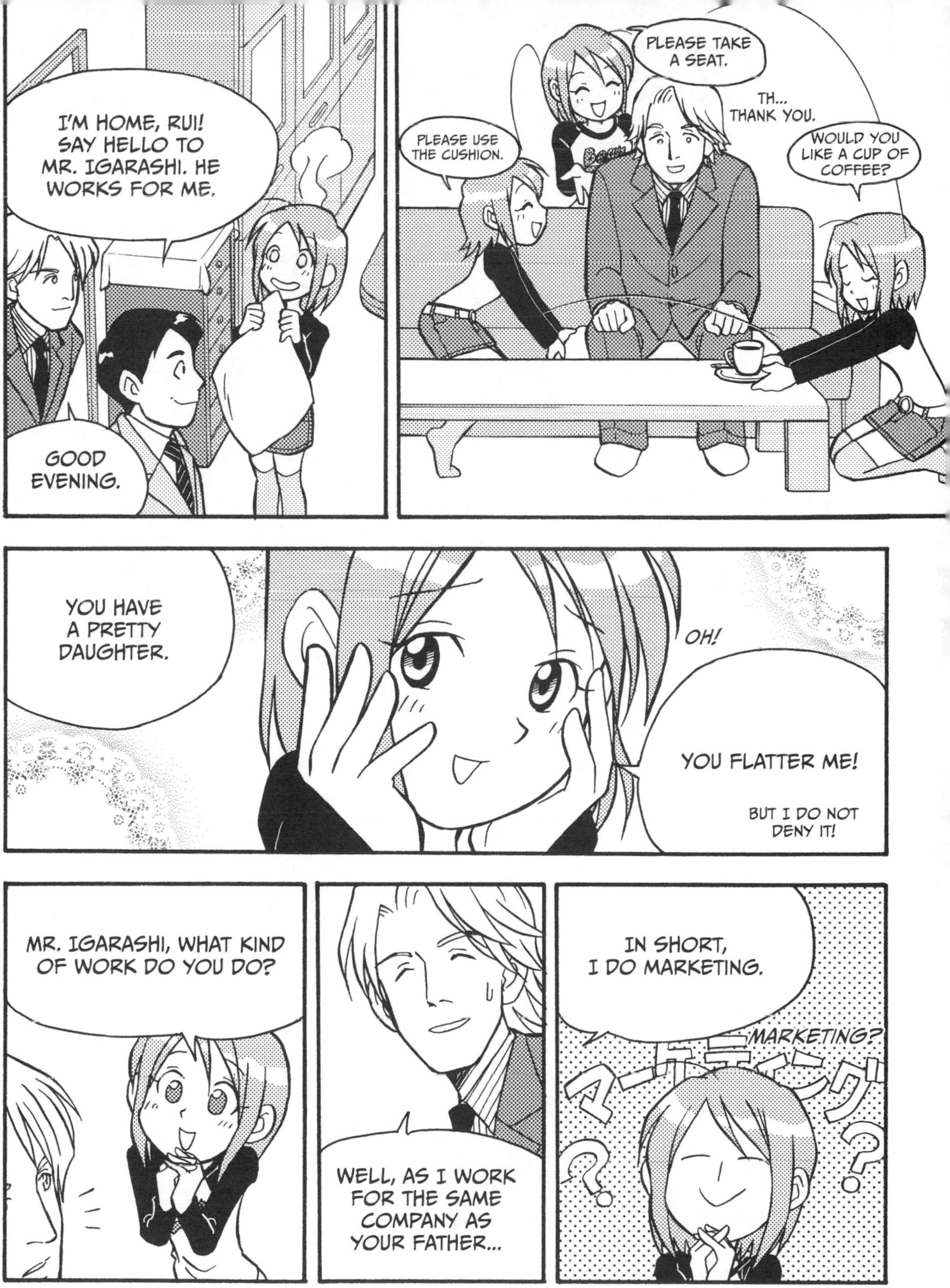
I'M HOME, RUI! SAY HELLO TO MR. IGARASHI. HE WORKS FOR ME.
GOOD EVENING.
PLEASE TAKE A SEAT.
TH... THANK YOU.
PLEASE USE THE CUSHION.
WOULD YOU LIKE A CUP OF COFFEE?
YOU HAVE A PRETTY DAUGHTER.
OH!
YOU FLATTER ME!
BUT I DO NOT DENY IT!
MR. IGARASHI, WHAT KIND OF WORK DO YOU DO?
WELL, AS I WORK FOR THE SAME COMPANY AS YOUR FATHER...
IN SHORT, I DO MARKETING.
MARKETING?
マーケティング

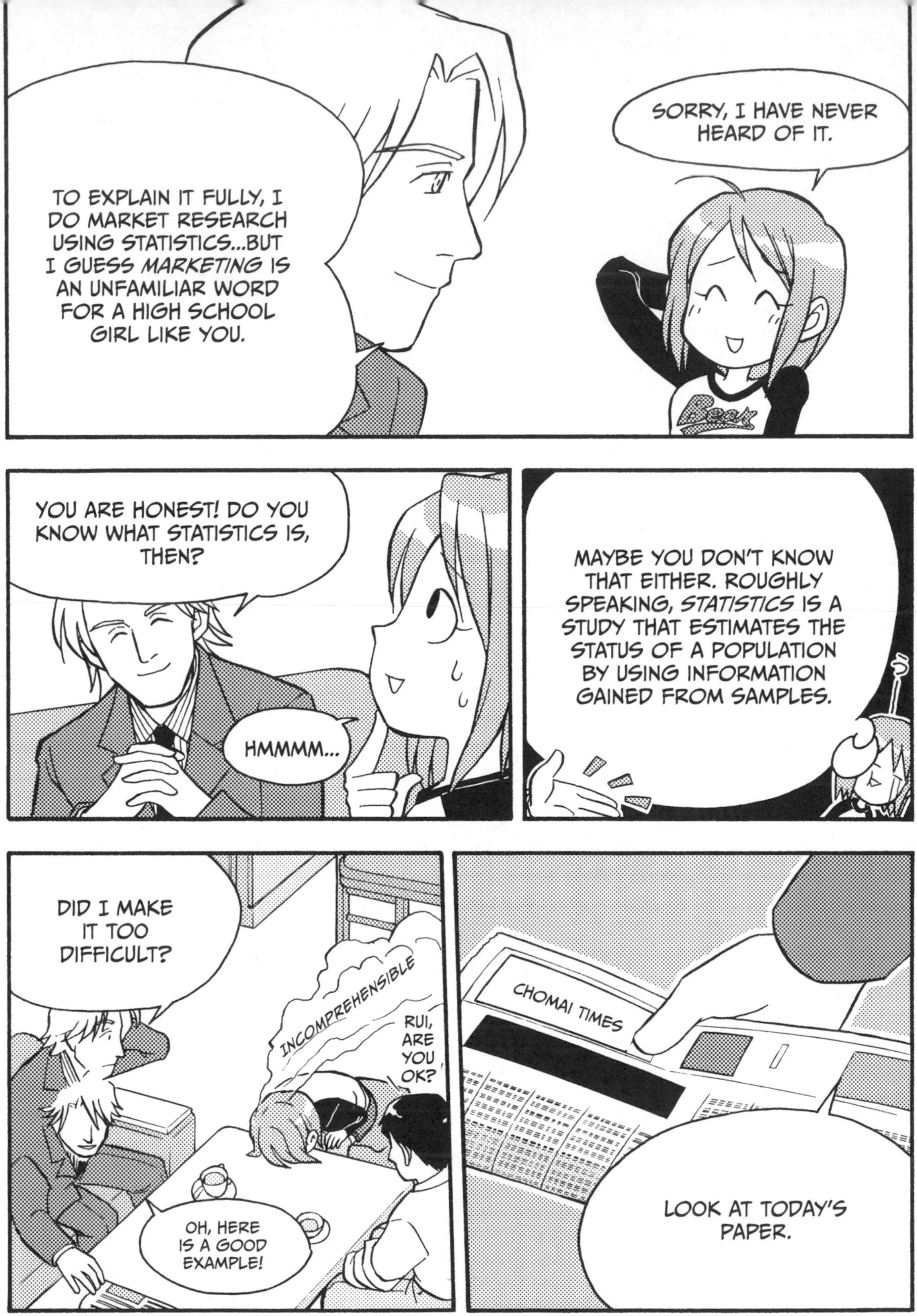
TO EXPLAIN IT FULLY, I DO MARKET RESEARCH USING STATISTICS...BUT I GUESS *MARKETING* IS AN UNFAMILIAR WORD FOR A HIGH SCHOOL GIRL LIKE YOU.
SORRY, I HAVE NEVER HEARD OF IT.
YOU ARE HONEST! DO YOU KNOW WHAT STATISTICS IS, THEN?
HMMMM...
MAYBE YOU DON'T KNOW THAT EITHER. ROUGHLY SPEAKING, *STATISTICS* IS A STUDY THAT ESTIMATES THE STATUS OF A POPULATION BY USING INFORMATION GAINED FROM SAMPLES.
DID I MAKE IT TOO DIFFICULT?
INCOMPREHENSIBLE
RUI, ARE YOU OK?
OH, HERE IS A GOOD EXAMPLE!
CHOMAI TIMES
LOOK AT TODAY'S PAPER.

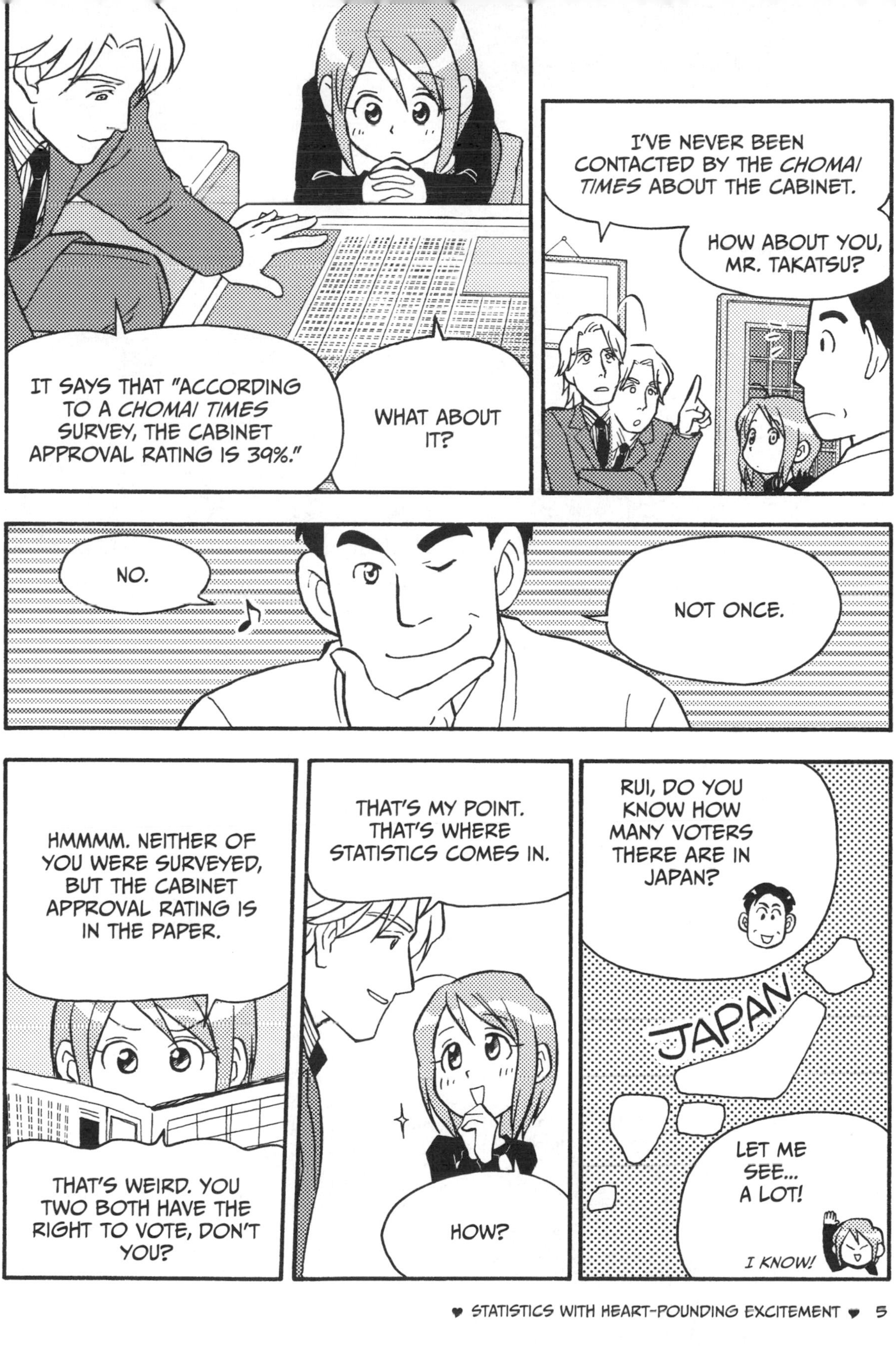
IT SAYS THAT "ACCORDING TO A *CHOMAI TIMES* SURVEY, THE CABINET APPROVAL RATING IS 39%."
WHAT ABOUT IT?
I'VE NEVER BEEN CONTACTED BY THE *CHOMAI TIMES* ABOUT THE CABINET.
HOW ABOUT YOU, MR. TAKATSU?
NO.
NOT ONCE.
HMMMM. NEITHER OF YOU WERE SURVEYED, BUT THE CABINET APPROVAL RATING IS IN THE PAPER.
THAT'S WEIRD. YOU TWO BOTH HAVE THE RIGHT TO VOTE, DON'T YOU?
THAT'S MY POINT. THAT'S WHERE STATISTICS COMES IN.
HOW?
RUI, DO YOU KNOW HOW MANY VOTERS THERE ARE IN JAPAN?
JAPAN
LET ME SEE... A LOT!
I KNOW!

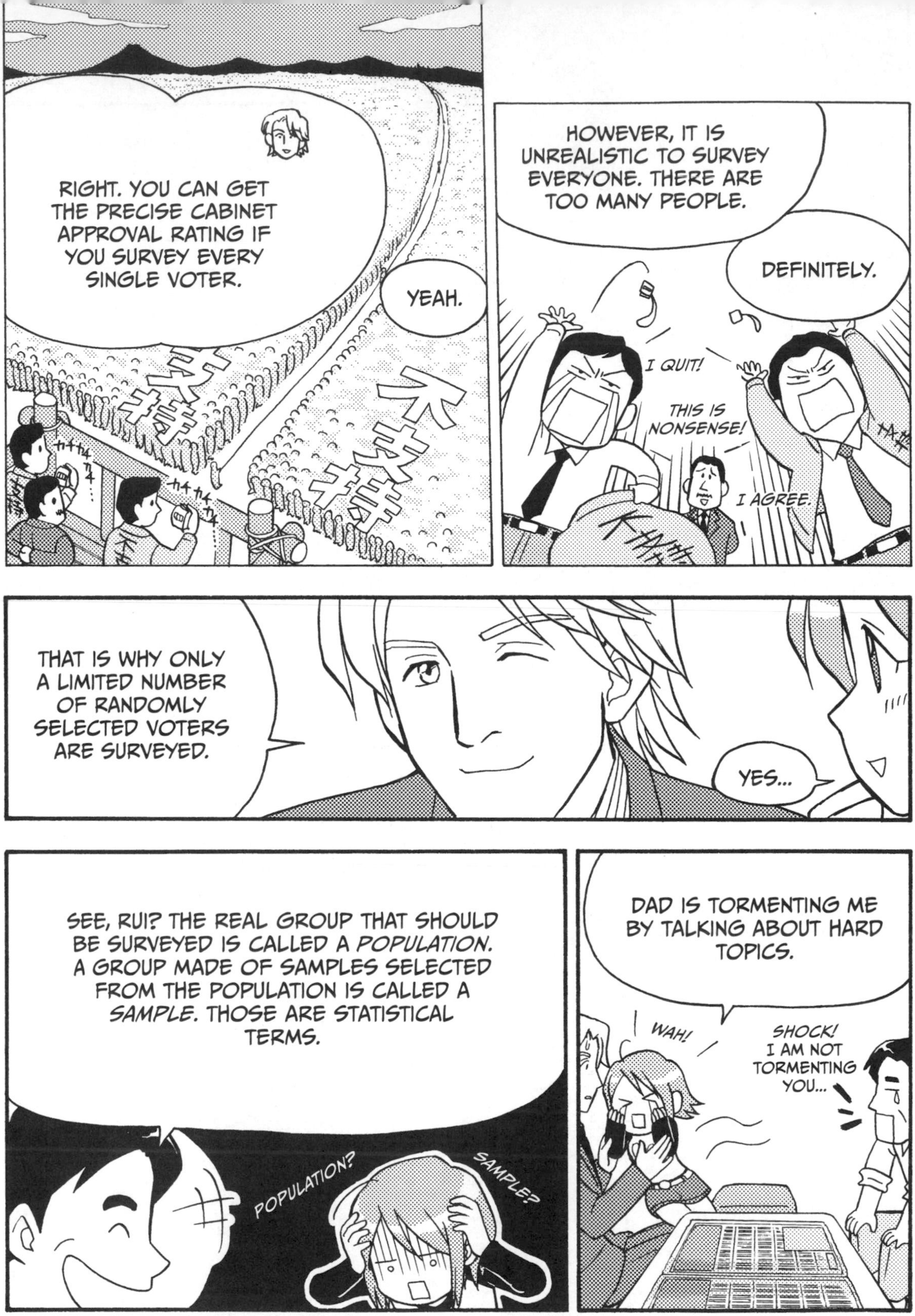
RIGHT. YOU CAN GET THE PRECISE CABINET APPROVAL RATING IF YOU SURVEY EVERY SINGLE VOTER.
YEAH.
支持
不支持
HOWEVER, IT IS UNREALISTIC TO SURVEY EVERYONE. THERE ARE TOO MANY PEOPLE.
DEFINITELY.
I QUIT!
THIS IS NONSENSE!
I AGREE.
THAT IS WHY ONLY A LIMITED NUMBER OF RANDOMLY SELECTED VOTERS ARE SURVEYED.
YES...
SEE, RUI? THE REAL GROUP THAT SHOULD BE SURVEYED IS CALLED A *POPULATION*. A GROUP MADE OF SAMPLES SELECTED FROM THE POPULATION IS CALLED A *SAMPLE*. THOSE ARE STATISTICAL TERMS.
POPULATION?
SAMPLE?
DAD IS TORMENTING ME BY TALKING ABOUT HARD TOPICS.
WAH!
SHOCK! I AM NOT TORMENTING YOU...

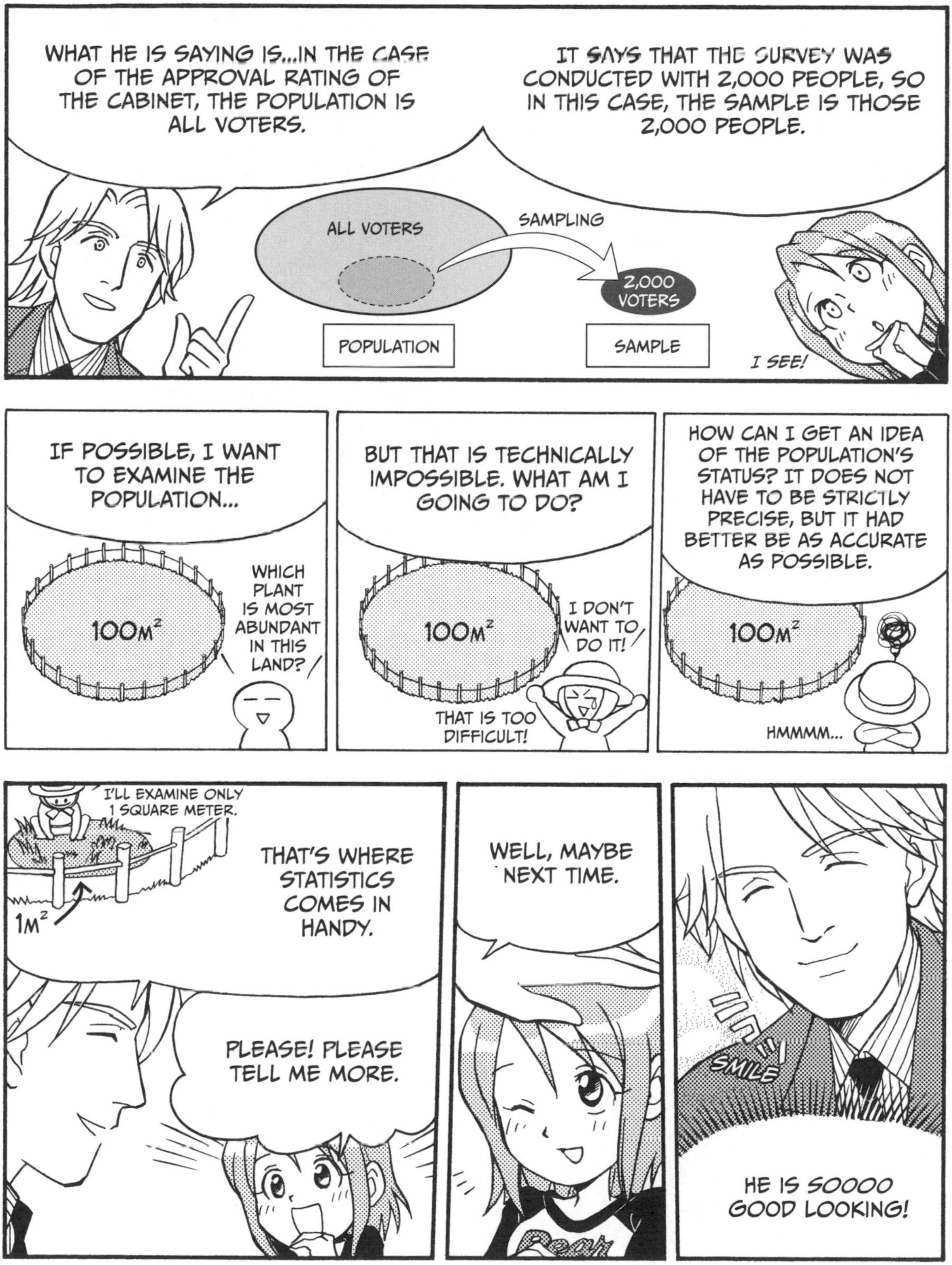

WHAT HE IS SAYING IS...IN THE CASE OF THE APPROVAL RATING OF THE CABINET, THE POPULATION IS ALL VOTERS.
IT SAYS THAT THE SURVEY WAS CONDUCTED WITH 2,000 PEOPLE, SO IN THIS CASE, THE SAMPLE IS THOSE 2,000 PEOPLE.
ALL VOTERS
SAMPLING
2,000 VOTERS
POPULATION
SAMPLE
I SEE!
IF POSSIBLE, I WANT TO EXAMINE THE POPULATION...
100M²
WHICH PLANT IS MOST ABUNDANT IN THIS LAND?
BUT THAT IS TECHNICALLY IMPOSSIBLE. WHAT AM I GOING TO DO?
100M²
I DON'T WANT TO DO IT!
THAT IS TOO DIFFICULT!
HOW CAN I GET AN IDEA OF THE POPULATION'S STATUS? IT DOES NOT HAVE TO BE STRICTLY PRECISE, BUT IT HAD BETTER BE AS ACCURATE AS POSSIBLE.
100M²
HMMMM...
I'LL EXAMINE ONLY 1 SQUARE METER.
1M²
THAT'S WHERE STATISTICS COMES IN HANDY.
PLEASE! PLEASE TELL ME MORE.
WELL, MAYBE NEXT TIME.
SMILE
HE IS SOOOO GOOD LOOKING!

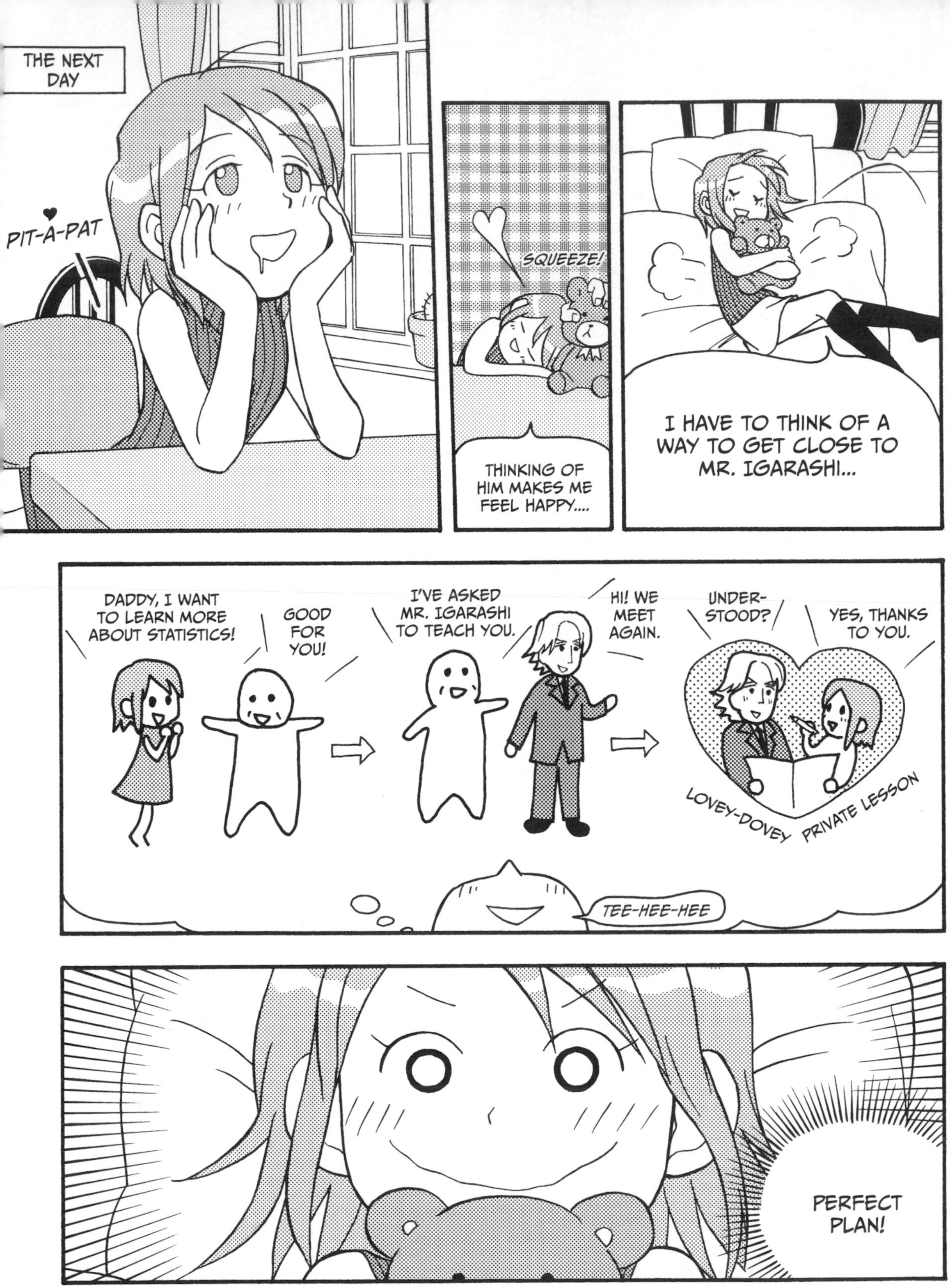
THE NEXT DAY
PIT-A-PAT
SQUEEZE!
THINKING OF HIM MAKES ME FEEL HAPPY....
I HAVE TO THINK OF A WAY TO GET CLOSE TO MR. IGARASHI...
DADDY, I WANT TO LEARN MORE ABOUT STATISTICS!
GOOD FOR YOU!
I'VE ASKED MR. IGARASHI TO TEACH YOU.
HI! WE MEET AGAIN.
UNDER-STOOD?
YES, THANKS TO YOU.
LOVEY-DOVEY PRIVATE LESSON
TEE-HEE-HEE
PERFECT PLAN!

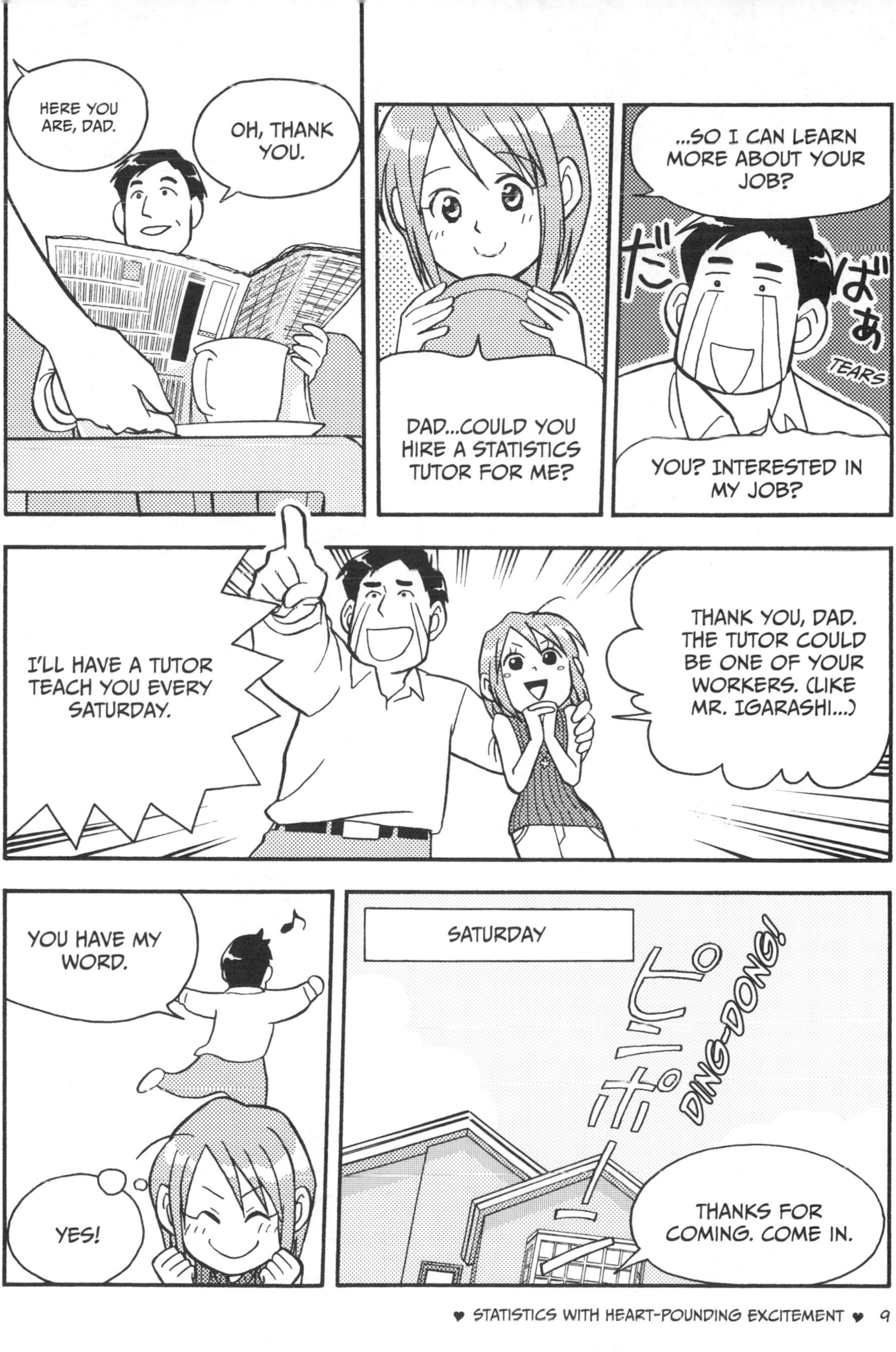
HERE YOU ARE, DAD.
OH, THANK YOU.
DAD...COULD YOU HIRE A STATISTICS TUTOR FOR ME?
...SO I CAN LEARN MORE ABOUT YOUR JOB?
だばぁ
TEARS
YOU? INTERESTED IN MY JOB?
I'LL HAVE A TUTOR TEACH YOU EVERY SATURDAY.
THANK YOU, DAD. THE TUTOR COULD BE ONE OF YOUR WORKERS. (LIKE MR. IGARASHI...)
YOU HAVE MY WORD.
YES!
SATURDAY
ピンポー
DING-DONG!
THANKS FOR COMING. COME IN.

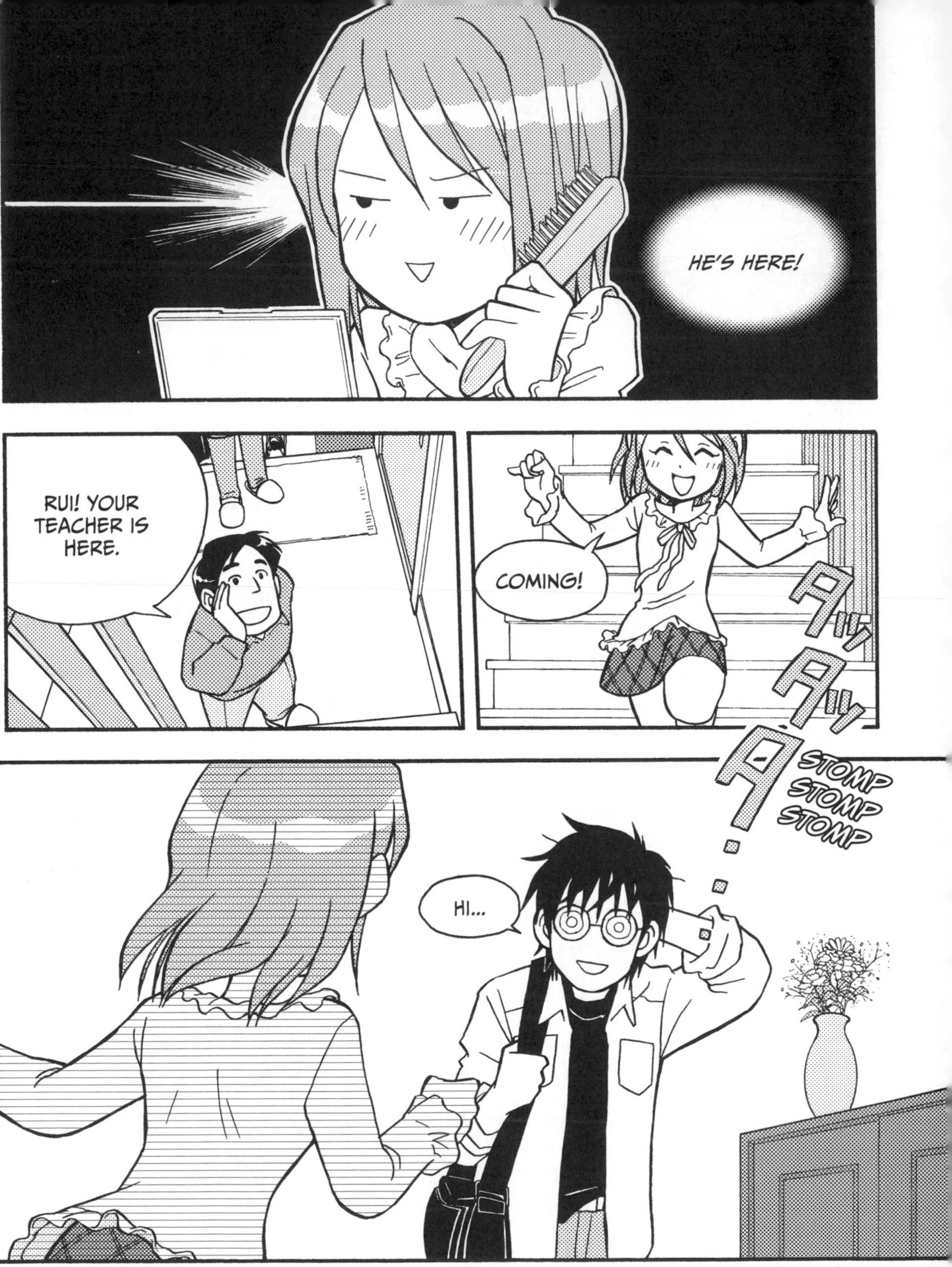
HE'S HERE!
RUI! YOUR TEACHER IS HERE.
COMING!
STOMP STOMP STOMP
HI...

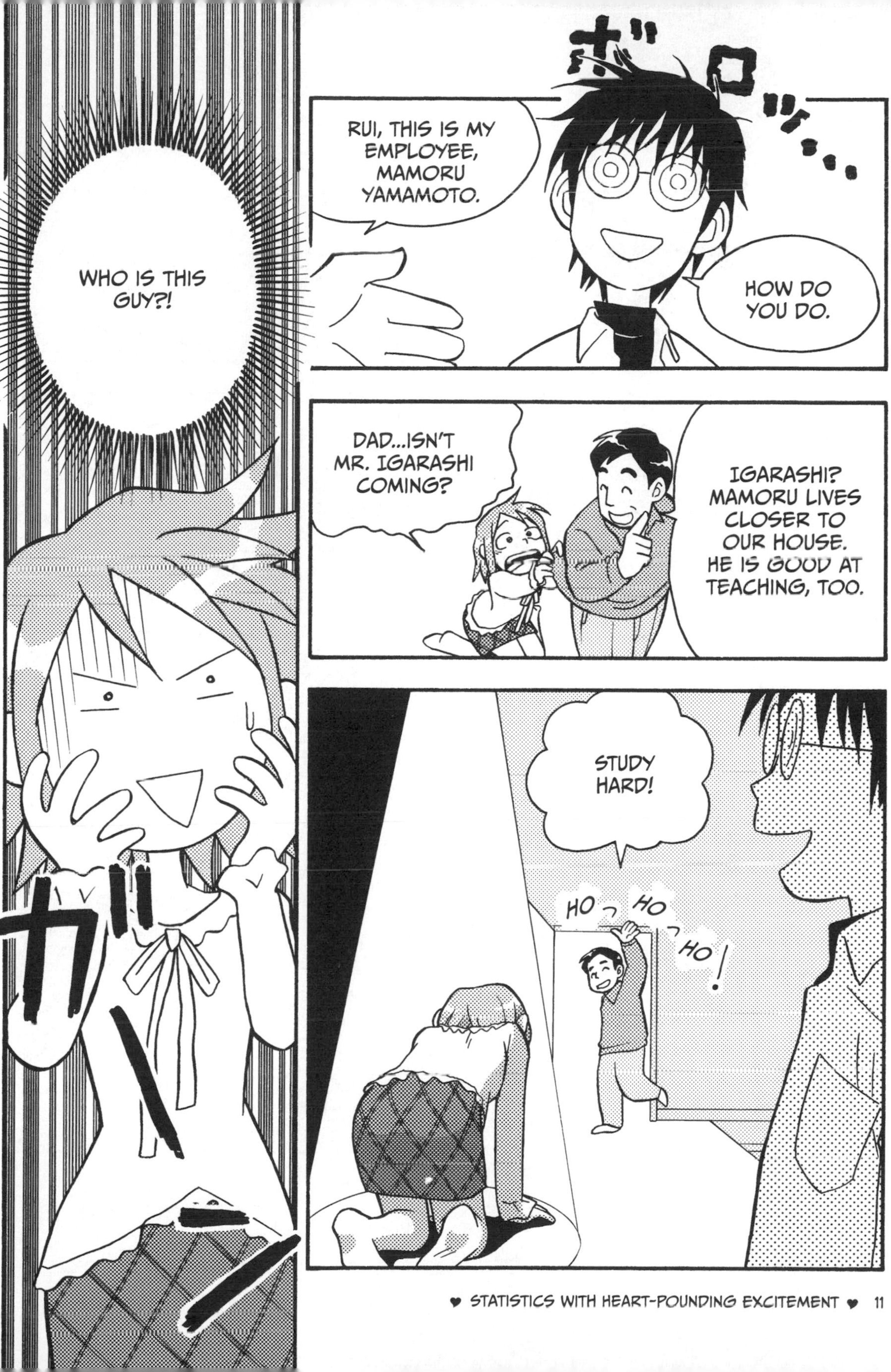
WHO IS THIS GUY?!
ボロッ...
RUI, THIS IS MY EMPLOYEE, MAMORU YAMAMOTO.
HOW DO YOU DO.
DAD...ISN'T MR. IGARASHI COMING?
IGARASHI? MAMORU LIVES CLOSER TO OUR HOUSE. HE IS GOOD AT TEACHING, TOO.
STUDY HARD!
HO HO HO!

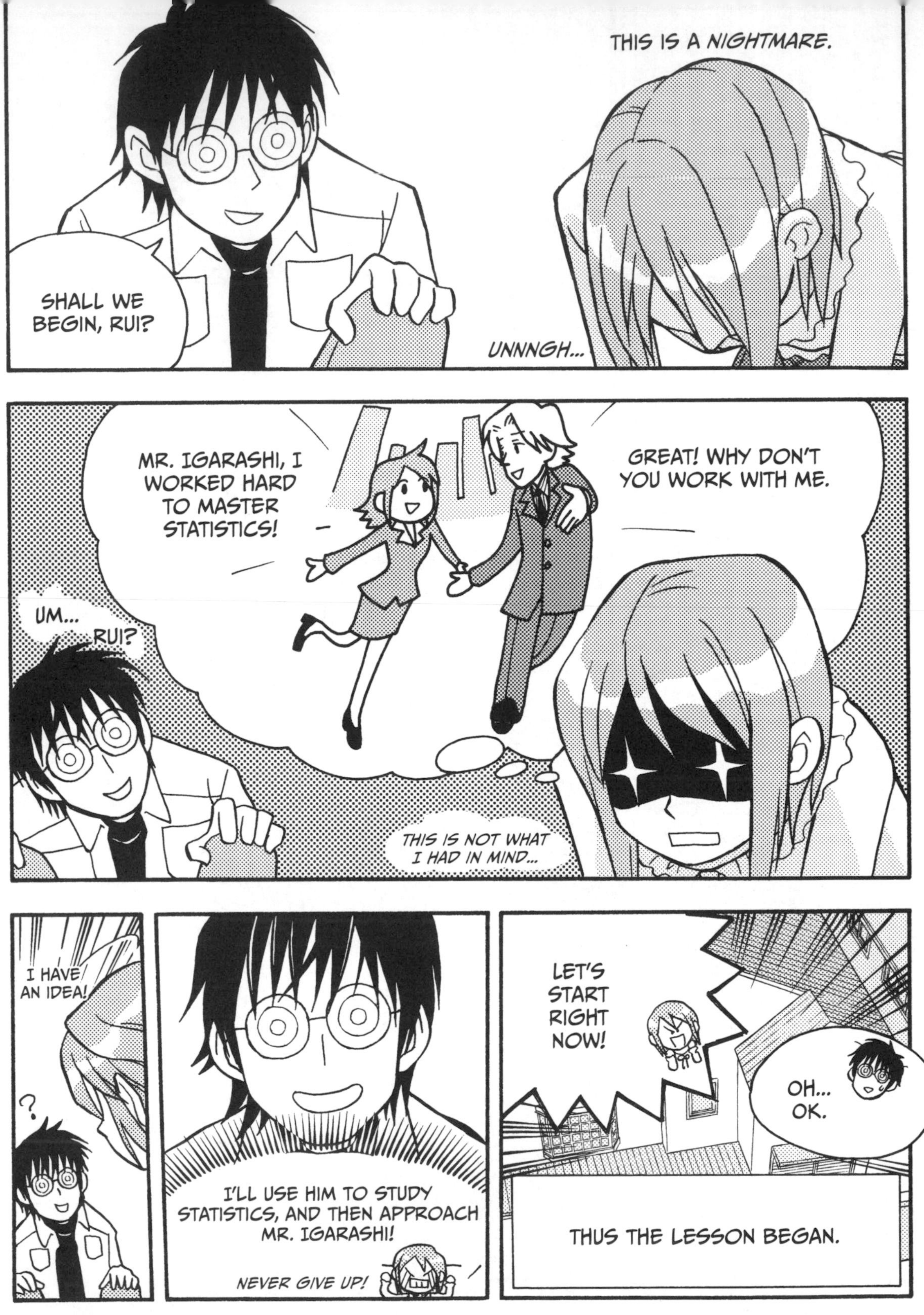
THIS IS A NIGHTMARE.
SHALL WE BEGIN, RUI?
UNNNGH...
MR. IGARASHI, I WORKED HARD TO MASTER STATISTICS!
GREAT! WHY DON'T YOU WORK WITH ME.
UM... RUI?
THIS IS NOT WHAT I HAD IN MIND...
I HAVE AN IDEA!
?
I'LL USE HIM TO STUDY STATISTICS, AND THEN APPROACH MR. IGARASHI!
NEVER GIVE UP!
LET'S START RIGHT NOW!
OH... OK.
THUS THE LESSON BEGAN.

1

DETERMINING DATA TYPES

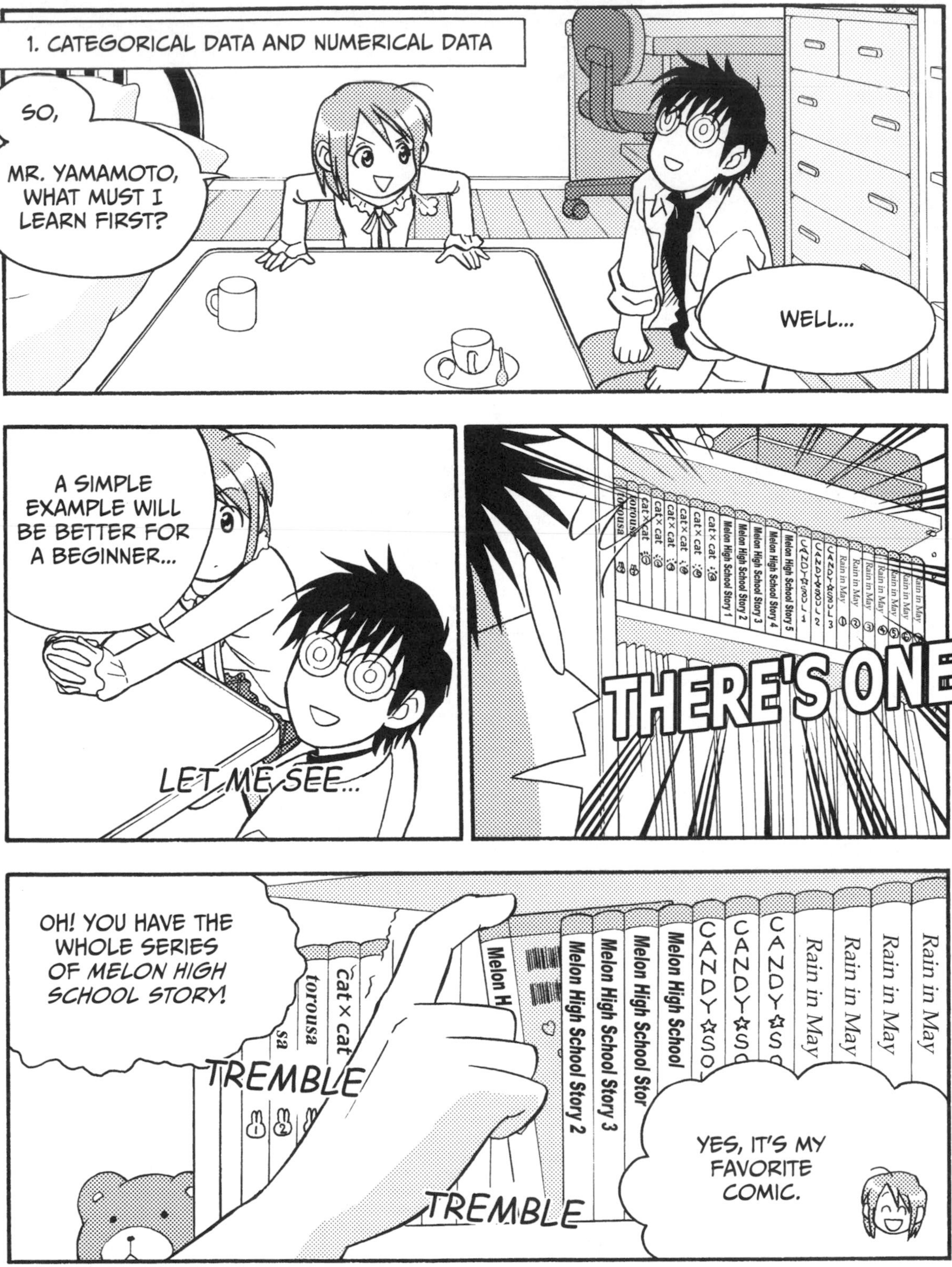
1. CATEGORICAL DATA AND NUMERICAL DATA
SO,
MR. YAMAMOTO, WHAT MUST I LEARN FIRST?
WELL...
A SIMPLE EXAMPLE WILL BE BETTER FOR A BEGINNER...
LET ME SEE...
THERE'S ONE!
OH! YOU HAVE THE WHOLE SERIES OF MELON HIGH SCHOOL STORY!
TREMBLE
TREMBLE
YES, IT'S MY FAVORITE COMIC.

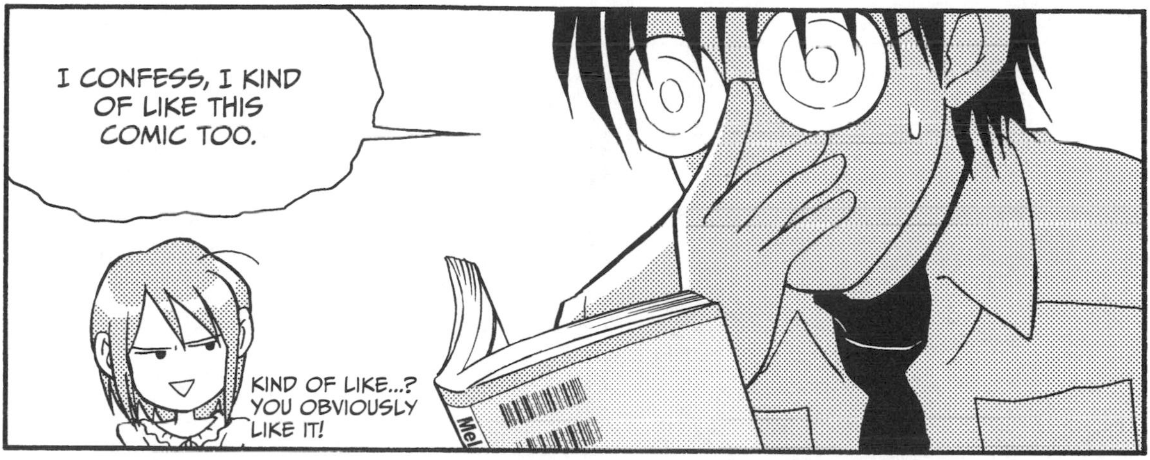
I CONFESS, I KIND OF LIKE THIS COMIC TOO.
KIND OF LIKE...? YOU OBVIOUSLY LIKE IT!
Mel

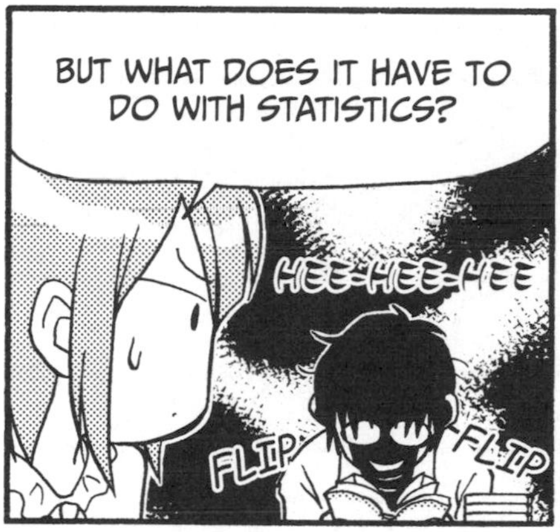
BUT WHAT DOES IT HAVE TO DO WITH STATISTICS?
HEE-HEE-HEE
FLIP
FLIP

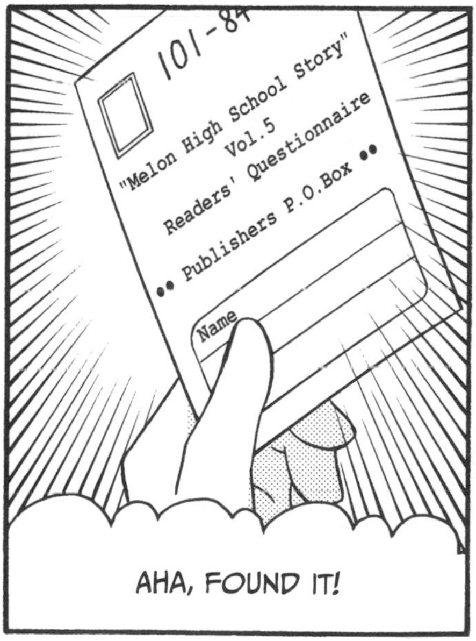
101-84
"Melon High School Story" Vol.5
Readers' Questionnaire
•• Publishers P.O.Box ••
Name
AHA, FOUND IT!

Melon High School Story Vol. 5 Reader Questionnaire

Q1. What is your impression of *Melon High School Story* Vol. 5?

1. Very fun
2. Rather fun
3. Average
4. Rather boring
5. Very boring

Q2. Sex

1. Female
2. Male

Q3. Age ________ years old

Q4. How many comics do you purchase per month? ________ titles

A Rina keychain will be given away to 30 lucky winners among those who send back this questionnaire!

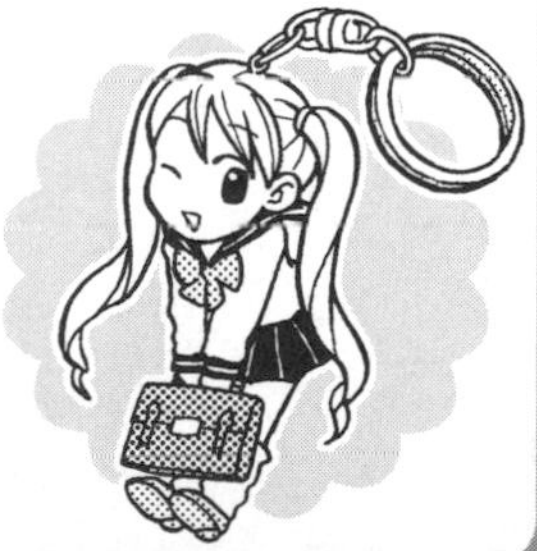

THANK YOU FOR YOUR CONTRIBUTION. YOUR VALUABLE OPINION WILL BE REFLECTED IN OUR FUTURE PUBLICATIONS.

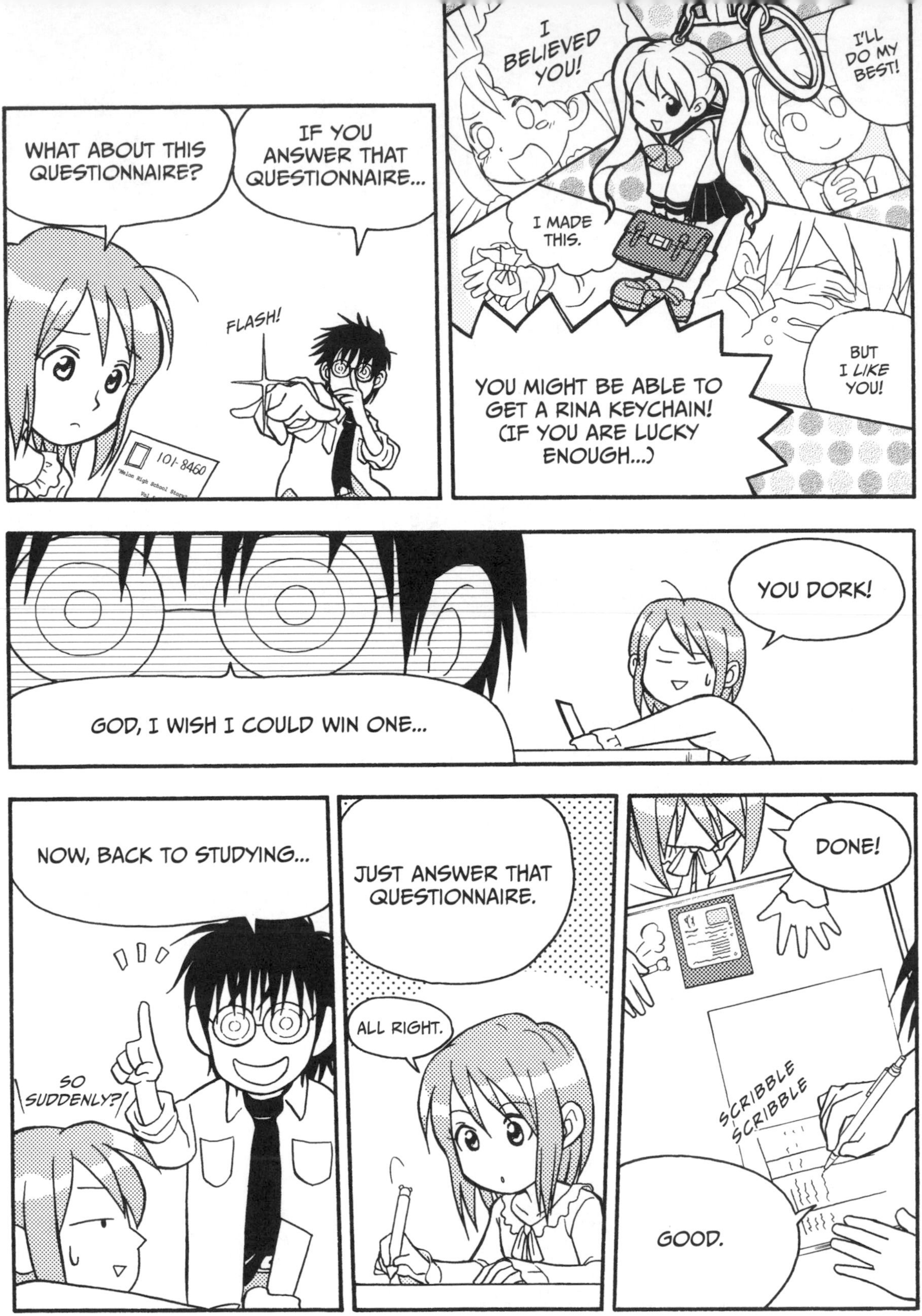
WHAT ABOUT THIS QUESTIONNAIRE?
IF YOU ANSWER THAT QUESTIONNAIRE...
FLASH!
101-8460
"Melon High School Story"
I BELIEVED YOU!
I'LL DO MY BEST!
I MADE THIS.
BUT I LIKE YOU!
YOU MIGHT BE ABLE TO GET A RINA KEYCHAIN! (IF YOU ARE LUCKY ENOUGH...)
GOD, I WISH I COULD WIN ONE...
YOU DORK!
NOW, BACK TO STUDYING...
SO SUDDENLY?
JUST ANSWER THAT QUESTIONNAIRE.
ALL RIGHT.
DONE!
SCRIBBLE SCRIBBLE
GOOD.

QUESTIONNAIRE RESULTS

RESPONDENT	Q1 YOUR IMPRESSION OF *MELON HIGH SCHOOL STORY*	Q2 SEX	Q3 AGE	Q4 COMIC BOOK PURCHASES PER MONTH
RUI	VERY FUN	FEMALE	17	2
A	RATHER FUN	FEMALE	17	1
B	AVERAGE	MALE	18	5
C	RATHER BORING	MALE	22	7
D	RATHER FUN	FEMALE	25	4
E	VERY BORING	MALE	20	3
F	VERY FUN	FEMALE	16	1
G	RATHER FUN	FEMALE	17	2
H	AVERAGE	MALE	18	0
I	AVERAGE	FEMALE	21	3
...	...	...	...	...

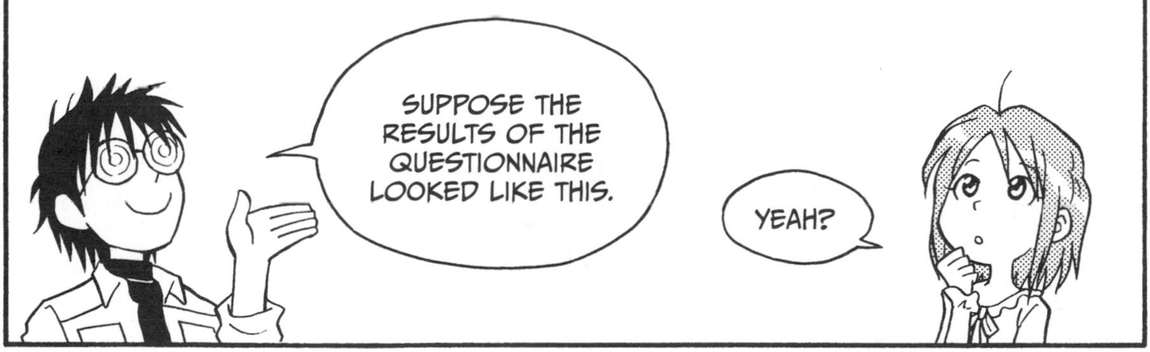
SUPPOSE THE RESULTS OF THE QUESTIONNAIRE LOOKED LIKE THIS.
YEAH?

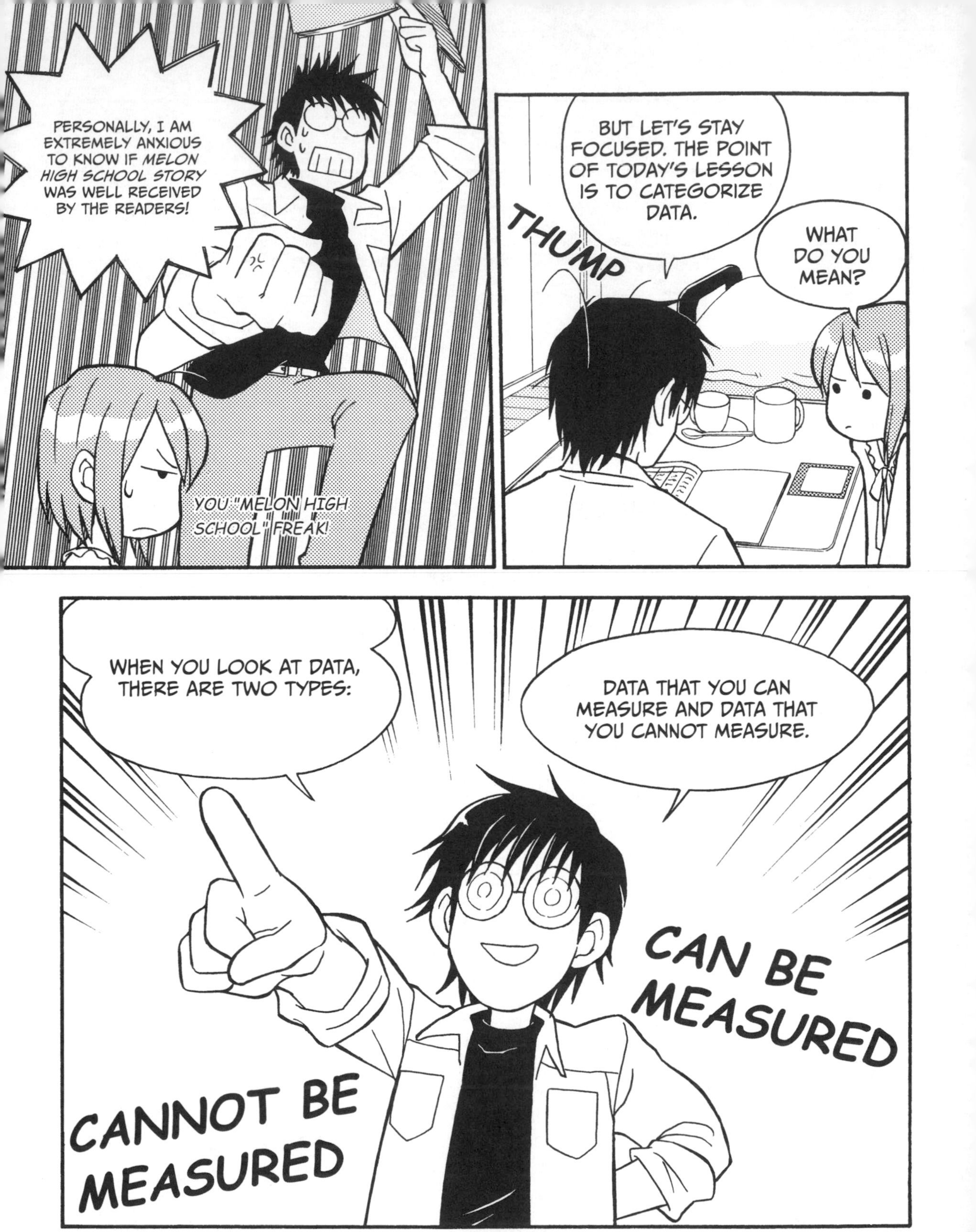
PERSONALLY, I AM EXTREMELY ANXIOUS TO KNOW IF *MELON HIGH SCHOOL STORY* WAS WELL RECEIVED BY THE READERS!
YOU "MELON HIGH SCHOOL" FREAK!
BUT LET'S STAY FOCUSED. THE POINT OF TODAY'S LESSON IS TO CATEGORIZE DATA.
THUMP
WHAT DO YOU MEAN?
WHEN YOU LOOK AT DATA, THERE ARE TWO TYPES:
DATA THAT YOU CAN MEASURE AND DATA THAT YOU CANNOT MEASURE.
CAN BE MEASURED
CANNOT BE MEASURED

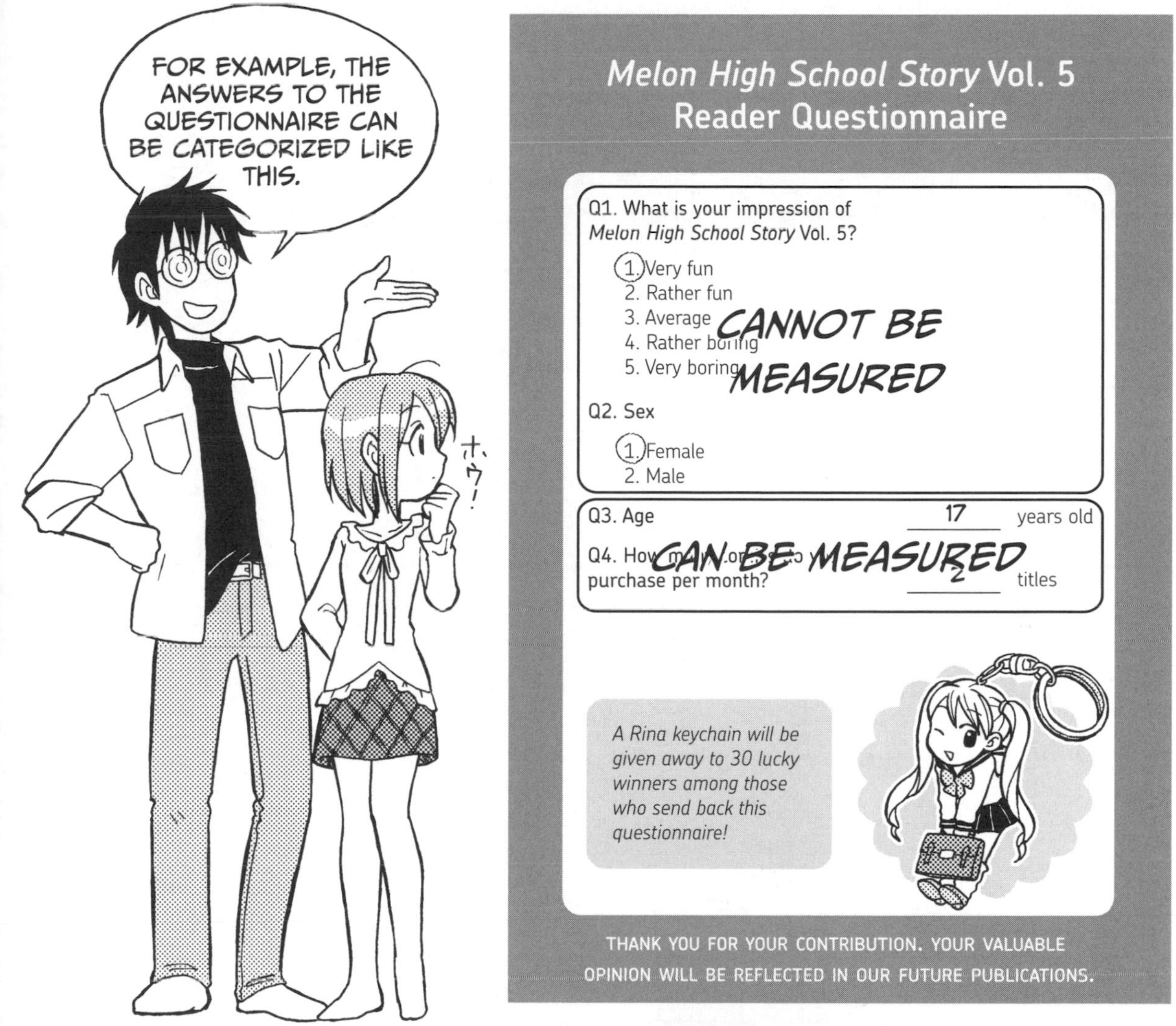

FOR EXAMPLE, THE ANSWERS TO THE QUESTIONNAIRE CAN BE CATEGORIZED LIKE THIS.
ホウ!
Melon High School Story Vol. 5
Reader Questionnaire
Q1. What is your impression of *Melon High School Story* Vol. 5?
1. Very fun
2. Rather fun
3. Average
4. Rather boring
5. Very boring
CANNOT BE MEASURED
Q2. Sex
1. Female
2. Male
Q3. Age 17 years old
Q4. How
purchase per month? 2 titles
CAN BE MEASURED
A Rina keychain will be given away to 30 lucky winners among those who send back this questionnaire!
THANK YOU FOR YOUR CONTRIBUTION. YOUR VALUABLE OPINION WILL BE REFLECTED IN OUR FUTURE PUBLICATIONS.

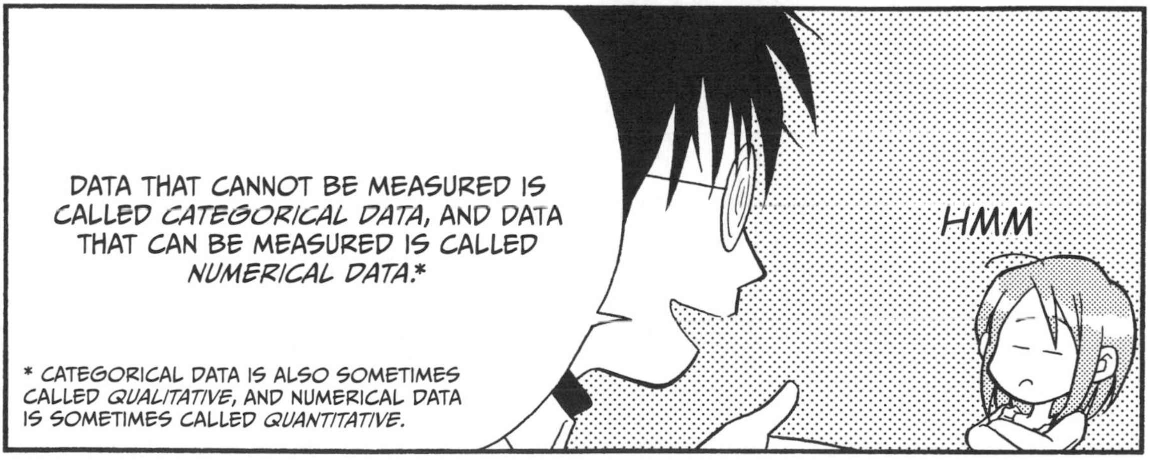

DATA THAT CANNOT BE MEASURED IS CALLED *CATEGORICAL DATA*, AND DATA THAT CAN BE MEASURED IS CALLED *NUMERICAL DATA*.*
HMM
* CATEGORICAL DATA IS ALSO SOMETIMES CALLED *QUALITATIVE*, AND NUMERICAL DATA IS SOMETIMES CALLED *QUANTITATIVE*.

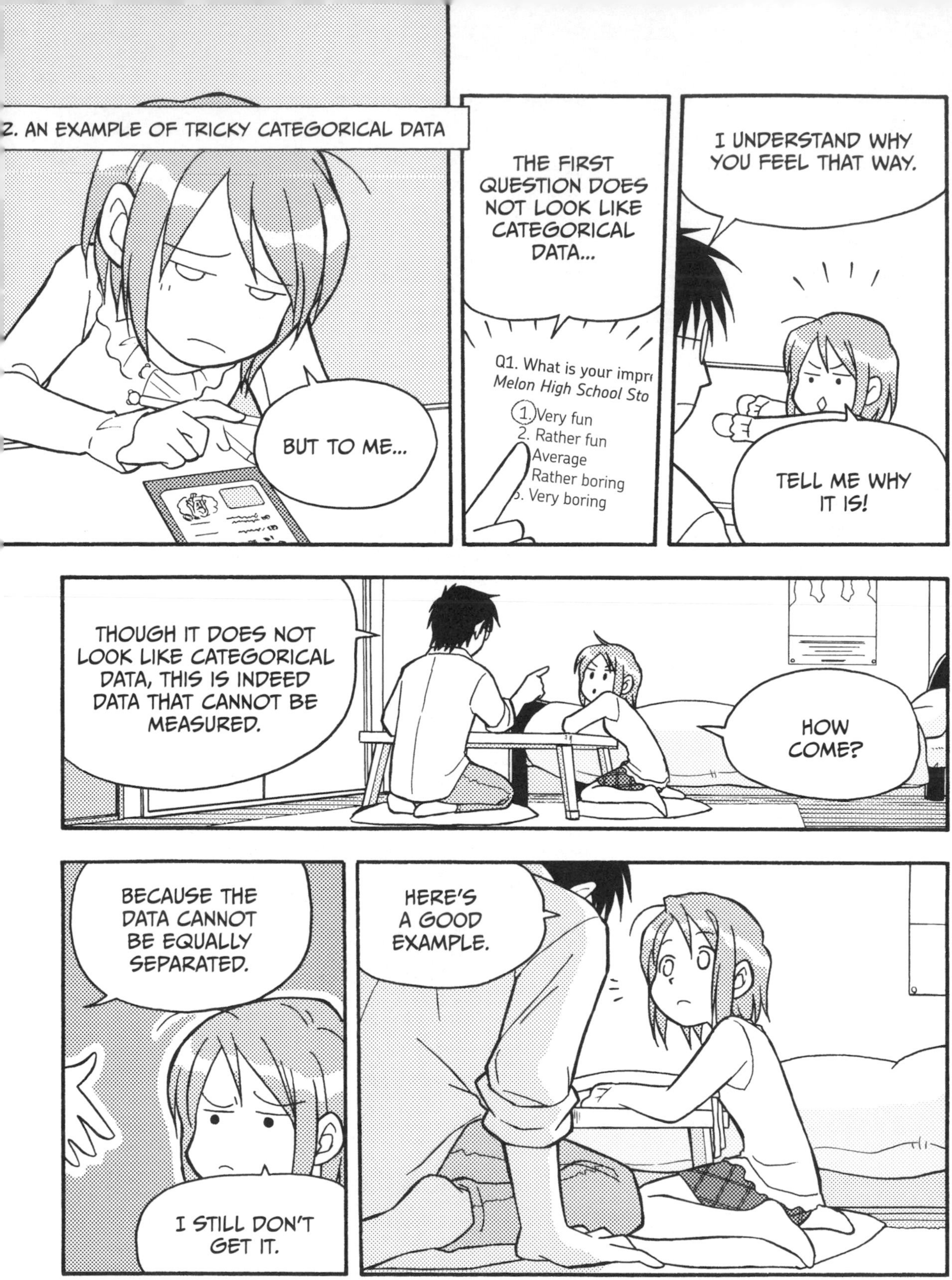
2. AN EXAMPLE OF TRICKY CATEGORICAL DATA
BUT TO ME...
THE FIRST QUESTION DOES NOT LOOK LIKE CATEGORICAL DATA...
Q1. What is your
Melon High School
1. Very fun
2. Rather fun
Average
Rather boring
Very boring
I UNDERSTAND WHY YOU FEEL THAT WAY.
TELL ME WHY IT IS!
THOUGH IT DOES NOT LOOK LIKE CATEGORICAL DATA, THIS IS INDEED DATA THAT CANNOT BE MEASURED.
HOW COME?
BECAUSE THE DATA CANNOT BE EQUALLY SEPARATED.
I STILL DON'T GET IT.
HERE'S A GOOD EXAMPLE.

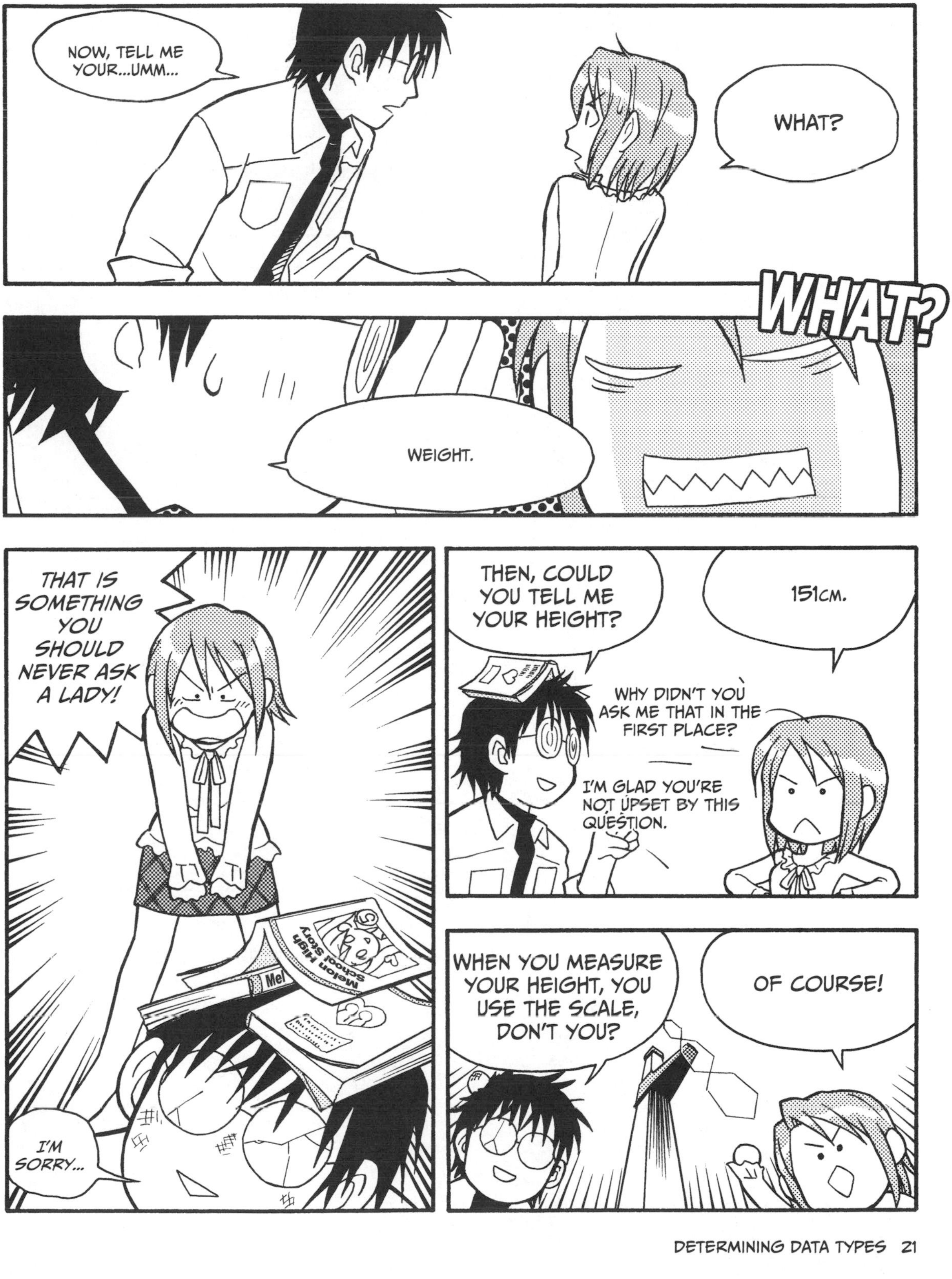
NOW, TELL ME YOUR...UMM...
WHAT?
WHAT?
WEIGHT.
THAT IS SOMETHING YOU SHOULD NEVER ASK A LADY!
Melon High School Story
I'M SORRY...
THEN, COULD YOU TELL ME YOUR HEIGHT?
151cm.
WHY DIDN'T YOU ASK ME THAT IN THE FIRST PLACE?
I'M GLAD YOU'RE NOT UPSET BY THIS QUESTION.
WHEN YOU MEASURE YOUR HEIGHT, YOU USE THE SCALE, DON'T YOU?
OF COURSE!

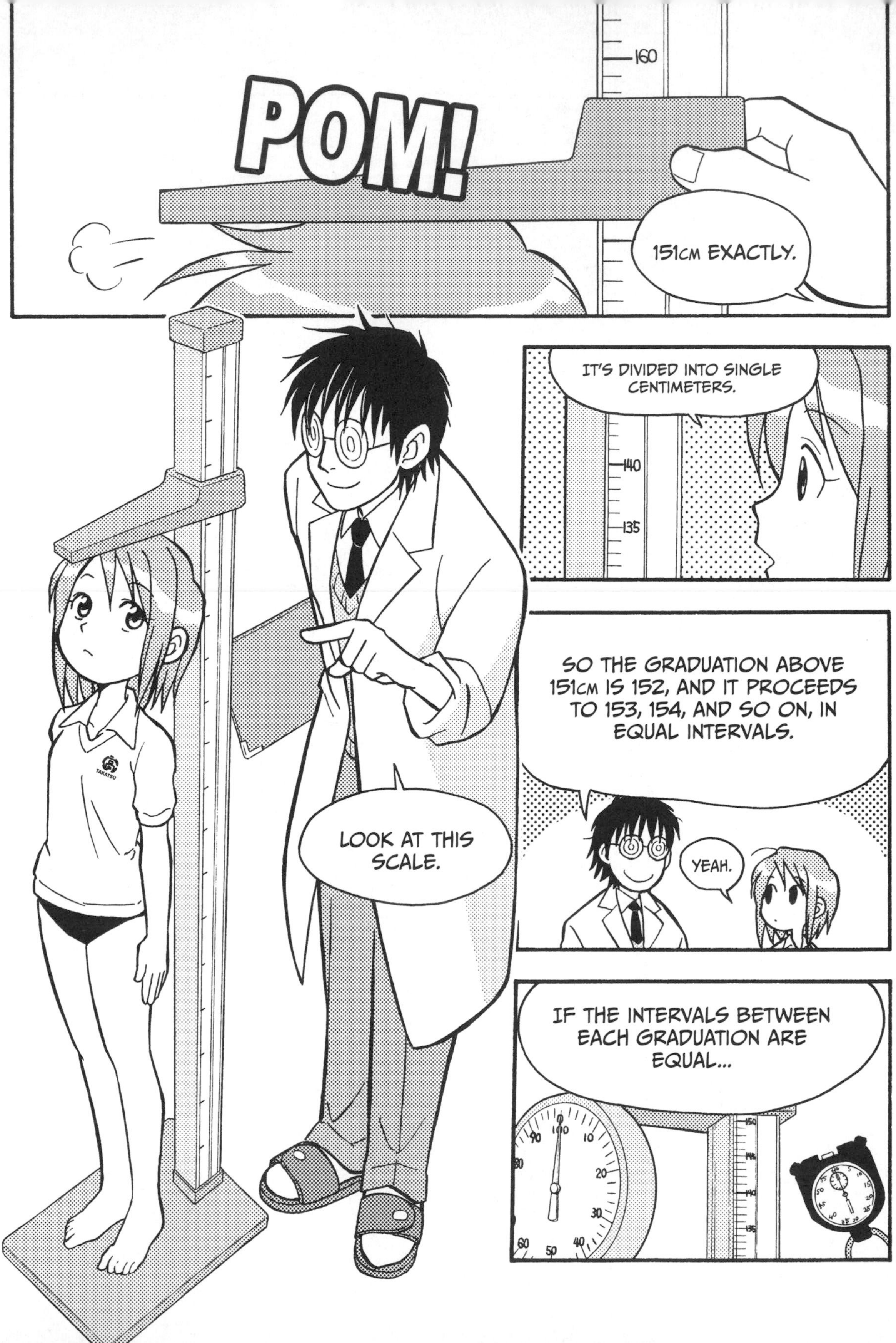
POM!
151cm EXACTLY.
LOOK AT THIS SCALE.
IT'S DIVIDED INTO SINGLE CENTIMETERS.
SO THE GRADUATION ABOVE 151cm IS 152, AND IT PROCEEDS TO 153, 154, AND SO ON, IN EQUAL INTERVALS.
YEAH.
IF THE INTERVALS BETWEEN EACH GRADUATION ARE EQUAL...

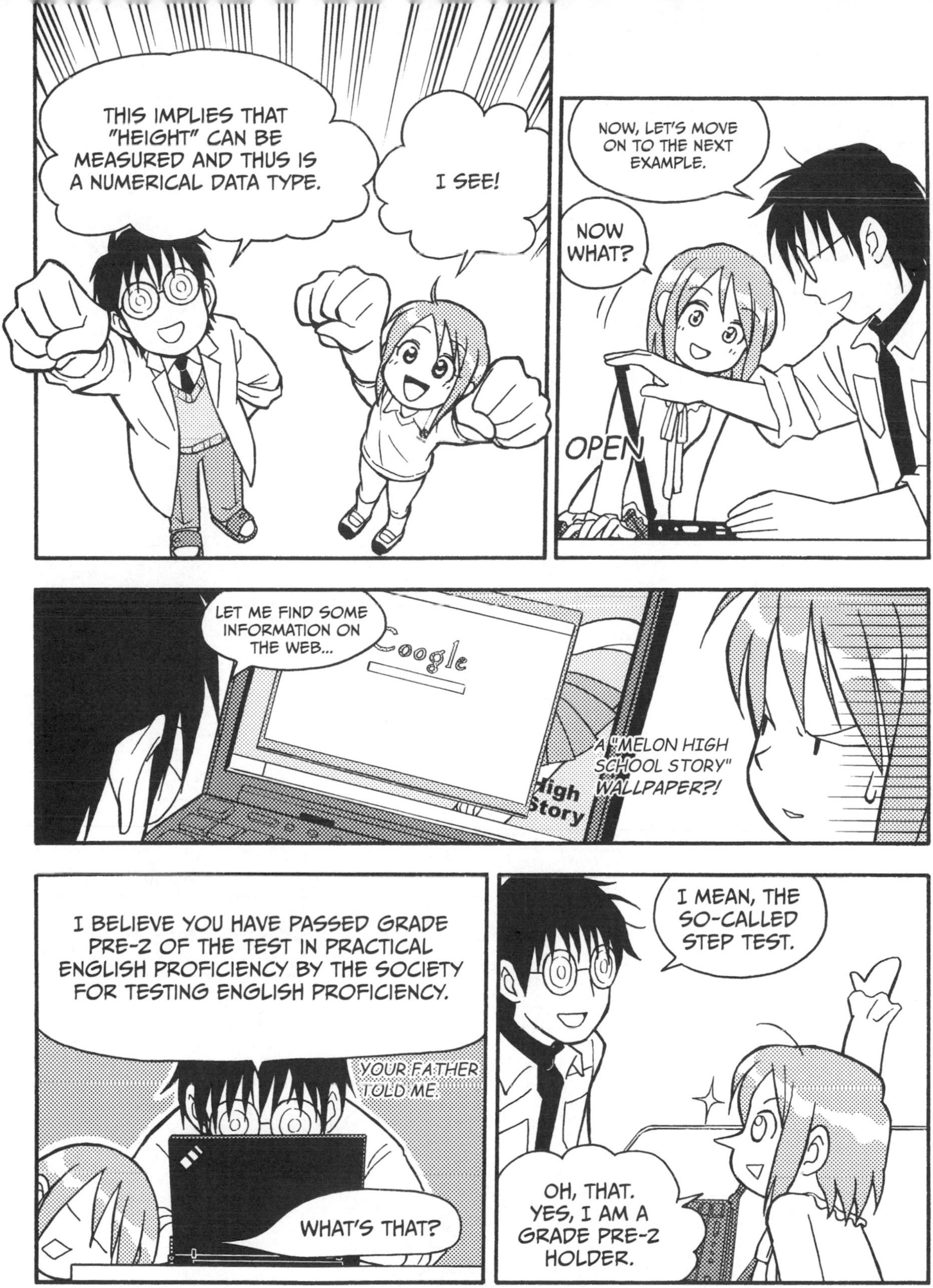
THIS IMPLIES THAT "HEIGHT" CAN BE MEASURED AND THUS IS A NUMERICAL DATA TYPE.
I SEE!
NOW, LET'S MOVE ON TO THE NEXT EXAMPLE.
NOW WHAT?
OPEN
LET ME FIND SOME INFORMATION ON THE WEB...
Google
A "MELON HIGH SCHOOL STORY" WALLPAPER?!
I BELIEVE YOU HAVE PASSED GRADE PRE-2 OF THE TEST IN PRACTICAL ENGLISH PROFICIENCY BY THE SOCIETY FOR TESTING ENGLISH PROFICIENCY.
YOUR FATHER TOLD ME.
WHAT'S THAT?
I MEAN, THE SO-CALLED STEP TEST.
OH, THAT. YES, I AM A GRADE PRE-2 HOLDER.

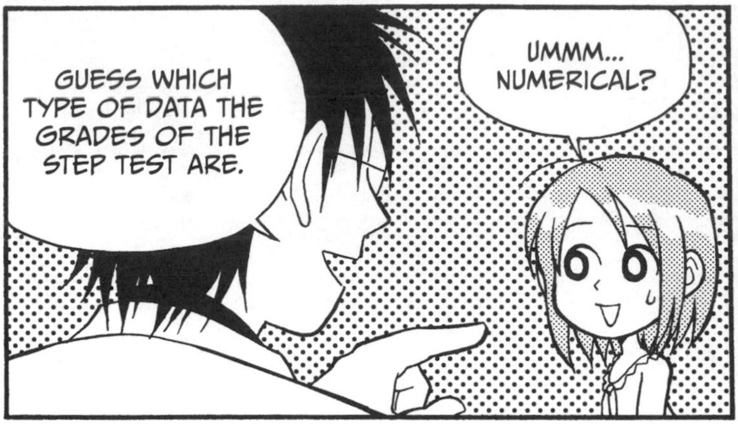
GUESS WHICH TYPE OF DATA THE GRADES OF THE STEP TEST ARE.
UMMM... NUMERICAL?

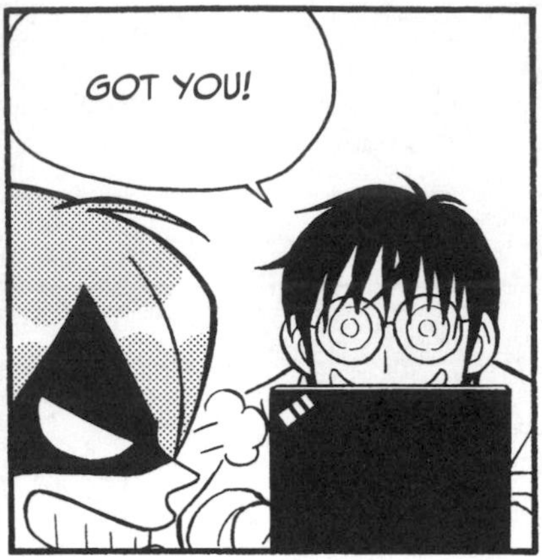
GOT YOU!

THE STEP TEST GRADES

Grade	Requirements
Grade 1	Advanced university graduate level, vocabulary 10,000–15,000 words
Grade 2	High school graduate level, vocabulary 5,100 words
Grade 3	Junior high school graduate level, vocabulary 2,100 words
Grade 4	Intermediate junior high school level, vocabulary 1,300 words
Grade 5	Beginner junior high school level, vocabulary 600 words

(from the Society for Testing English Proficiency, http://www.eiken.or.jp/)

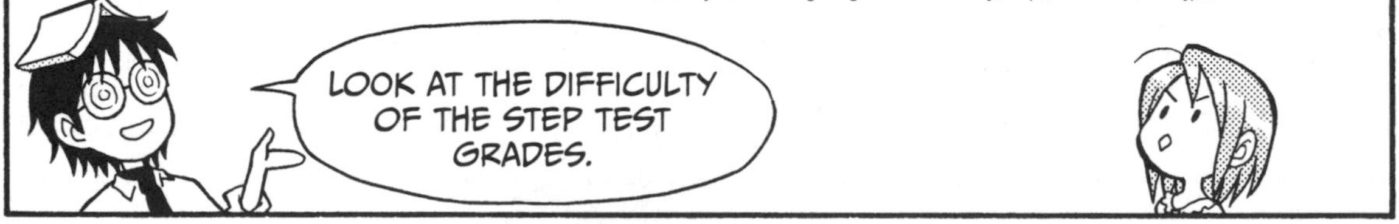
LOOK AT THE DIFFICULTY OF THE STEP TEST GRADES.

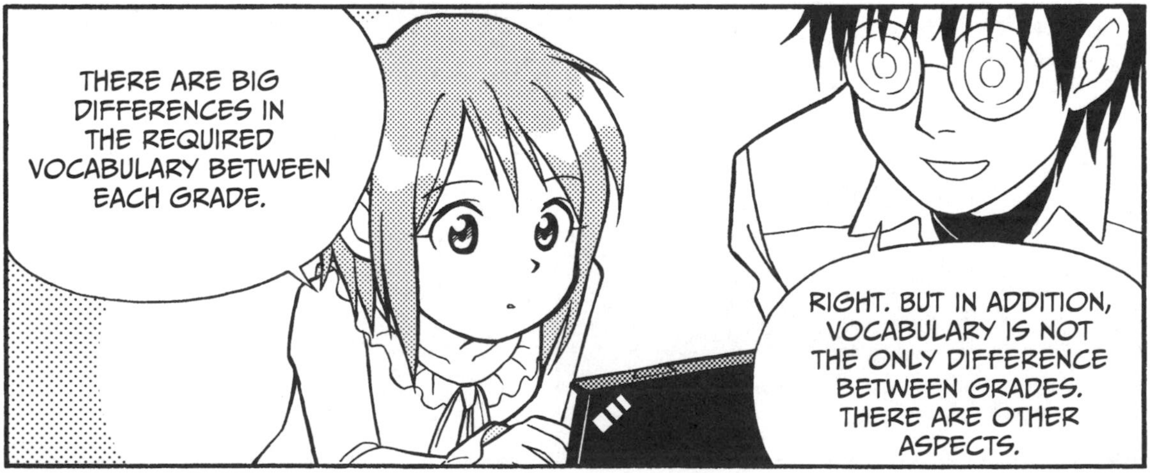
THERE ARE BIG DIFFERENCES IN THE REQUIRED VOCABULARY BETWEEN EACH GRADE.
RIGHT. BUT IN ADDITION, VOCABULARY IS NOT THE ONLY DIFFERENCE BETWEEN GRADES. THERE ARE OTHER ASPECTS.

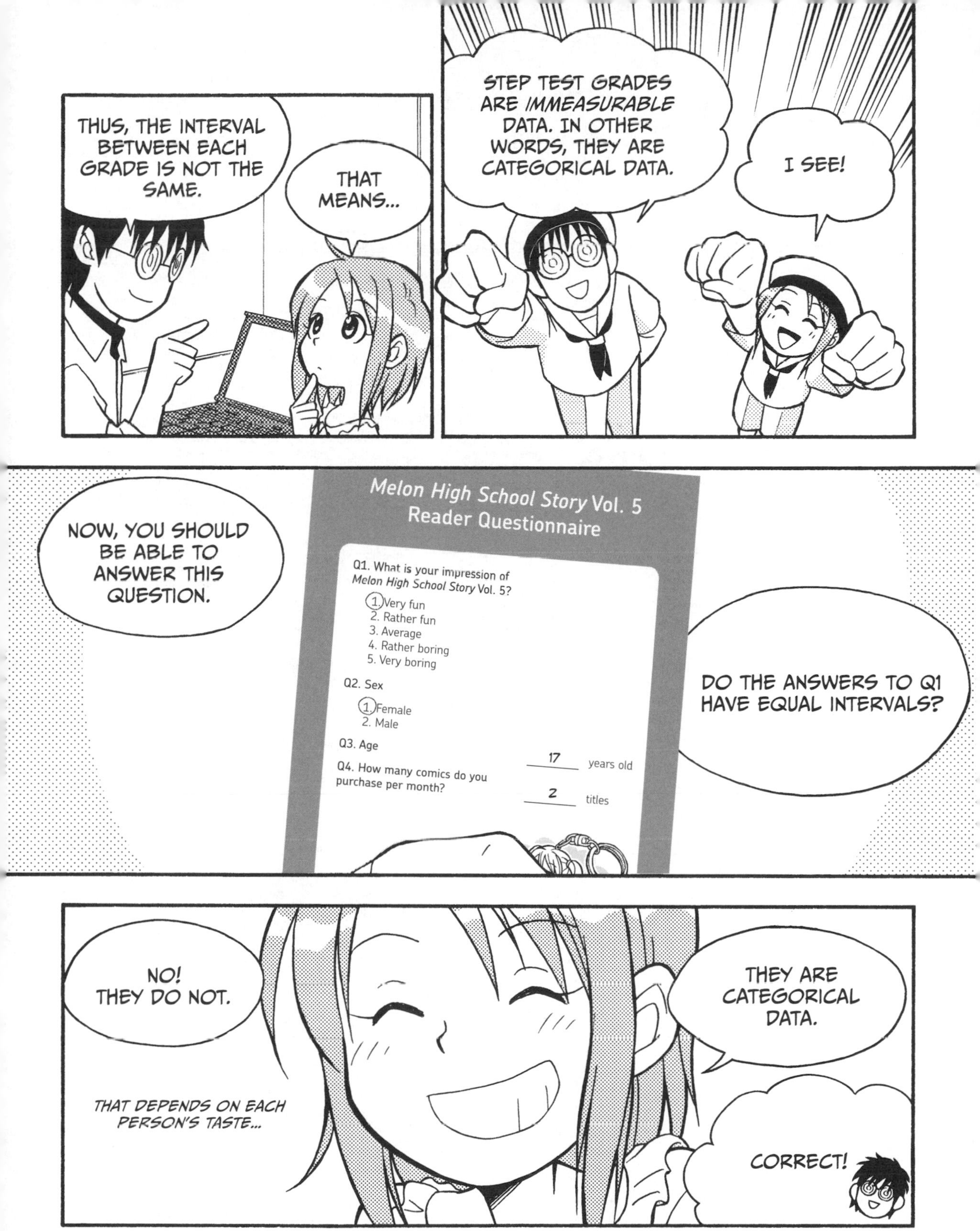
THUS, THE INTERVAL BETWEEN EACH GRADE IS NOT THE SAME.
THAT MEANS...
STEP TEST GRADES ARE *IMMEASURABLE* DATA. IN OTHER WORDS, THEY ARE CATEGORICAL DATA.
I SEE!
NOW, YOU SHOULD BE ABLE TO ANSWER THIS QUESTION.
Melon High School Story Vol. 5
Reader Questionnaire
Q1. What is your impression of *Melon High School Story* Vol. 5?
1. Very fun
2. Rather fun
3. Average
4. Rather boring
5. Very boring
Q2. Sex
1. Female
2. Male
Q3. Age
17 years old
Q4. How many comics do you purchase per month?
2 titles
DO THE ANSWERS TO Q1 HAVE EQUAL INTERVALS?
NO! THEY DO NOT.
THEY ARE CATEGORICAL DATA.
THAT DEPENDS ON EACH PERSON'S TASTE...
CORRECT!

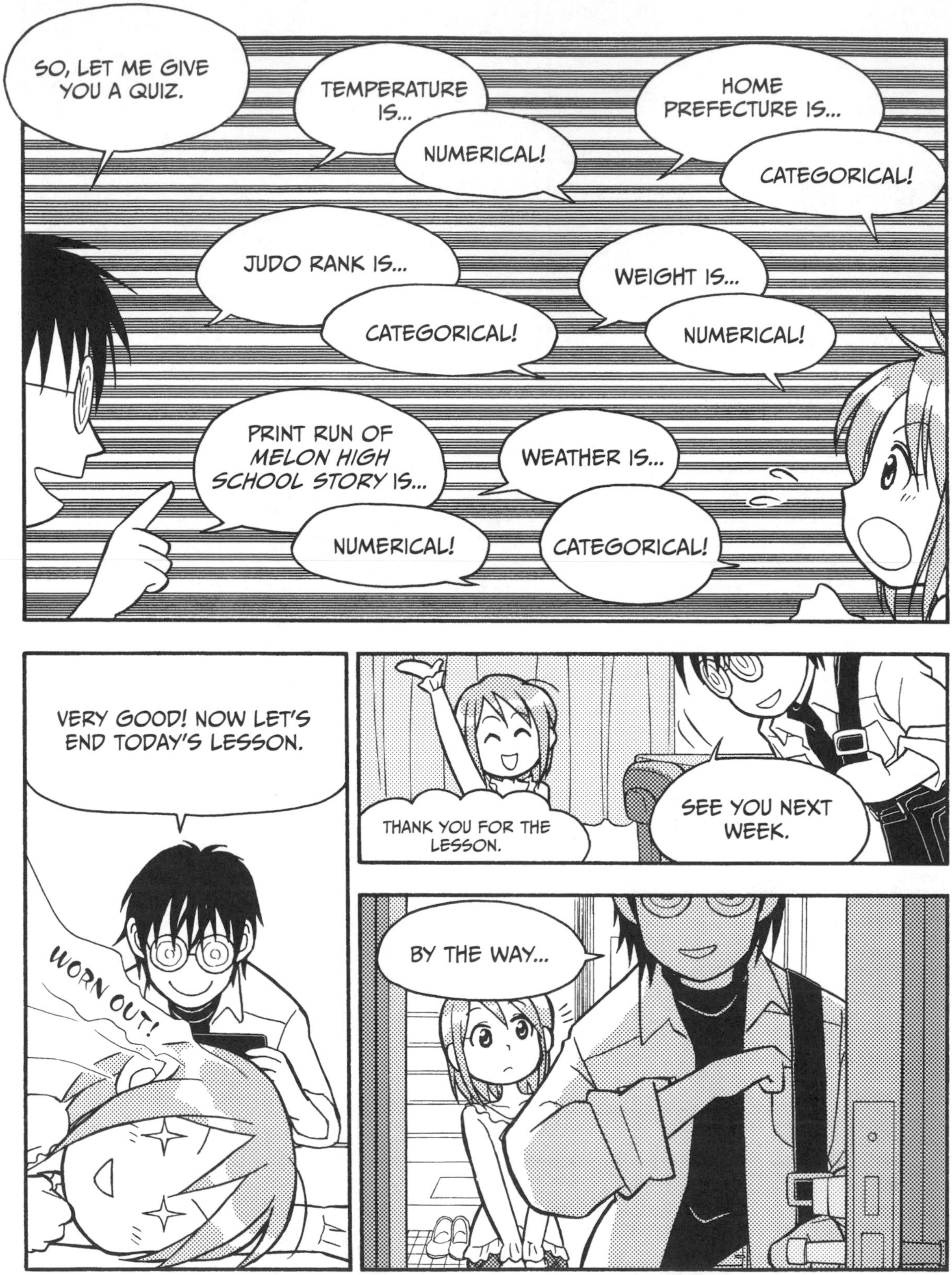
SO, LET ME GIVE YOU A QUIZ.
TEMPERATURE IS...
NUMERICAL!
HOME PREFECTURE IS...
CATEGORICAL!
JUDO RANK IS...
CATEGORICAL!
WEIGHT IS...
NUMERICAL!
PRINT RUN OF MELON HIGH SCHOOL STORY IS...
NUMERICAL!
WEATHER IS...
CATEGORICAL!
VERY GOOD! NOW LET'S END TODAY'S LESSON.
WORN OUT!
THANK YOU FOR THE LESSON.
SEE YOU NEXT WEEK.
BY THE WAY...

3. HOW MULTIPLE-CHOICE ANSWERS ARE HANDLED IN PRACTICE

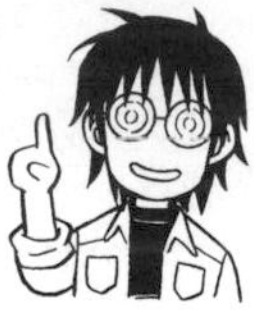

As mentioned on page 25, the multiple-choice answers for the first question of the readers' questionnaire are categorical data. However, in practice, it is possible to handle such data as numerical data when processing consumer questionnaires and so on. Some examples are below.

Very fun	⇨	5 points
Rather fun	⇨	4 points
Average	⇨	3 points
Rather boring	⇨	2 points
Very boring	⇨	1 point

Very fun	⇨	2 points
Rather fun	⇨	1 point
Average	⇨	0 points
Rather boring	⇨	–1 points
Very boring	⇨	–2 points

The same data is handled differently in theory and in practice. Keep in mind that data may be categorized differently in different situations.

EXERCISE

Determine whether the data in the following table is categorical data or numerical data.

Respondent	Blood type	Opinion on sports drink X	Comfortable air conditioning temperature (°C)	100m track race record (seconds)
Mr./Ms. A	B	Not good	25	14.1
Mr./Ms. B	A	Good	24	12.2
Mr./Ms. C	AB	Good	25	17.0
Mr./Ms. D	O	Average	27	15.6
Mr./Ms. E	A	Not good	24	18.4
...	...	...	...	...

ANSWER

Blood type and opinion on sports drink X are examples of categorical data. Comfortable air conditioning temperature and 100m track race record are examples of numerical data.

SUMMARY

- Data is classified as categorical data or numerical data.
- Some data, such as "very fun" or "very boring," is theoretically categorical data. However, in practice, it is possible to treat it as numerical data.

2

GETTING THE BIG PICTURE: UNDERSTANDING NUMERICAL DATA

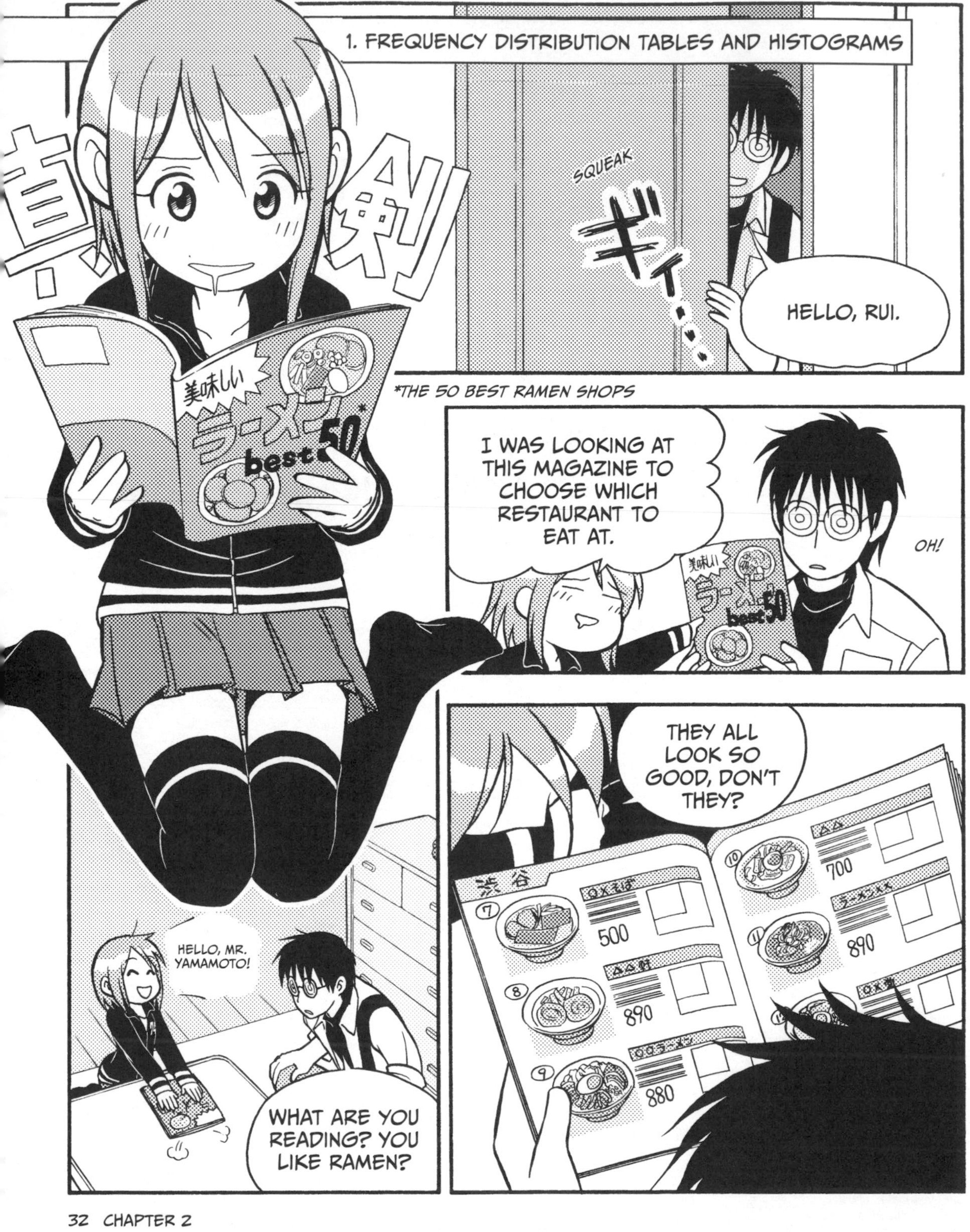
1. FREQUENCY DISTRIBUTION TABLES AND HISTOGRAMS
真剣
美味しい ラーメン best 50*
HELLO, MR. YAMAMOTO!
WHAT ARE YOU READING? YOU LIKE RAMEN?
SQUEAK
ギィ・・・
HELLO, RUI.
*THE 50 BEST RAMEN SHOPS
I WAS LOOKING AT THIS MAGAZINE TO CHOOSE WHICH RESTAURANT TO EAT AT.
OH!
THEY ALL LOOK SO GOOD, DON'T THEY?
渋谷
500
890
880
700
890

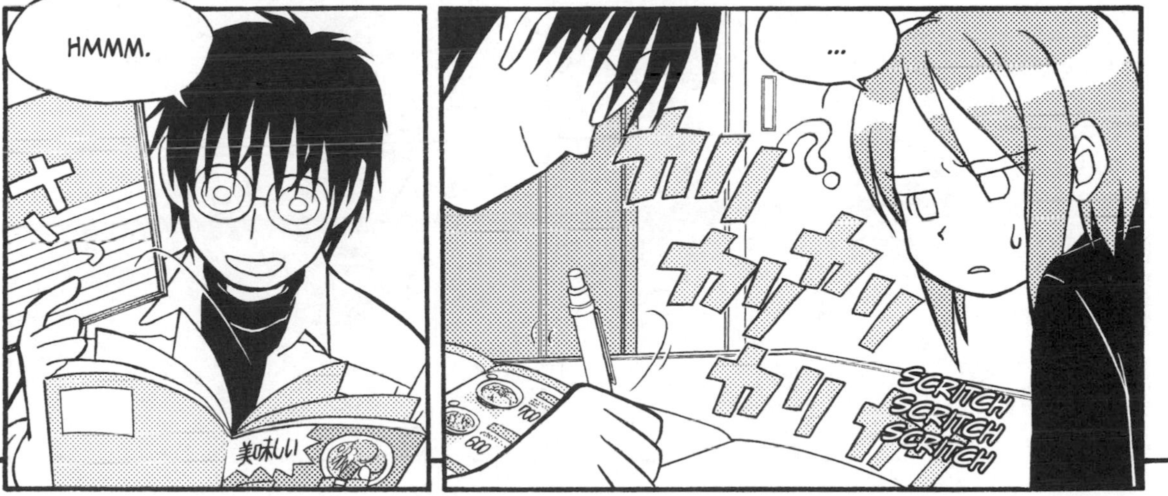
HMMM.
美味しい
...
SCRITCH
SCRITCH
SCRITCH

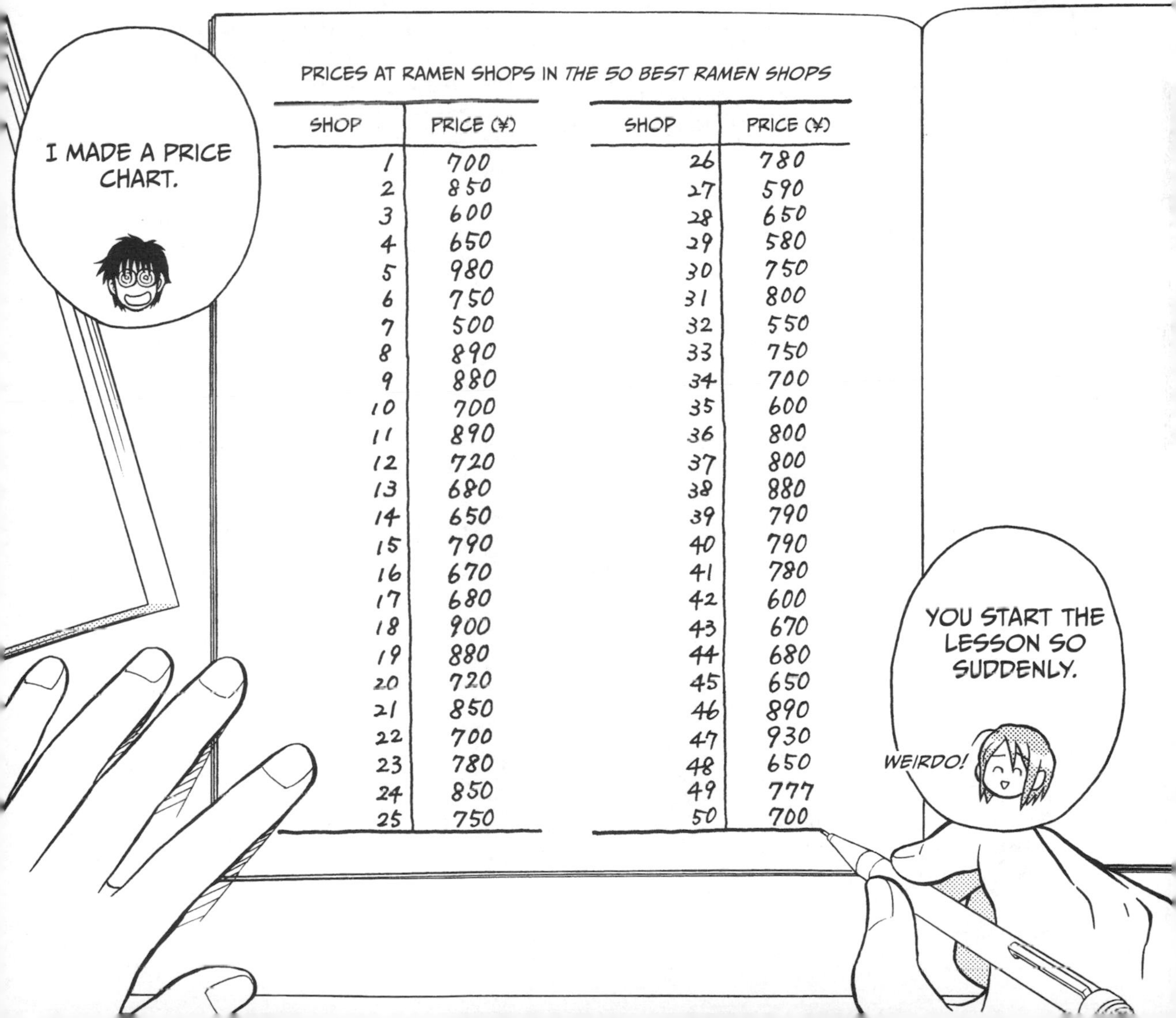
I MADE A PRICE CHART.
YOU START THE LESSON SO SUDDENLY.
WEIRDO!

PRICES AT RAMEN SHOPS IN *THE 50 BEST RAMEN SHOPS*

SHOP	PRICE (¥)	SHOP	PRICE (¥)
1	700	26	780
2	850	27	590
3	600	28	650
4	650	29	580
5	980	30	750
6	750	31	800
7	500	32	550
8	890	33	750
9	880	34	700
10	700	35	600
11	890	36	800
12	720	37	800
13	680	38	880
14	650	39	790
15	790	40	790
16	670	41	780
17	680	42	600
18	900	43	670
19	880	44	680
20	720	45	650
21	850	46	890
22	700	47	930
23	780	48	650
24	850	49	777
25	750	50	700

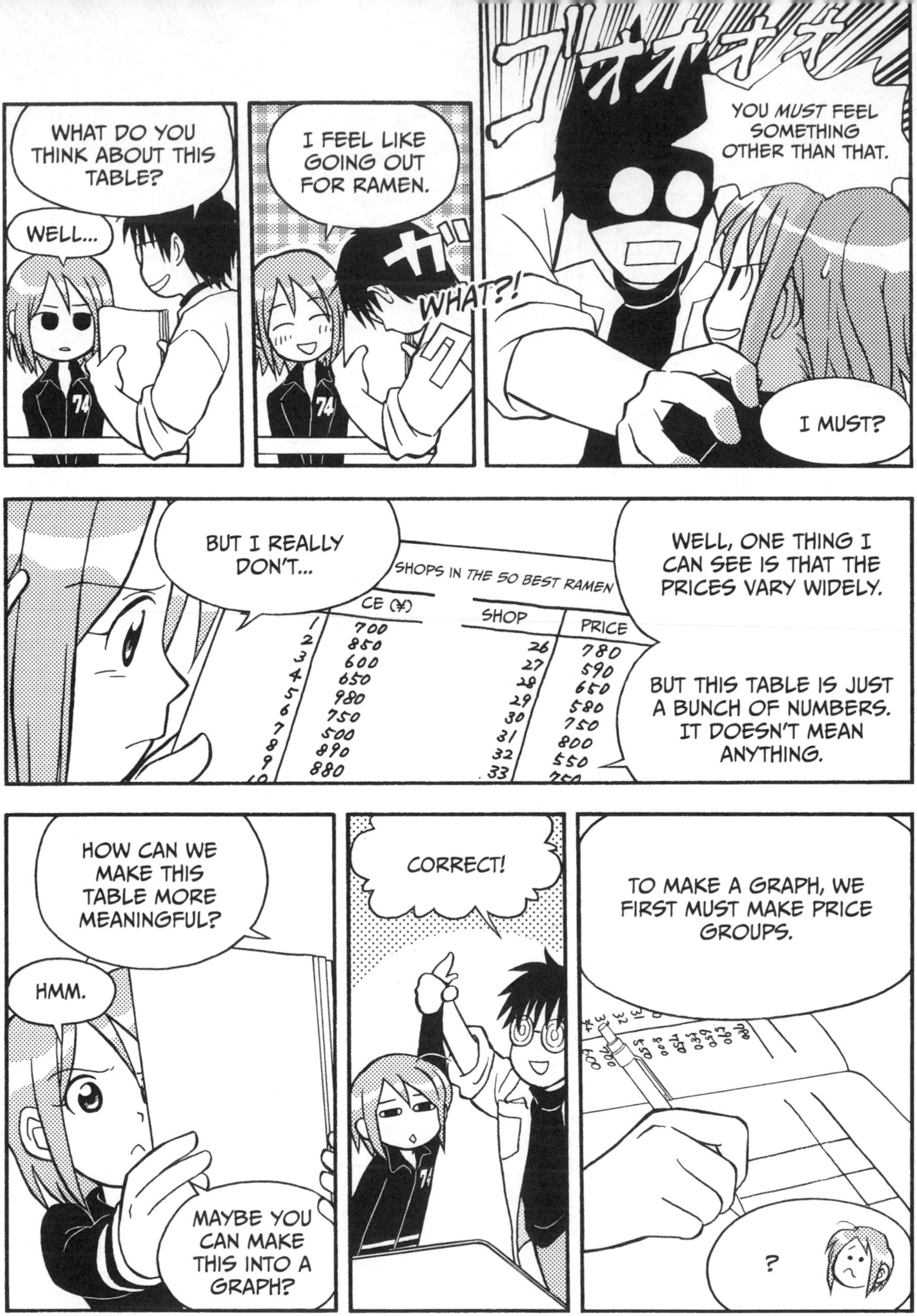
WHAT DO YOU THINK ABOUT THIS TABLE?
WELL...
I FEEL LIKE GOING OUT FOR RAMEN.
ガ
WHAT?!
ゴオオオオ
YOU *MUST* FEEL SOMETHING OTHER THAN THAT.
I MUST?
BUT I REALLY DON'T...
SHOPS IN *THE 50 BEST RAMEN*
CE (¥)
SHOP
PRICE
1 700
2 850
3 600
4 650
5 980
6 750
7 500
8 890
9 880
26 780
27 590
28 650
29 580
30 750
31 800
32 550
WELL, ONE THING I CAN SEE IS THAT THE PRICES VARY WIDELY.
BUT THIS TABLE IS JUST A BUNCH OF NUMBERS. IT DOESN'T MEAN ANYTHING.
HOW CAN WE MAKE THIS TABLE MORE MEANINGFUL?
HMM.
MAYBE YOU CAN MAKE THIS INTO A GRAPH?
CORRECT!
TO MAKE A GRAPH, WE FIRST MUST MAKE PRICE GROUPS.
?

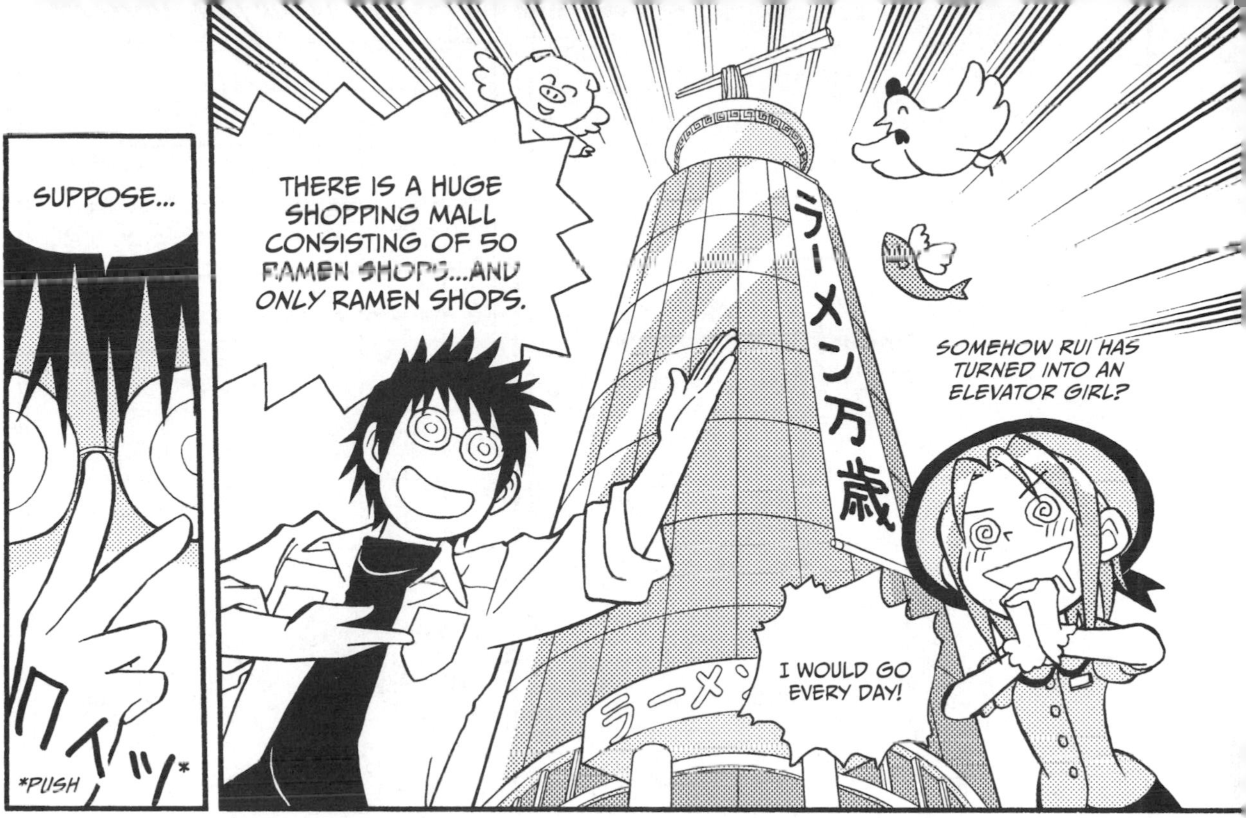
SUPPOSE...
クイッ*
*PUSH
THERE IS A HUGE SHOPPING MALL CONSISTING OF 50 RAMEN SHOPS...AND ONLY RAMEN SHOPS.
ラーメン万歳
SOMEHOW RUI HAS TURNED INTO AN ELEVATOR GIRL?
I WOULD GO EVERY DAY!

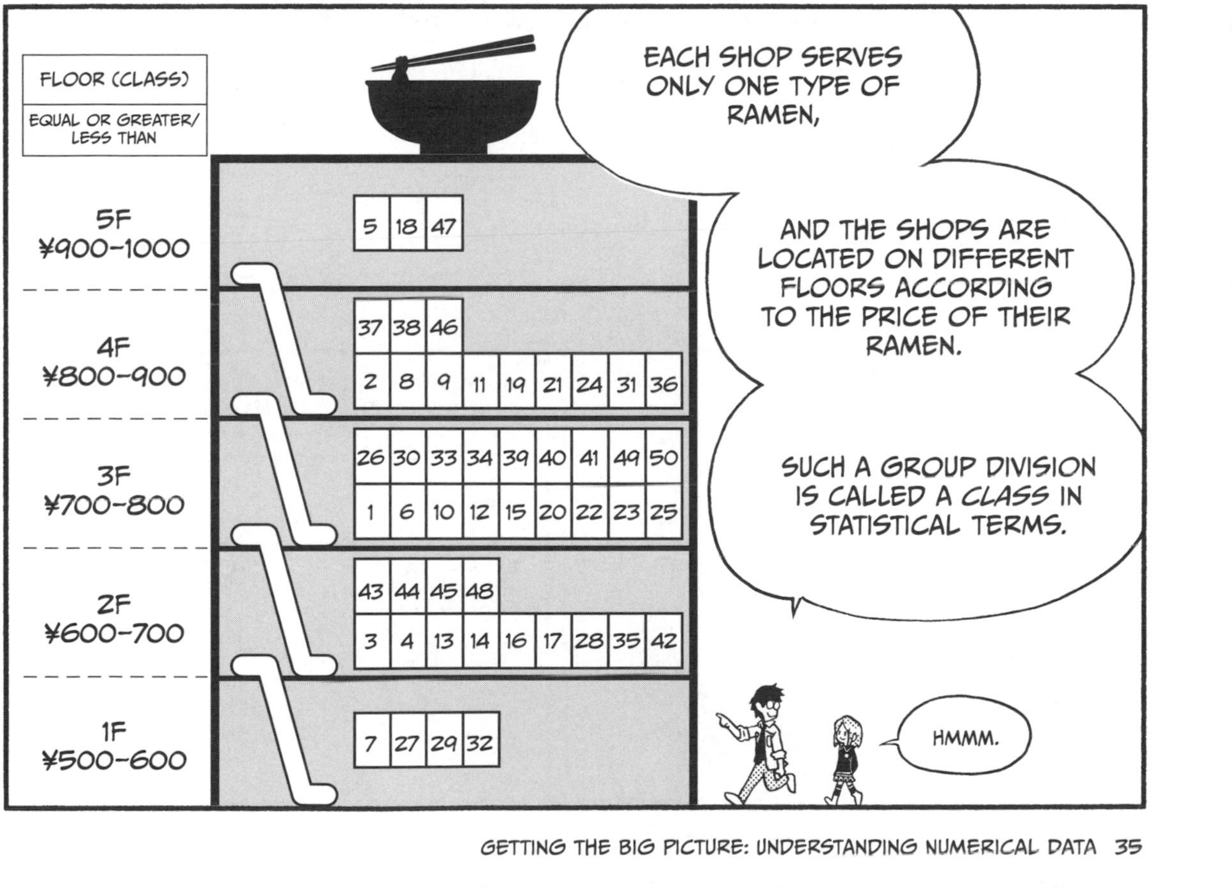
FLOOR (CLASS)
EQUAL OR GREATER/ LESS THAN
5F ¥900–1000
5 18 47
4F ¥800–900
37 38 46
2 8 9 11 19 21 24 31 36
3F ¥700–800
26 30 33 34 39 40 41 49 50
1 6 10 12 15 20 22 23 25
2F ¥600–700
43 44 45 48
3 4 13 14 16 17 28 35 42
1F ¥500–600
7 27 29 32
EACH SHOP SERVES ONLY ONE TYPE OF RAMEN,
AND THE SHOPS ARE LOCATED ON DIFFERENT FLOORS ACCORDING TO THE PRICE OF THEIR RAMEN.
SUCH A GROUP DIVISION IS CALLED A CLASS IN STATISTICAL TERMS.
HMMM.

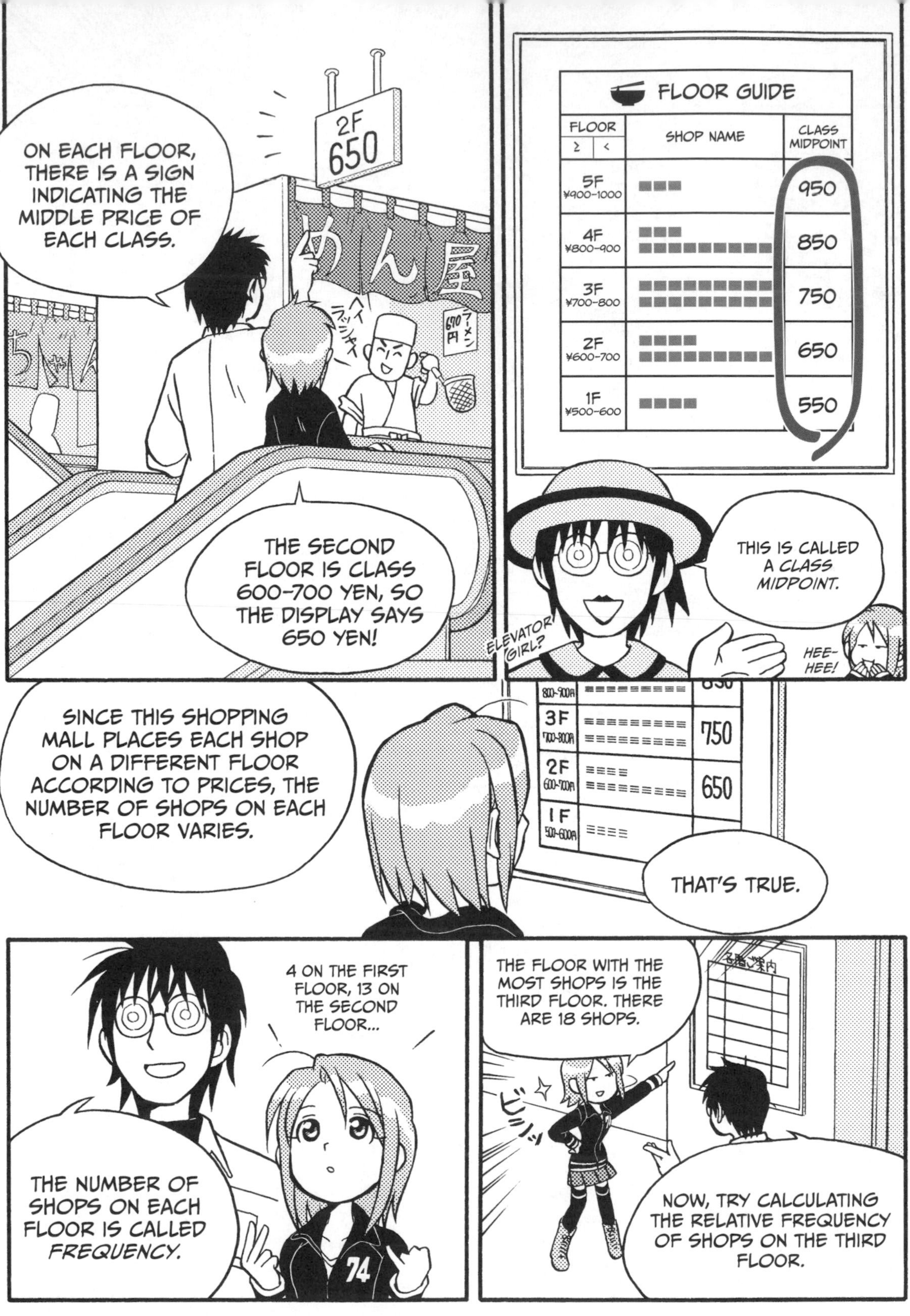
ON EACH FLOOR, THERE IS A SIGN INDICATING THE MIDDLE PRICE OF EACH CLASS.
2F 650
めん屋
ヘイ ラッシャイ
ラーメン 670円
THE SECOND FLOOR IS CLASS 600–700 YEN, SO THE DISPLAY SAYS 650 YEN!
FLOOR GUIDE
FLOOR
≥ <
SHOP NAME
CLASS MIDPOINT
5F ¥900–1000 950
4F ¥800–900 850
3F ¥700–800 750
2F ¥600–700 650
1F ¥500–600 550
THIS IS CALLED A *CLASS MIDPOINT.*
ELEVATOR GIRL?
HEE-HEE!
SINCE THIS SHOPPING MALL PLACES EACH SHOP ON A DIFFERENT FLOOR ACCORDING TO PRICES, THE NUMBER OF SHOPS ON EACH FLOOR VARIES.
THAT'S TRUE.
4 ON THE FIRST FLOOR, 13 ON THE SECOND FLOOR...
THE NUMBER OF SHOPS ON EACH FLOOR IS CALLED *FREQUENCY.*
74
THE FLOOR WITH THE MOST SHOPS IS THE THIRD FLOOR. THERE ARE 18 SHOPS.
ビシッ
NOW, TRY CALCULATING THE RELATIVE FREQUENCY OF SHOPS ON THE THIRD FLOOR.

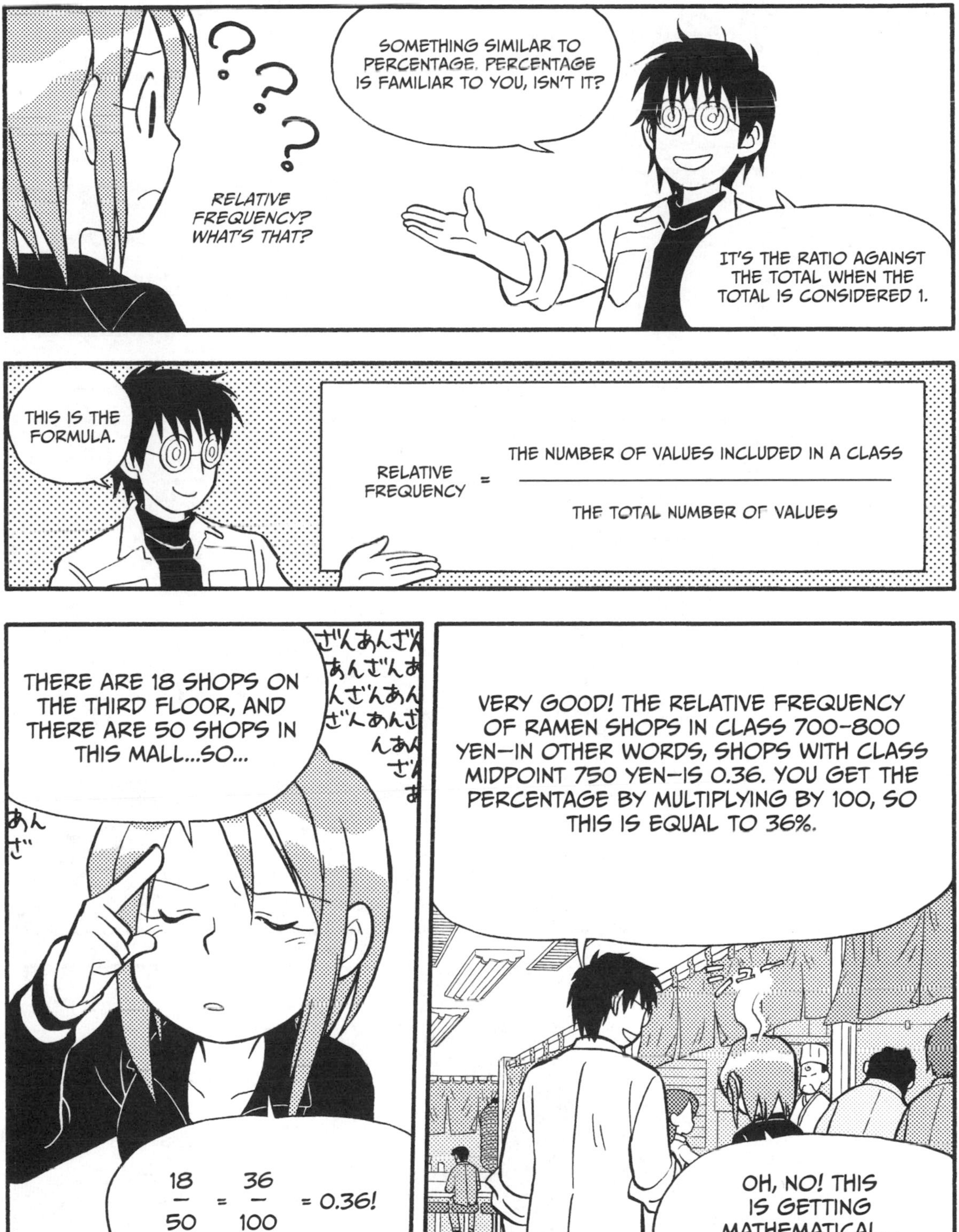
RELATIVE FREQUENCY? WHAT'S THAT?
SOMETHING SIMILAR TO PERCENTAGE. PERCENTAGE IS FAMILIAR TO YOU, ISN'T IT?
IT'S THE RATIO AGAINST THE TOTAL WHEN THE TOTAL IS CONSIDERED 1.
THIS IS THE FORMULA.
RELATIVE FREQUENCY = THE NUMBER OF VALUES INCLUDED IN A CLASS / THE TOTAL NUMBER OF VALUES
THERE ARE 18 SHOPS ON THE THIRD FLOOR, AND THERE ARE 50 SHOPS IN THIS MALL...SO...
18/50 = 36/100 = 0.36!
VERY GOOD! THE RELATIVE FREQUENCY OF RAMEN SHOPS IN CLASS 700–800 YEN—IN OTHER WORDS, SHOPS WITH CLASS MIDPOINT 750 YEN—IS 0.36. YOU GET THE PERCENTAGE BY MULTIPLYING BY 100, SO THIS IS EQUAL TO 36%.
OH, NO! THIS IS GETTING MATHEMATICAL...

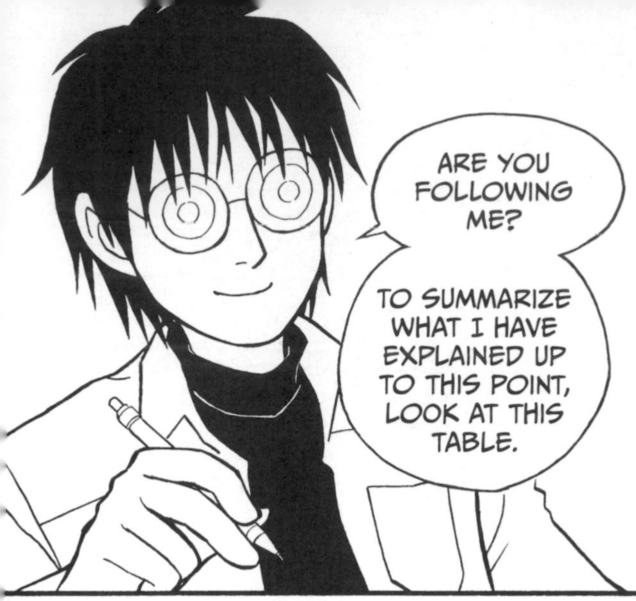
ARE YOU FOLLOWING ME?
TO SUMMARIZE WHAT I HAVE EXPLAINED UP TO THIS POINT, LOOK AT THIS TABLE.

50 BEST RAMEN SHOPS FREQUENCY TABLE

CLASS (EQUAL OR GREATER/ LESS THAN)	CLASS MIDPOINT	FRE-QUENCY	RELATIVE FREQUENCY
500–600	550	4	0.08
600–700	650	13	0.26
700–800	750	18	0.36
800–900	850	12	0.24
900–1000	950	3	0.06
SUM		50	1.00

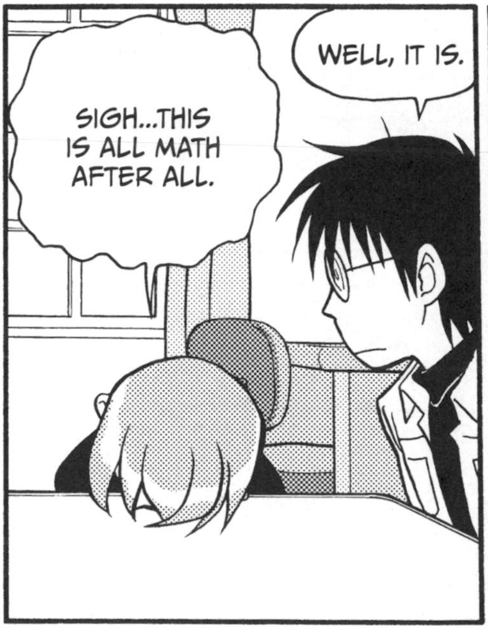
SIGH...THIS IS ALL MATH AFTER ALL.
WELL, IT IS.

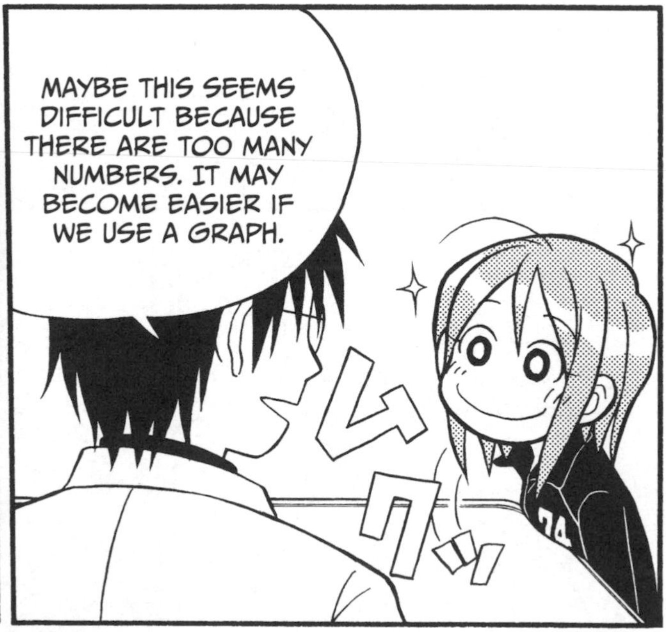
MAYBE THIS SEEMS DIFFICULT BECAUSE THERE ARE TOO MANY NUMBERS. IT MAY BECOME EASIER IF WE USE A GRAPH.

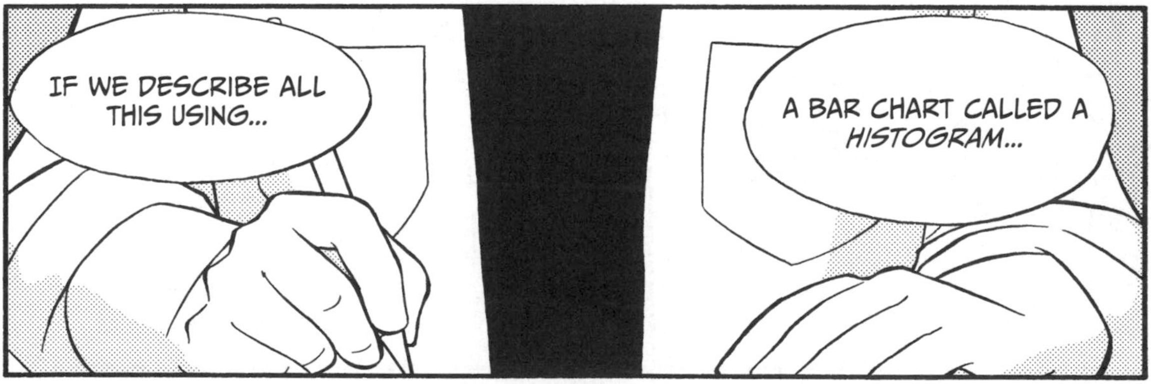
IF WE DESCRIBE ALL THIS USING...
A BAR CHART CALLED A *HISTOGRAM*...

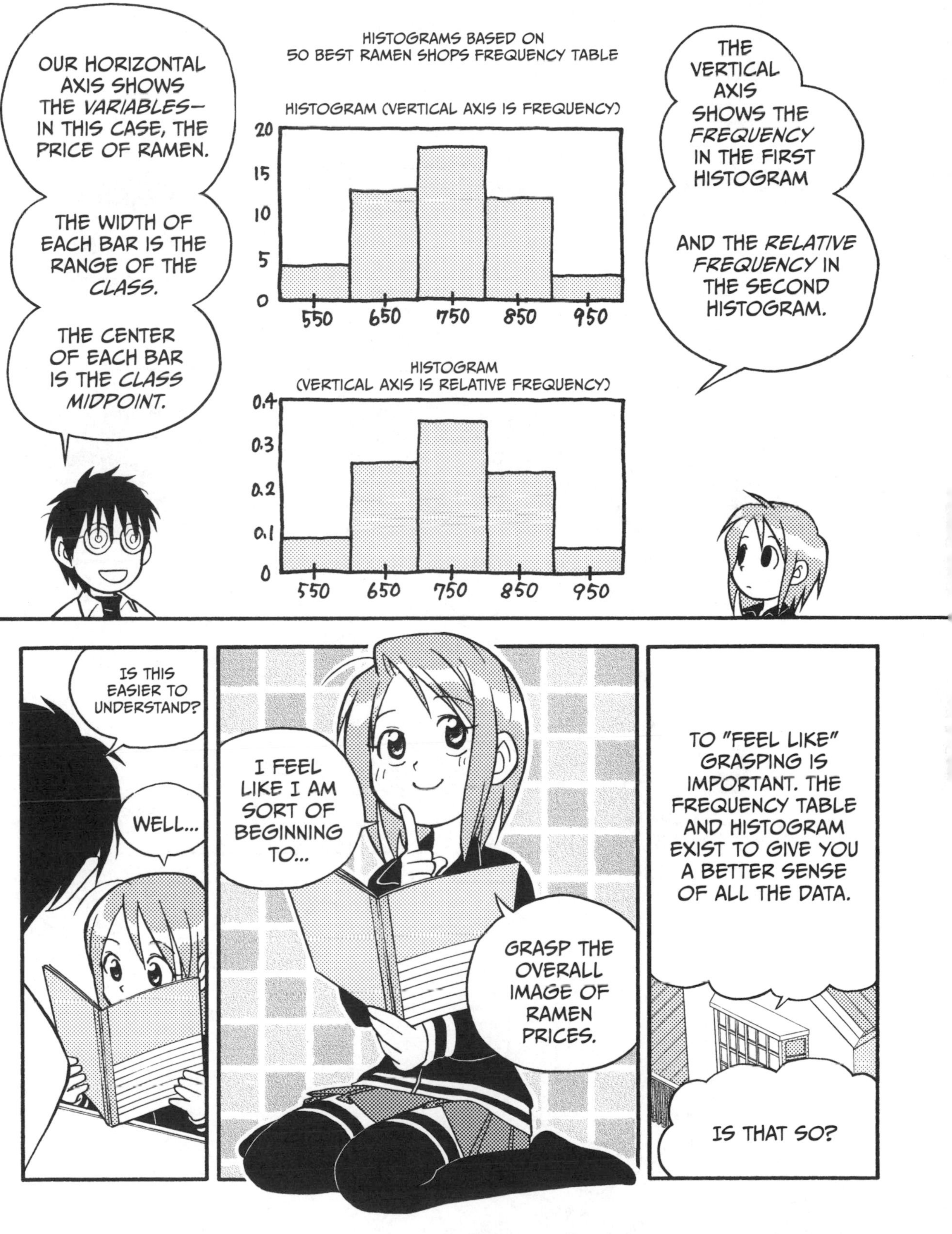

OUR HORIZONTAL AXIS SHOWS THE *VARIABLES*— IN THIS CASE, THE PRICE OF RAMEN.
THE WIDTH OF EACH BAR IS THE RANGE OF THE *CLASS.*
THE CENTER OF EACH BAR IS THE *CLASS MIDPOINT.*
HISTOGRAMS BASED ON 50 BEST RAMEN SHOPS FREQUENCY TABLE
HISTOGRAM (VERTICAL AXIS IS FREQUENCY)
20
15
10
5
0
550 650 750 850 950
HISTOGRAM (VERTICAL AXIS IS RELATIVE FREQUENCY)
0.4
0.3
0.2
0.1
0
550 650 750 850 950
THE VERTICAL AXIS SHOWS THE *FREQUENCY* IN THE FIRST HISTOGRAM
AND THE *RELATIVE FREQUENCY* IN THE SECOND HISTOGRAM.
IS THIS EASIER TO UNDERSTAND?
WELL...
I FEEL LIKE I AM SORT OF BEGINNING TO...
GRASP THE OVERALL IMAGE OF RAMEN PRICES.
TO "FEEL LIKE" GRASPING IS IMPORTANT. THE FREQUENCY TABLE AND HISTOGRAM EXIST TO GIVE YOU A BETTER SENSE OF ALL THE DATA.
IS THAT SO?

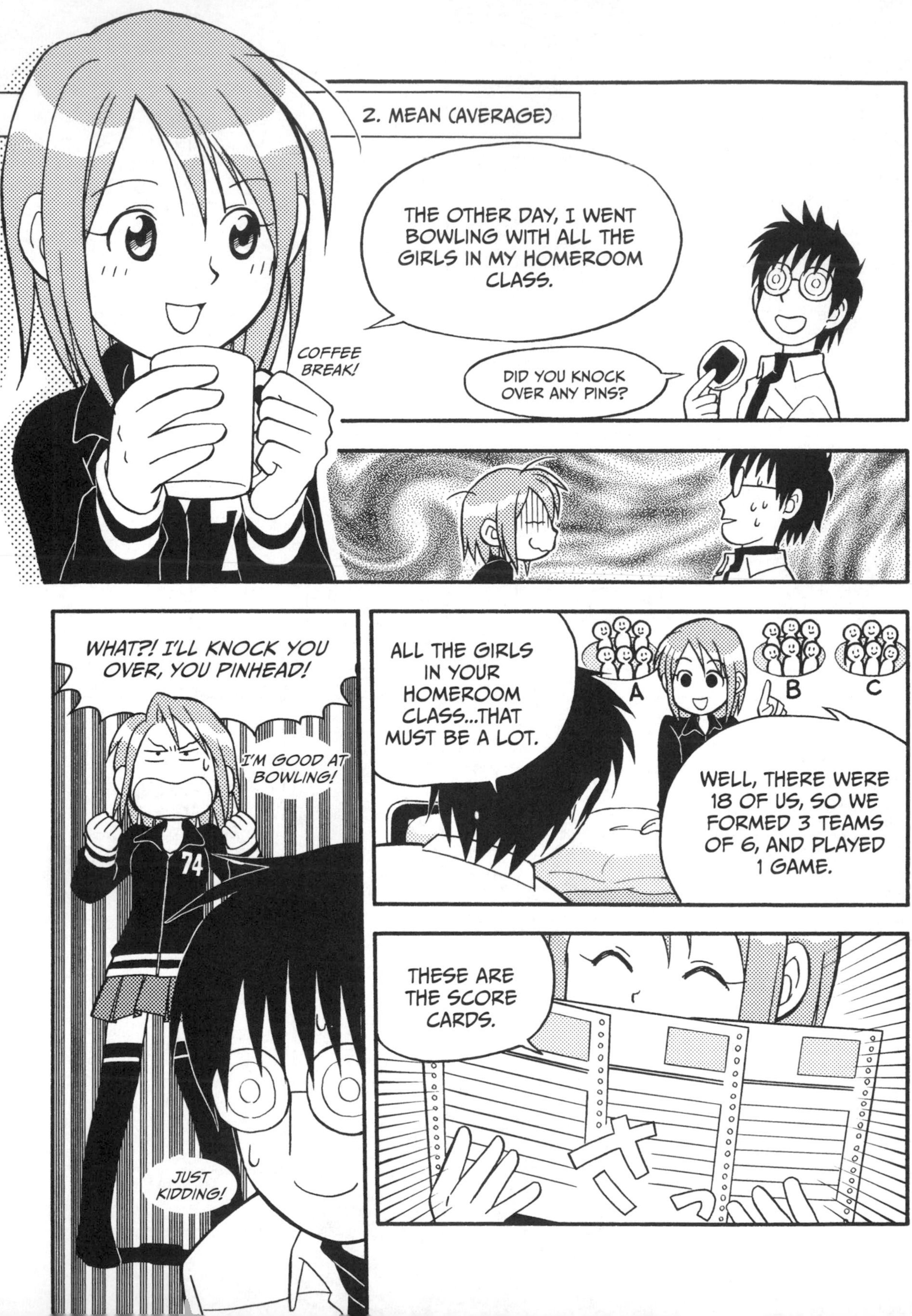
COFFEE BREAK!
2. MEAN (AVERAGE)
THE OTHER DAY, I WENT BOWLING WITH ALL THE GIRLS IN MY HOMEROOM CLASS.
DID YOU KNOCK OVER ANY PINS?
WHAT?! I'LL KNOCK YOU OVER, YOU PINHEAD!
I'M GOOD AT BOWLING!
JUST KIDDING!
ALL THE GIRLS IN YOUR HOMEROOM CLASS...THAT MUST BE A LOT.
A
B
C
WELL, THERE WERE 18 OF US, SO WE FORMED 3 TEAMS OF 6, AND PLAYED 1 GAME.
THESE ARE THE SCORE CARDS.

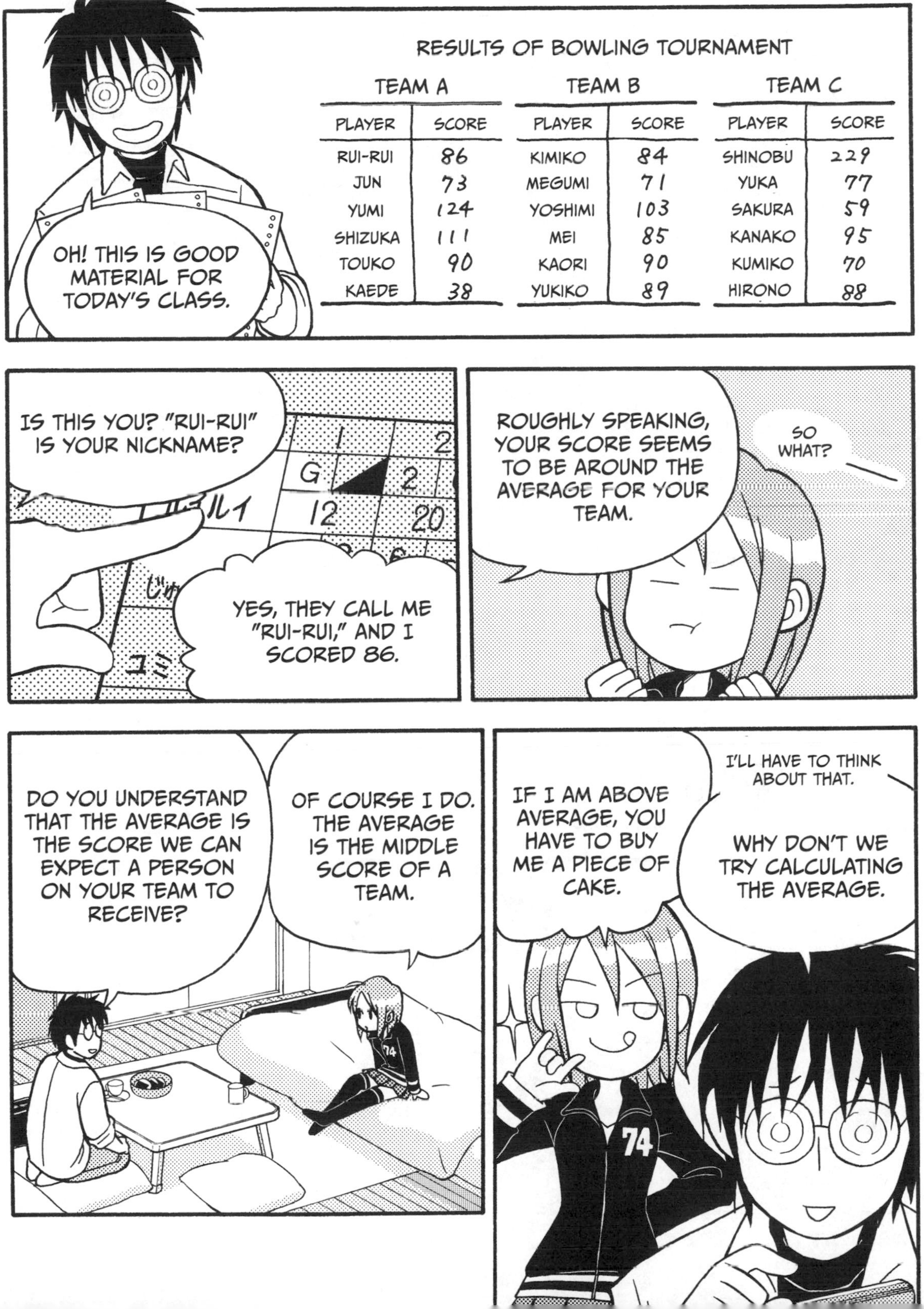
OH! THIS IS GOOD MATERIAL FOR TODAY'S CLASS.
IS THIS YOU? "RUI-RUI" IS YOUR NICKNAME?
YES, THEY CALL ME "RUI-RUI," AND I SCORED 86.
ROUGHLY SPEAKING, YOUR SCORE SEEMS TO BE AROUND THE AVERAGE FOR YOUR TEAM.
SO WHAT?
DO YOU UNDERSTAND THAT THE AVERAGE IS THE SCORE WE CAN EXPECT A PERSON ON YOUR TEAM TO RECEIVE?
OF COURSE I DO. THE AVERAGE IS THE MIDDLE SCORE OF A TEAM.
IF I AM ABOVE AVERAGE, YOU HAVE TO BUY ME A PIECE OF CAKE.
I'LL HAVE TO THINK ABOUT THAT.
WHY DON'T WE TRY CALCULATING THE AVERAGE.

RESULTS OF BOWLING TOURNAMENT

TEAM A		TEAM B		TEAM C	
PLAYER	SCORE	PLAYER	SCORE	PLAYER	SCORE
RUI-RUI	86	KIMIKO	84	SHINOBU	229
JUN	73	MEGUMI	71	YUKA	77
YUMI	124	YOSHIMI	103	SAKURA	59
SHIZUKA	111	MEI	85	KANAKO	95
TOUKO	90	KAORI	90	KUMIKO	70
KAEDE	38	YUKIKO	89	HIRONO	88

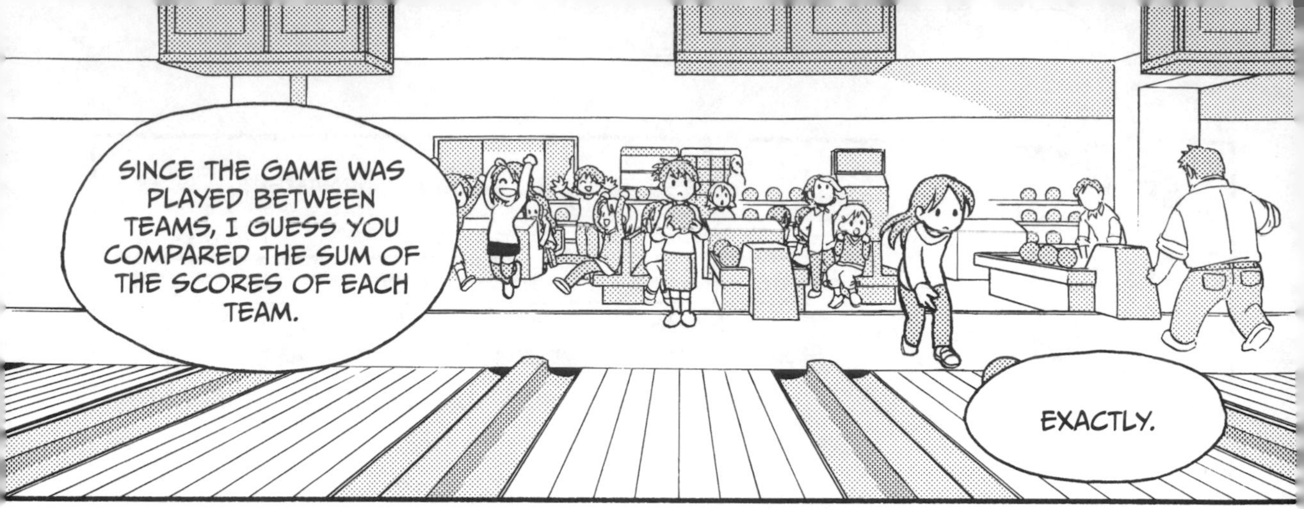
SINCE THE GAME WAS PLAYED BETWEEN TEAMS, I GUESS YOU COMPARED THE SUM OF THE SCORES OF EACH TEAM.
EXACTLY.

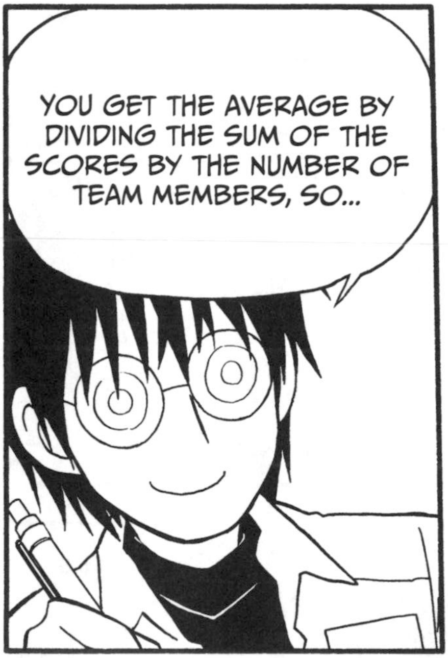
YOU GET THE AVERAGE BY DIVIDING THE SUM OF THE SCORES BY THE NUMBER OF TEAM MEMBERS, SO...

TEAM A

$$\frac{86+73+124+111+90+38}{6}=\frac{522}{6}=87$$

TEAM B

$$\frac{84+71+103+85+90+89}{6}=\frac{522}{6}=87$$

TEAM C

$$\frac{229+77+59+95+70+88}{6}=\frac{618}{6}=103$$

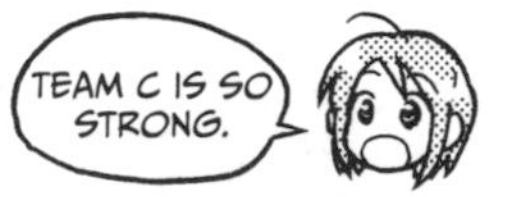
TEAM C IS SO STRONG.

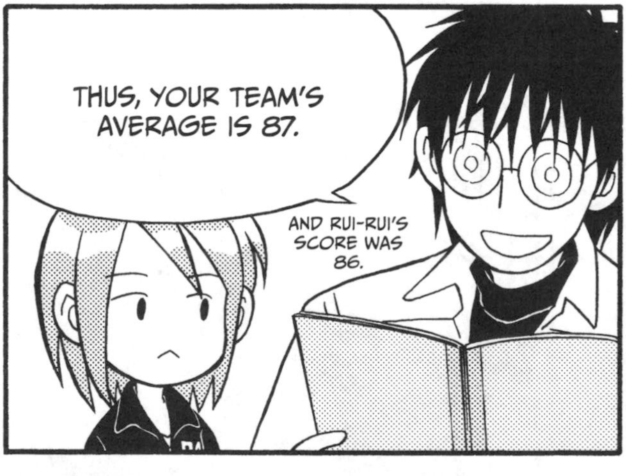
THUS, YOUR TEAM'S AVERAGE IS 87.
AND RUI-RUI'S SCORE WAS 86.

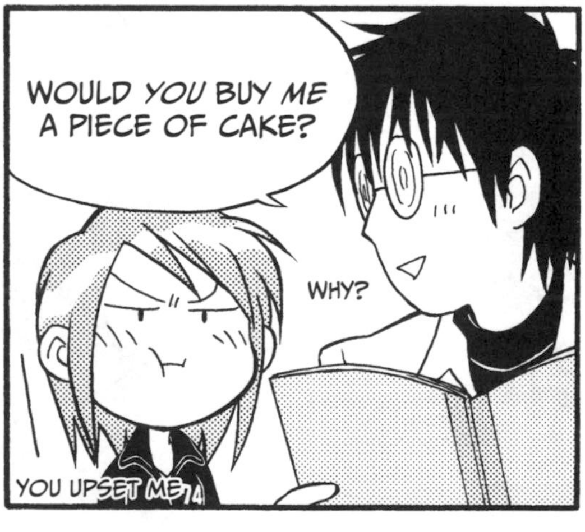
WOULD *YOU* BUY *ME* A PIECE OF CAKE?
WHY?
YOU UPSET ME.

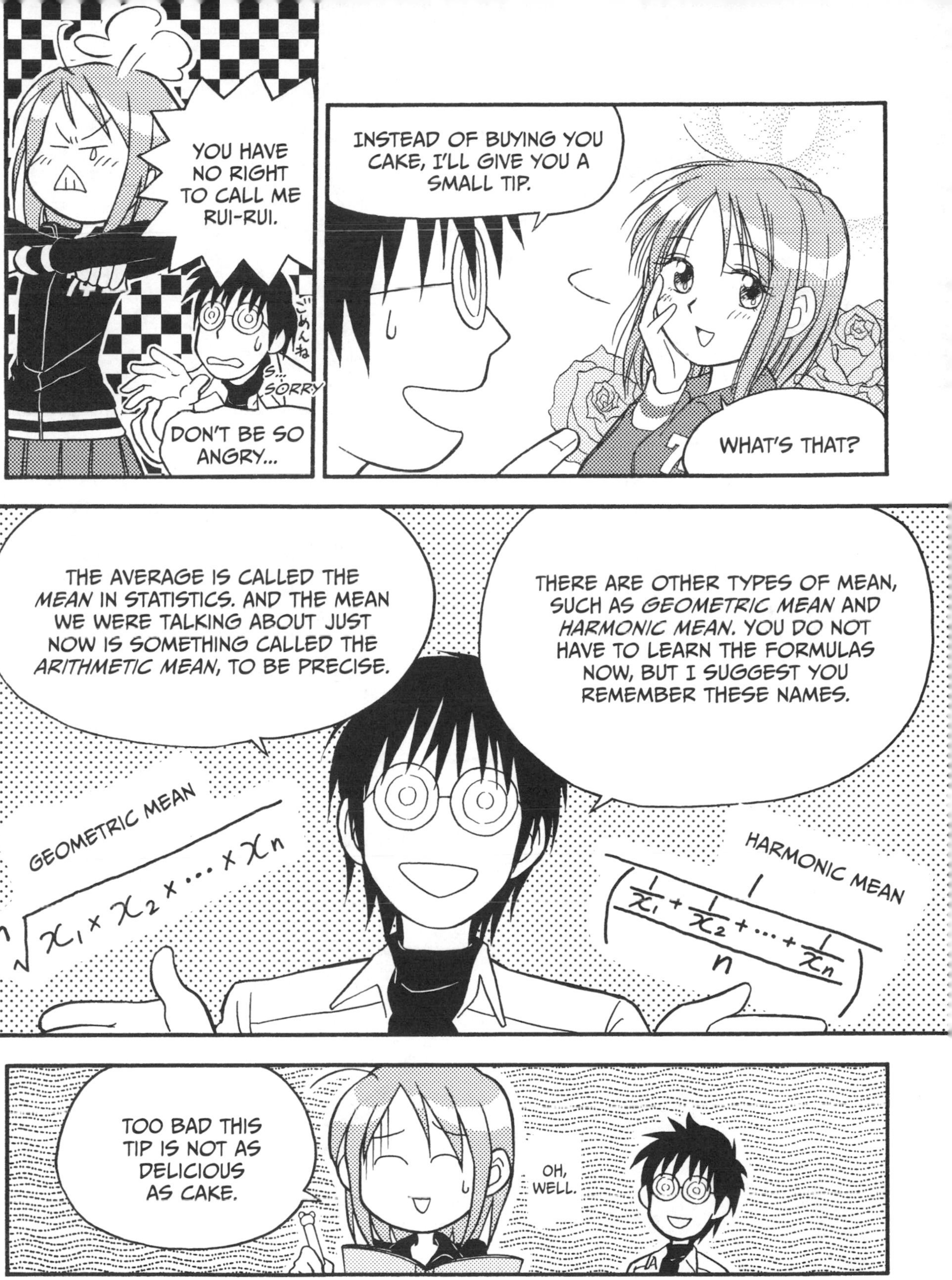
YOU HAVE NO RIGHT TO CALL ME RUI-RUI.
ごめんね
S... SORRY
DON'T BE SO ANGRY...
INSTEAD OF BUYING YOU CAKE, I'LL GIVE YOU A SMALL TIP.
WHAT'S THAT?
THE AVERAGE IS CALLED THE *MEAN* IN STATISTICS. AND THE MEAN WE WERE TALKING ABOUT JUST NOW IS SOMETHING CALLED THE *ARITHMETIC MEAN*, TO BE PRECISE.
THERE ARE OTHER TYPES OF MEAN, SUCH AS *GEOMETRIC MEAN* AND *HARMONIC MEAN*. YOU DO NOT HAVE TO LEARN THE FORMULAS NOW, BUT I SUGGEST YOU REMEMBER THESE NAMES.
GEOMETRIC MEAN
$\sqrt[n]{x_1 \times x_2 \times \cdots \times x_n}$
HARMONIC MEAN
$\dfrac{1}{\left(\dfrac{\frac{1}{x_1} + \frac{1}{x_2} + \cdots + \frac{1}{x_n}}{n}\right)}$
TOO BAD THIS TIP IS NOT AS DELICIOUS AS CAKE.
OH, WELL.

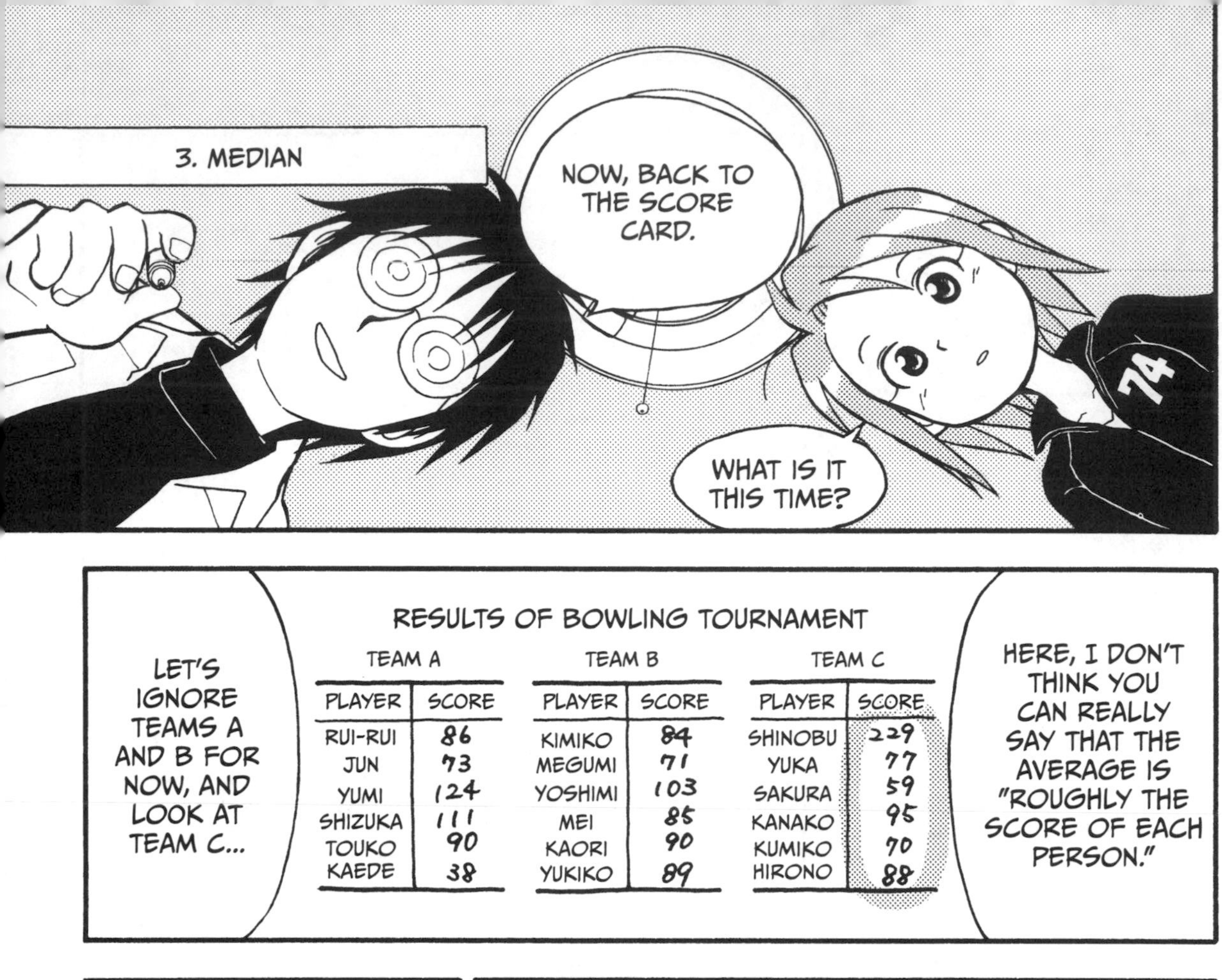
3. MEDIAN
NOW, BACK TO THE SCORE CARD.
WHAT IS IT THIS TIME?

LET'S IGNORE TEAMS A AND B FOR NOW, AND LOOK AT TEAM C...

RESULTS OF BOWLING TOURNAMENT

TEAM A		TEAM B		TEAM C	
PLAYER	SCORE	PLAYER	SCORE	PLAYER	SCORE
RUI-RUI	86	KIMIKO	84	SHINOBU	229
JUN	73	MEGUMI	71	YUKA	77
YUMI	124	YOSHIMI	103	SAKURA	59
SHIZUKA	111	MEI	85	KANAKO	95
TOUKO	90	KAORI	90	KUMIKO	70
KAEDE	38	YUKIKO	89	HIRONO	88

HERE, I DON'T THINK YOU CAN REALLY SAY THAT THE AVERAGE IS "ROUGHLY THE SCORE OF EACH PERSON."

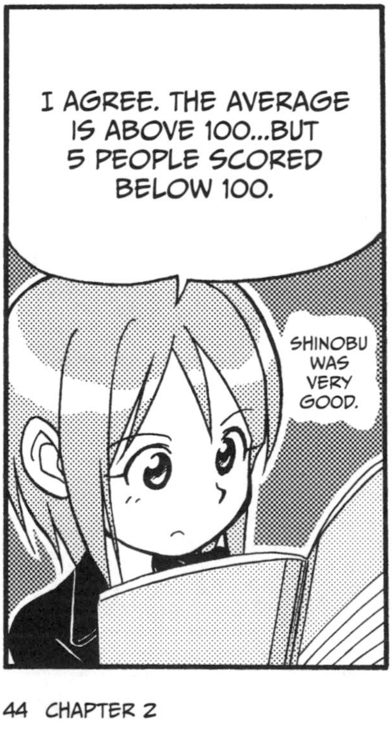
I AGREE. THE AVERAGE IS ABOVE 100...BUT 5 PEOPLE SCORED BELOW 100.
SHINOBU WAS VERY GOOD.

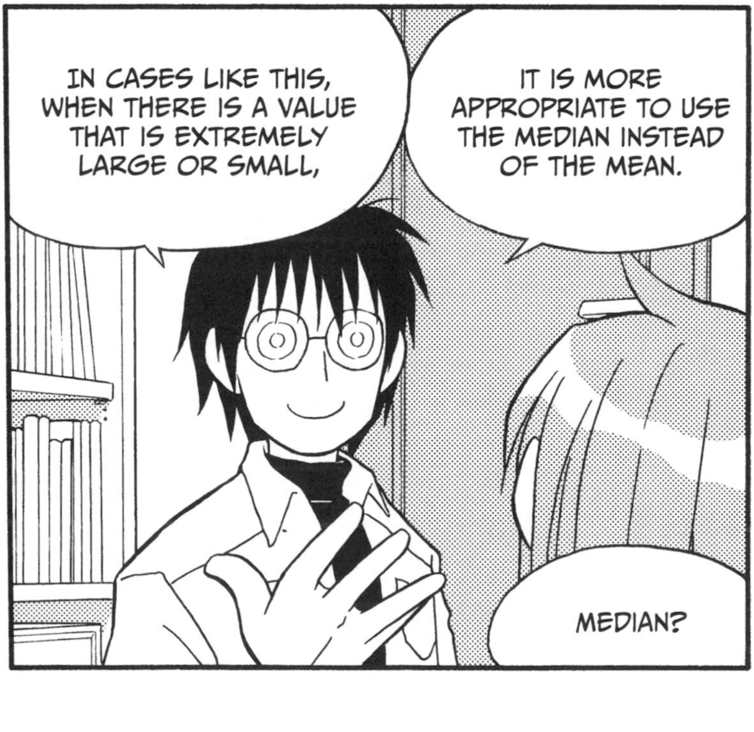
IN CASES LIKE THIS, WHEN THERE IS A VALUE THAT IS EXTREMELY LARGE OR SMALL,
IT IS MORE APPROPRIATE TO USE THE MEDIAN INSTEAD OF THE MEAN.
MEDIAN?

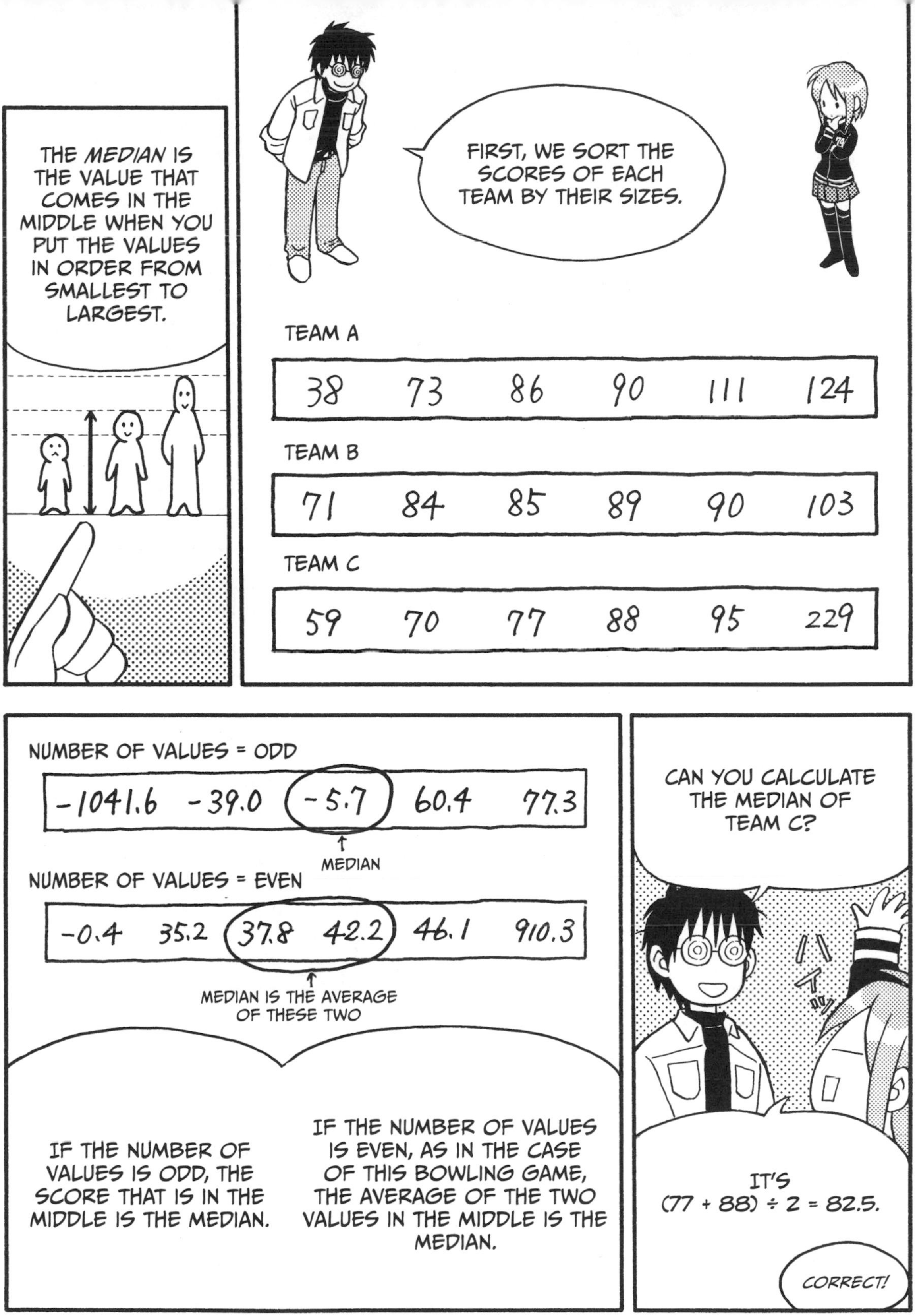

THE *MEDIAN* IS THE VALUE THAT COMES IN THE MIDDLE WHEN YOU PUT THE VALUES IN ORDER FROM SMALLEST TO LARGEST.
FIRST, WE SORT THE SCORES OF EACH TEAM BY THEIR SIZES.
TEAM A
38 73 86 90 111 124
TEAM B
71 84 85 89 90 103
TEAM C
59 70 77 88 95 229
NUMBER OF VALUES = ODD
-1041.6 -39.0 -5.7 60.4 77.3
MEDIAN
NUMBER OF VALUES = EVEN
-0.4 35.2 37.8 42.2 46.1 910.3
MEDIAN IS THE AVERAGE OF THESE TWO
IF THE NUMBER OF VALUES IS ODD, THE SCORE THAT IS IN THE MIDDLE IS THE MEDIAN.
IF THE NUMBER OF VALUES IS EVEN, AS IN THE CASE OF THIS BOWLING GAME, THE AVERAGE OF THE TWO VALUES IN THE MIDDLE IS THE MEDIAN.
CAN YOU CALCULATE THE MEDIAN OF TEAM C?
IT'S (77 + 88) ÷ 2 = 82.5.
CORRECT!

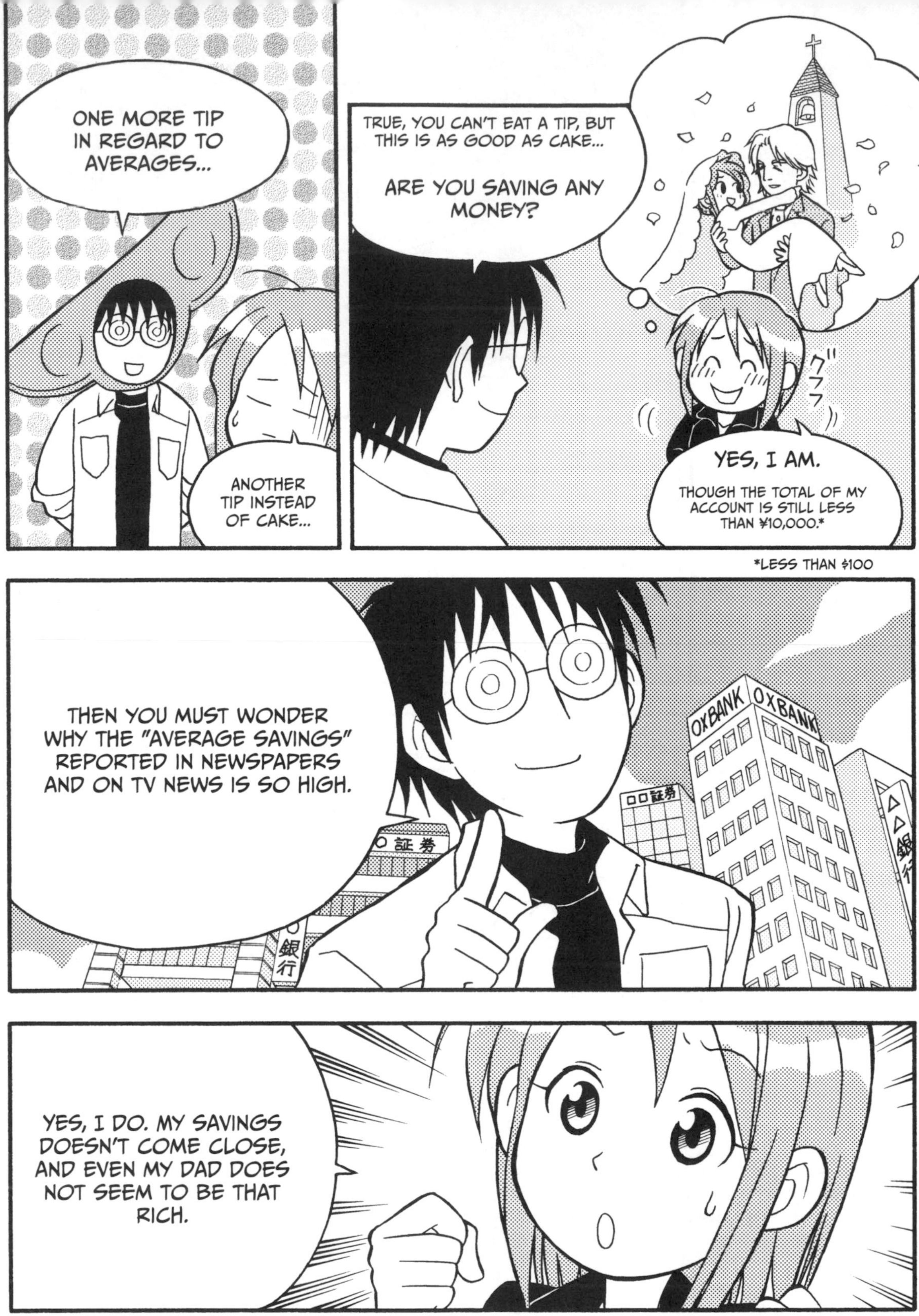
ONE MORE TIP IN REGARD TO AVERAGES...
ANOTHER TIP INSTEAD OF CAKE...
TRUE, YOU CAN'T EAT A TIP, BUT THIS IS AS GOOD AS CAKE...
ARE YOU SAVING ANY MONEY?
YES, I AM.
THOUGH THE TOTAL OF MY ACCOUNT IS STILL LESS THAN ¥10,000.*
*LESS THAN $100
THEN YOU MUST WONDER WHY THE "AVERAGE SAVINGS" REPORTED IN NEWSPAPERS AND ON TV NEWS IS SO HIGH.
OXBANK
OXBANK
YES, I DO. MY SAVINGS DOESN'T COME CLOSE, AND EVEN MY DAD DOES NOT SEEM TO BE THAT RICH.

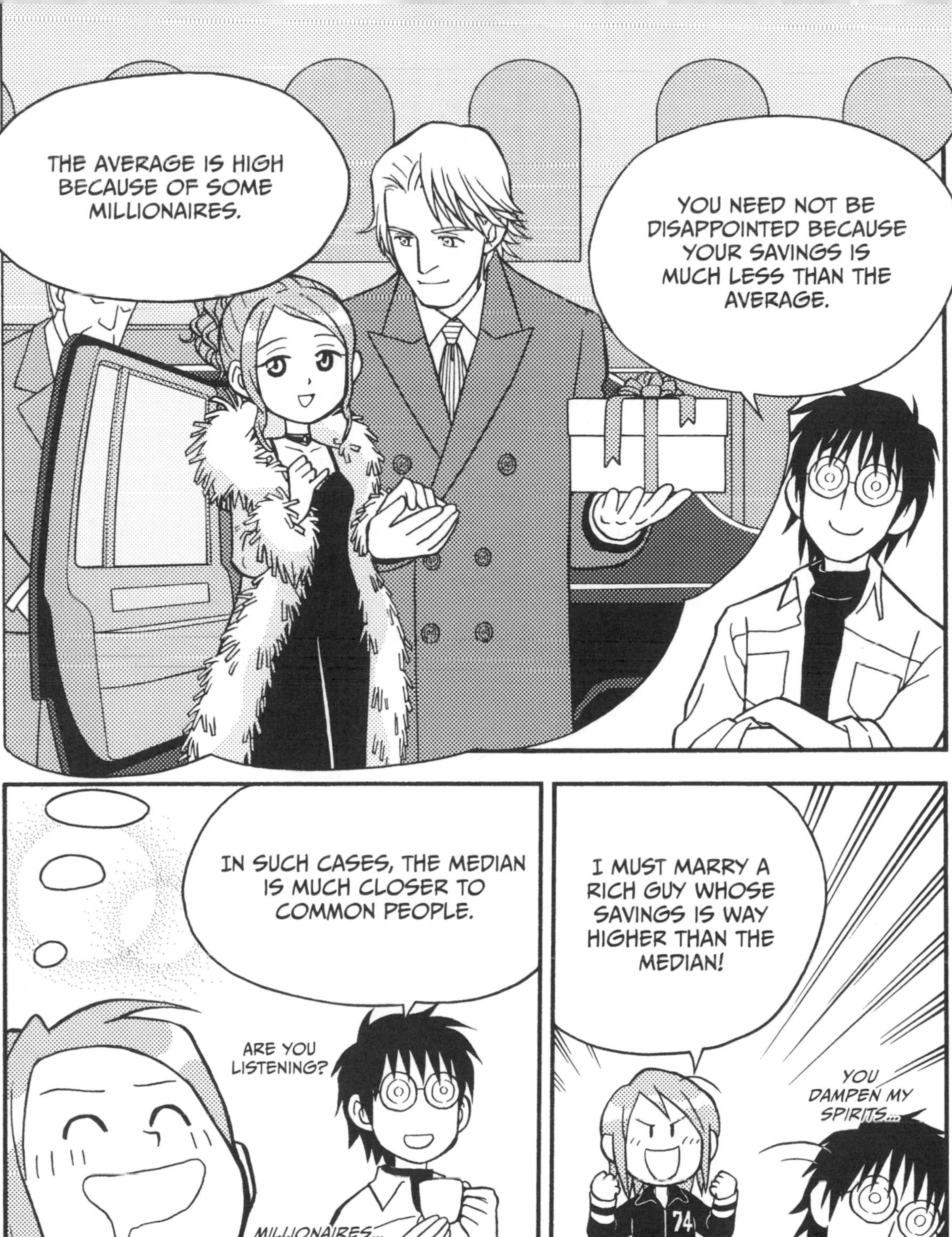

THE AVERAGE IS HIGH BECAUSE OF SOME MILLIONAIRES.
YOU NEED NOT BE DISAPPOINTED BECAUSE YOUR SAVINGS IS MUCH LESS THAN THE AVERAGE.
IN SUCH CASES, THE MEDIAN IS MUCH CLOSER TO COMMON PEOPLE.
ARE YOU LISTENING?
MILLIONAIRES...
I MUST MARRY A RICH GUY WHOSE SAVINGS IS WAY HIGHER THAN THE MEDIAN!
YOU DAMPEN MY SPIRITS...

4. STANDARD DEVIATION

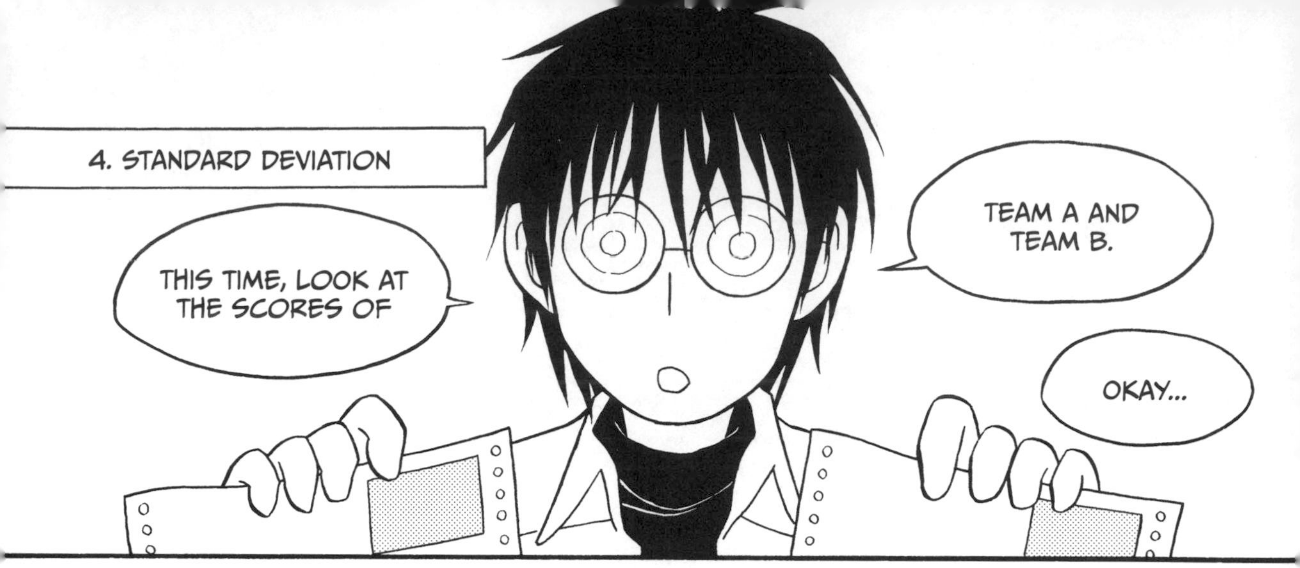
THIS TIME, LOOK AT THE SCORES OF
TEAM A AND TEAM B.
OKAY...

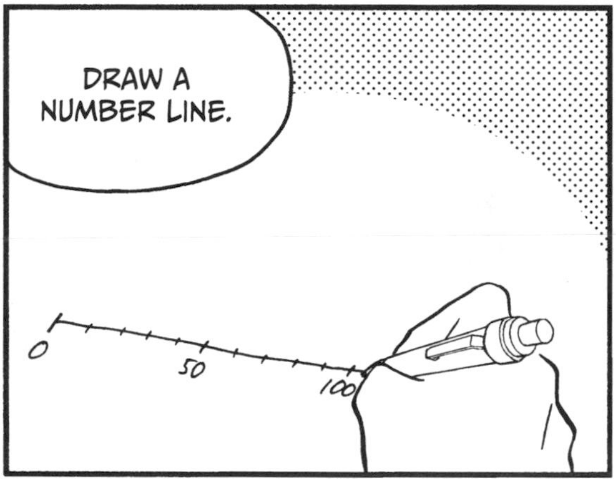
DRAW A NUMBER LINE.
0
50
100

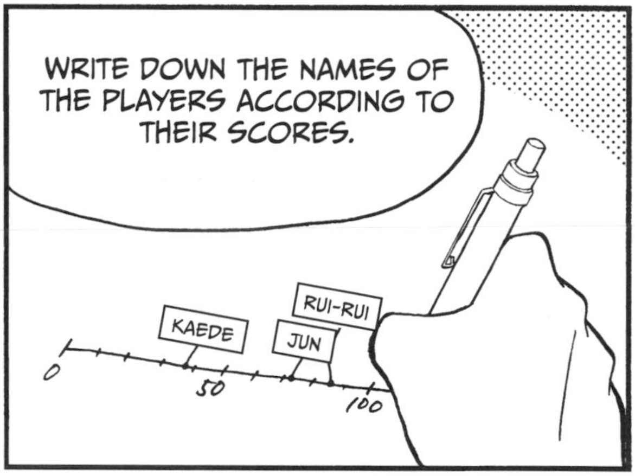
WRITE DOWN THE NAMES OF THE PLAYERS ACCORDING TO THEIR SCORES.
KAEDE
JUN
RUI-RUI
0
50
100

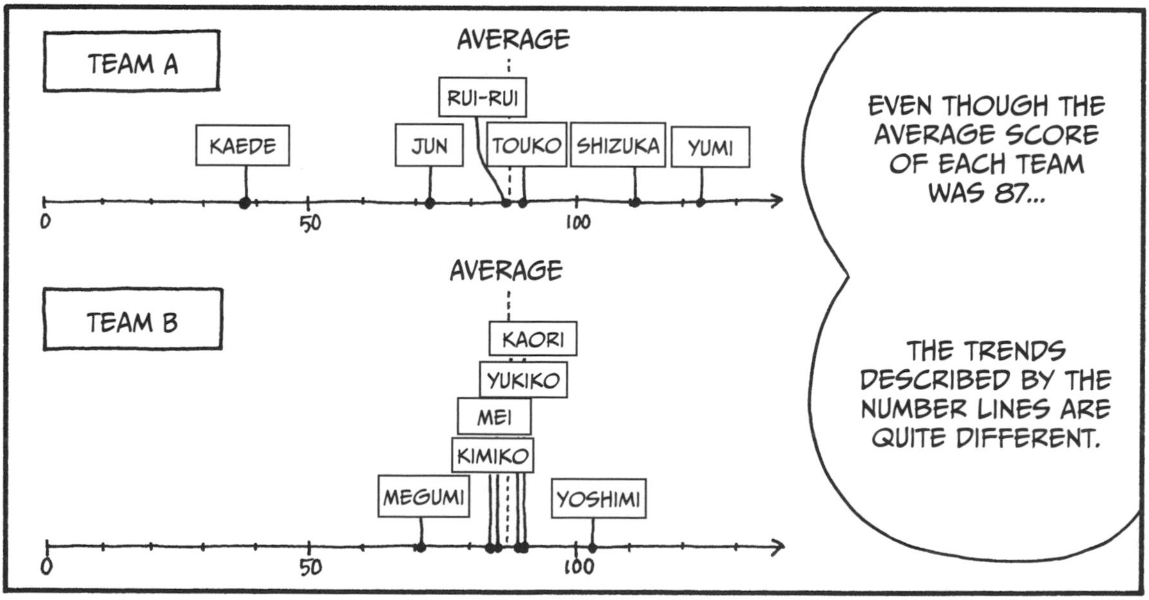
TEAM A
AVERAGE
RUI-RUI
KAEDE
JUN
TOUKO
SHIZUKA
YUMI
0
50
100
TEAM B
AVERAGE
KAORI
YUKIKO
MEI
KIMIKO
MEGUMI
YOSHIMI
0
50
100
EVEN THOUGH THE AVERAGE SCORE OF EACH TEAM WAS 87...
THE TRENDS DESCRIBED BY THE NUMBER LINES ARE QUITE DIFFERENT.

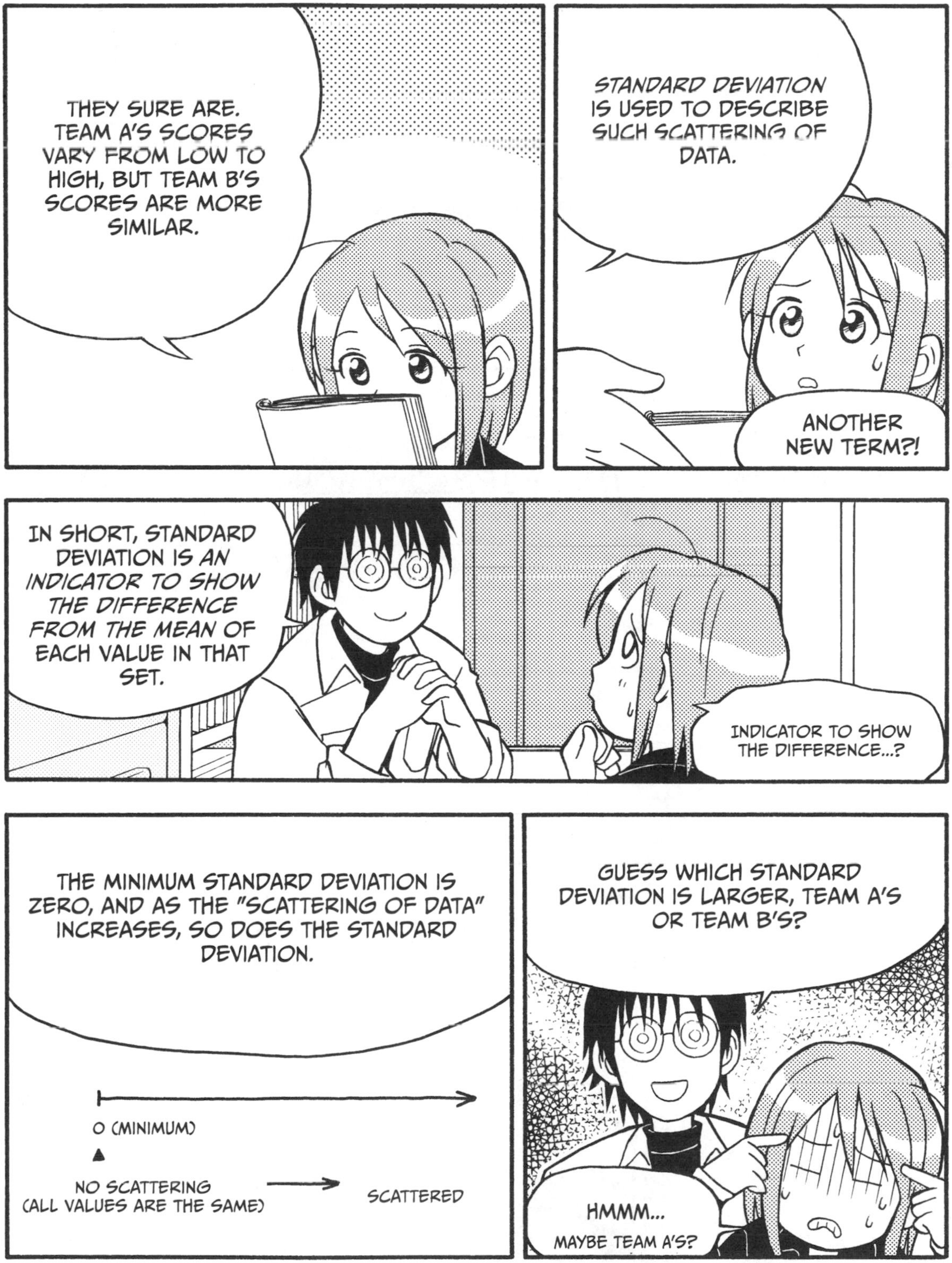
THEY SURE ARE. TEAM A'S SCORES VARY FROM LOW TO HIGH, BUT TEAM B'S SCORES ARE MORE SIMILAR.
STANDARD DEVIATION IS USED TO DESCRIBE SUCH SCATTERING OF DATA.
ANOTHER NEW TERM?!
IN SHORT, STANDARD DEVIATION IS AN INDICATOR TO SHOW THE DIFFERENCE FROM THE MEAN OF EACH VALUE IN THAT SET.
INDICATOR TO SHOW THE DIFFERENCE...?
THE MINIMUM STANDARD DEVIATION IS ZERO, AND AS THE "SCATTERING OF DATA" INCREASES, SO DOES THE STANDARD DEVIATION.
0 (MINIMUM)
NO SCATTERING (ALL VALUES ARE THE SAME)
SCATTERED
GUESS WHICH STANDARD DEVIATION IS LARGER, TEAM A'S OR TEAM B'S?
HMMM...
MAYBE TEAM A'S?

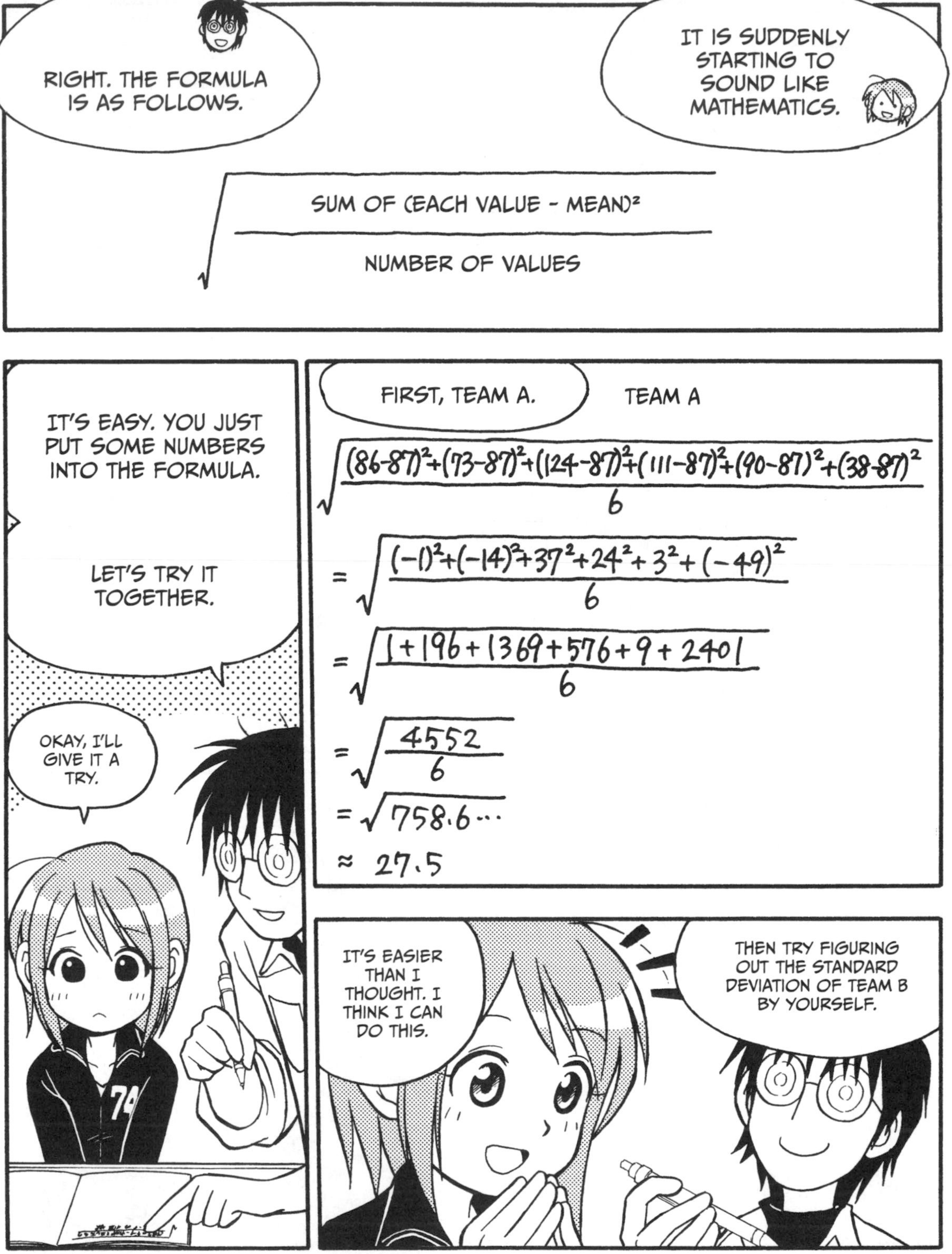
RIGHT. THE FORMULA IS AS FOLLOWS.
IT IS SUDDENLY STARTING TO SOUND LIKE MATHEMATICS.
√(SUM OF (EACH VALUE - MEAN)² / NUMBER OF VALUES)
IT'S EASY. YOU JUST PUT SOME NUMBERS INTO THE FORMULA.
LET'S TRY IT TOGETHER.
OKAY, I'LL GIVE IT A TRY.
FIRST, TEAM A.
TEAM A
√(((86-87)²+(73-87)²+(124-87)²+(111-87)²+(90-87)²+(38-87)²) / 6)
= √(((-1)²+(-14)²+37²+24²+3²+(-49)²) / 6)
= √((1+196+1369+576+9+2401) / 6)
= √(4552 / 6)
= √758.6…
≈ 27.5
IT'S EASIER THAN I THOUGHT. I THINK I CAN DO THIS.
THEN TRY FIGURING OUT THE STANDARD DEVIATION OF TEAM B BY YOURSELF.

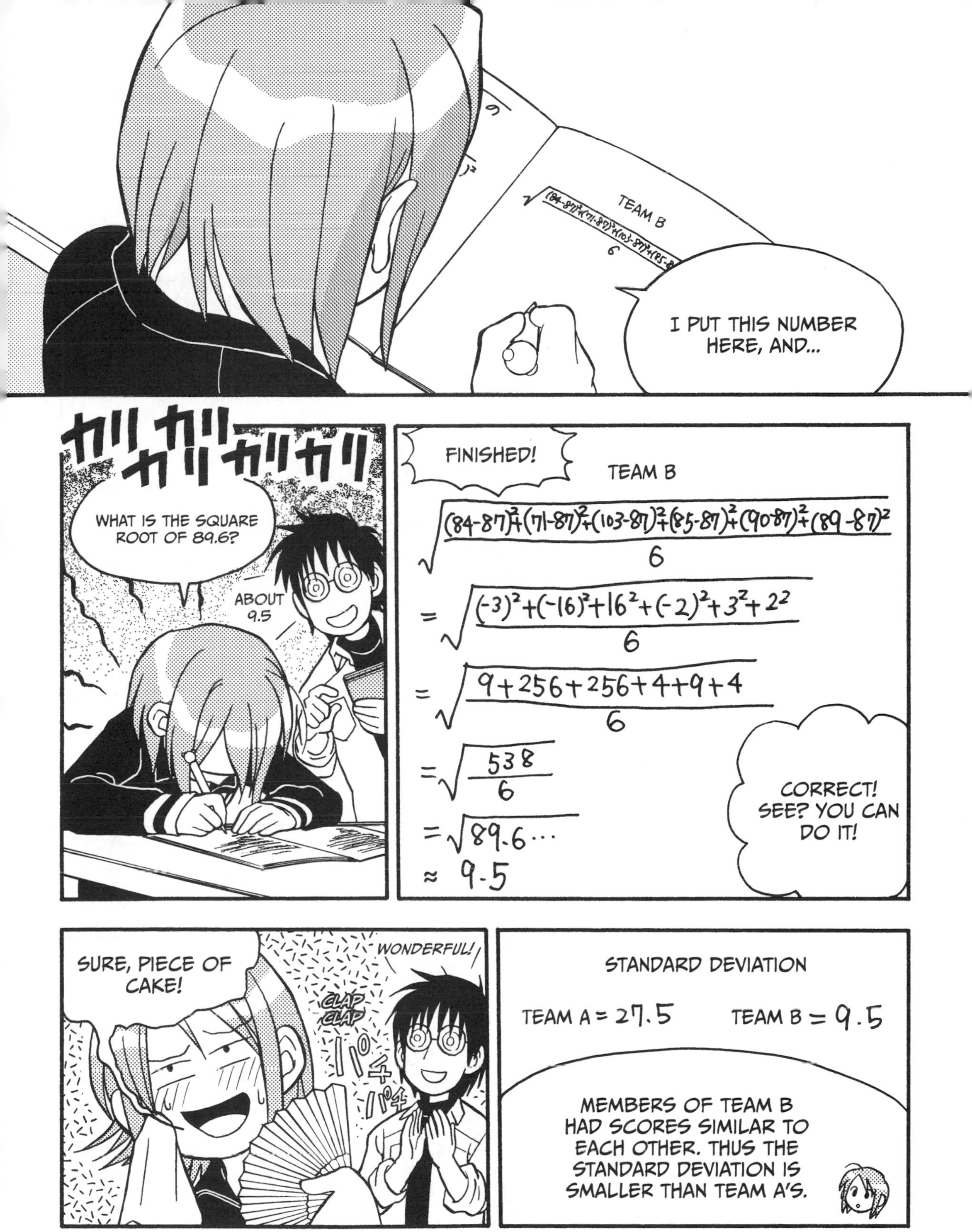
TEAM B
I PUT THIS NUMBER HERE, AND...
カリカリ カリカリカリ
WHAT IS THE SQUARE ROOT OF 89.6?
ABOUT 9.5
FINISHED!
TEAM B
√[((84-87)² + (71-87)² + (103-87)² + (85-87)² + (90-87)² + (89-87)²) / 6]
= √[((-3)² + (-16)² + 16² + (-2)² + 3² + 2²) / 6]
= √[(9 + 256 + 256 + 4 + 9 + 4) / 6]
= √(538 / 6)
= √89.6...
≈ 9.5
CORRECT! SEE? YOU CAN DO IT!
SURE, PIECE OF CAKE!
WONDERFUL!
CLAP CLAP
パチパチ
STANDARD DEVIATION
TEAM A = 27.5
TEAM B = 9.5
MEMBERS OF TEAM B HAD SCORES SIMILAR TO EACH OTHER. THUS THE STANDARD DEVIATION IS SMALLER THAN TEAM A'S.

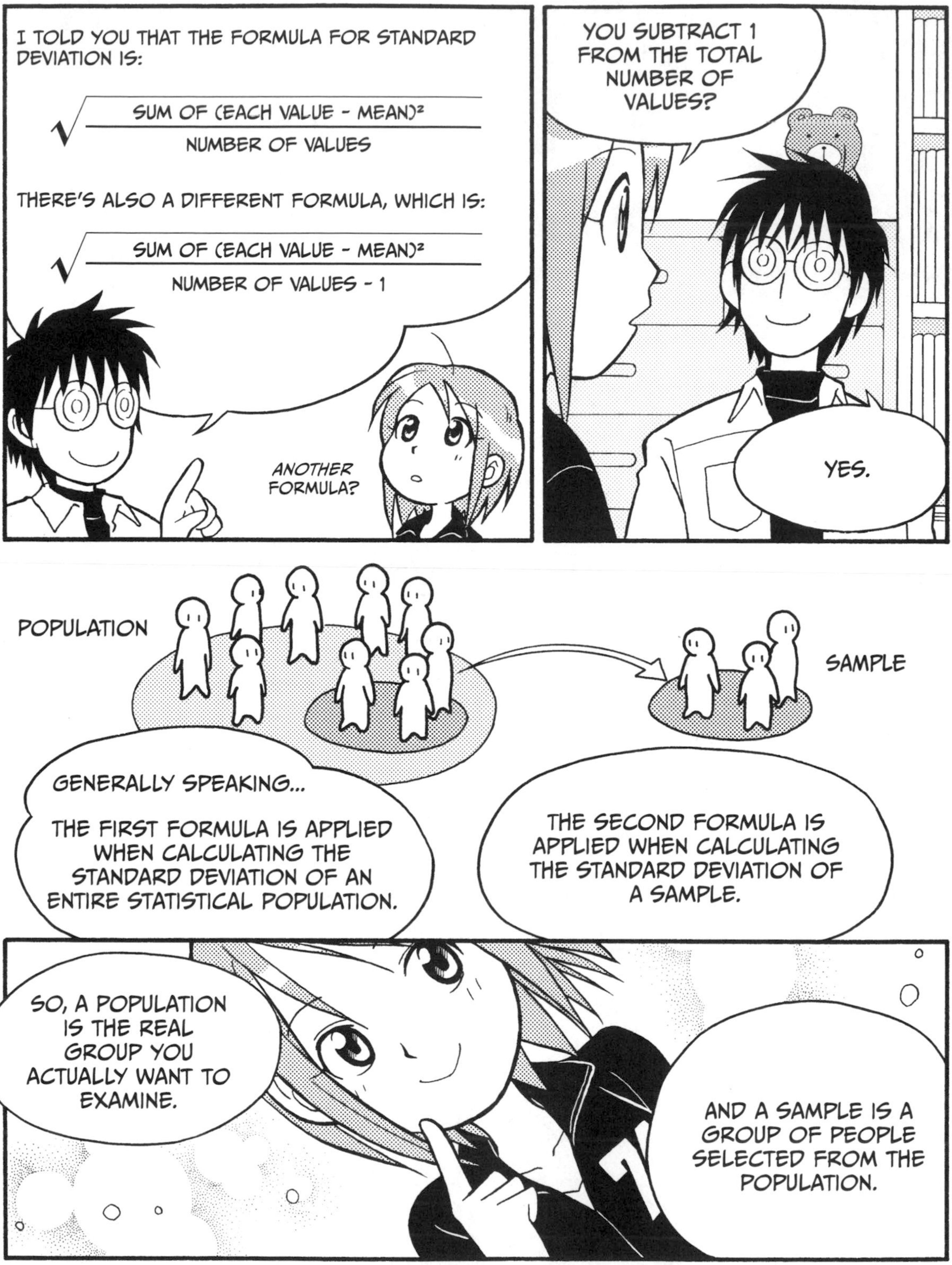
I TOLD YOU THAT THE FORMULA FOR STANDARD DEVIATION IS:
$\sqrt{\frac{\text{SUM OF (EACH VALUE - MEAN)}^2}{\text{NUMBER OF VALUES}}}$
THERE'S ALSO A DIFFERENT FORMULA, WHICH IS:
$\sqrt{\frac{\text{SUM OF (EACH VALUE - MEAN)}^2}{\text{NUMBER OF VALUES - 1}}}$
ANOTHER FORMULA?
YOU SUBTRACT 1 FROM THE TOTAL NUMBER OF VALUES?
YES.
POPULATION
SAMPLE
GENERALLY SPEAKING...
THE FIRST FORMULA IS APPLIED WHEN CALCULATING THE STANDARD DEVIATION OF AN ENTIRE STATISTICAL POPULATION.
THE SECOND FORMULA IS APPLIED WHEN CALCULATING THE STANDARD DEVIATION OF A SAMPLE.
SO, A POPULATION IS THE REAL GROUP YOU ACTUALLY WANT TO EXAMINE.
AND A SAMPLE IS A GROUP OF PEOPLE SELECTED FROM THE POPULATION.

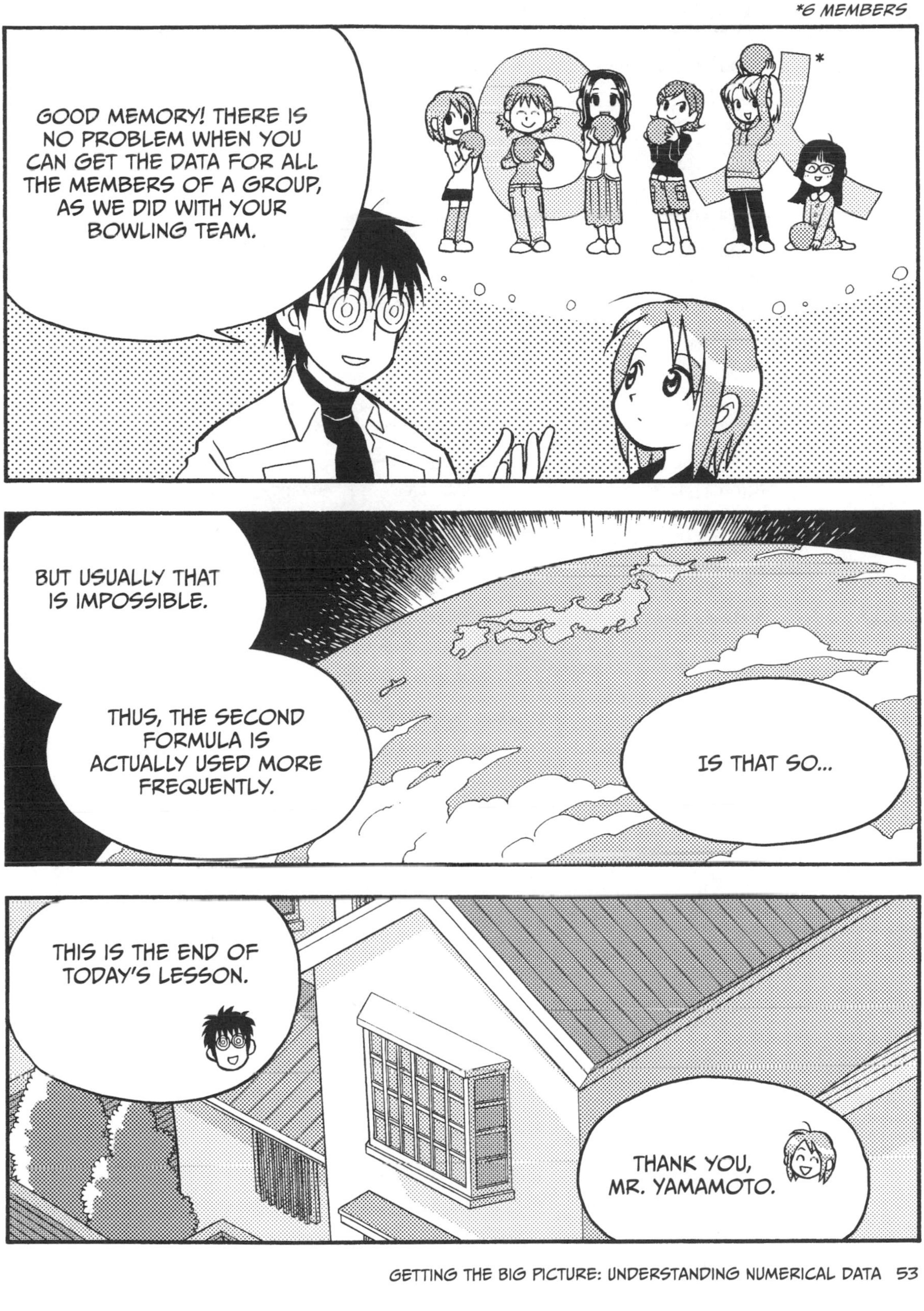
*6 MEMBERS
GOOD MEMORY! THERE IS NO PROBLEM WHEN YOU CAN GET THE DATA FOR ALL THE MEMBERS OF A GROUP, AS WE DID WITH YOUR BOWLING TEAM.
BUT USUALLY THAT IS IMPOSSIBLE.
THUS, THE SECOND FORMULA IS ACTUALLY USED MORE FREQUENTLY.
IS THAT SO...
THIS IS THE END OF TODAY'S LESSON.
THANK YOU, MR. YAMAMOTO.

5. THE RANGE OF CLASS OF A FREQUENCY TABLE

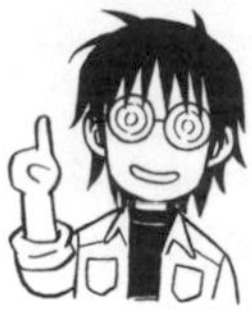

If you felt that something was unclear in "Frequency Distribution Tables and Histograms" on page 32, take another look here at the table introduced on page 38.

TABLE 2-1: 50 BEST RAMEN SHOPS FREQUENCY TABLE

Class (equal or greater/less than)	Class midpoint	Frequency	Relative frequency
500–600	550	4	0.08
600–700	650	13	0.26
700–800	750	18	0.36
800–900	850	12	0.24
900–1000	950	3	0.06
Sum		50	1.00

As you can see, the range of class in this table is 100. The range was not determined according to any kind of mathematical standard—I set the range subjectively. Determining the range of class is up to the person who is analyzing the data.

But shouldn't there be a way to set the range of class mathematically? A frequency table may seem invalid if its range is determined subjectively.

There is a way to figure out the range of class mathematically. This is explained on the following pages. You'll also find a sample calculation using the data in Table 2-1.

Step 1

Calculate the number of classes using the Sturges' Rule below:

$$1 + \frac{\log_{10}(\text{number of values})}{\log_{10}2}$$

$$1 + \frac{\log_{10}50}{\log_{10}2} = 1 + 5.6438... = 6.6438... \approx 7$$

Step 2

Calculate the range of class using the formula below:

$$\frac{(\text{the maximum value}) - (\text{the minimum value})}{\text{the number of classes calculated from the Sturges' Rule}}$$

$$\frac{980 - 500}{7} = \frac{480}{7} = 68.5714... \approx 69$$

Below is a frequency chart organized according to the range of class as calculated by the formula in step 2.

TABLE 2-2: 50 BEST RAMEN SHOPS FREQUENCY TABLE
(RANGE OF CLASS DETERMINED MATHEMATICALLY)

Class (equal or greater/less than)	Class midpoint	Frequency	Relative frequency
500–569	534.5	2	0.04
569–638	603.5	5	0.10
638–707	672.5	15	0.30
707–776	741.5	6	0.12
776–845	810.5	10	0.20
845–914	879.5	10	0.20
914–983	948.5	2	0.04
Sum		50	1.00

What do you think of this? Does this table seem even less convincing compared to Table 2-1? And why is the interval 69 yen?

If you try to explain to people that "this was calculated by a formula called the Sturges' Rule," they will only get mad and say, "Who cares about Stur . . . whatever! Why did you set the interval to a weird amount like 69 yen?"

To summarize, some people may hesitate to set the range of class subjectively. However, as the table above indicates, determining the range of class with the Sturges' Rule does not necessarily provide a convincing table. A frequency table is, after all, a tool to help you visualize data. The analyst should set the range of class to any amount he or she thinks is appropriate.

6. ESTIMATION THEORY AND DESCRIPTIVE STATISTICS

In the prologue, we explain that statistics can make an estimate about the situation of the population based on information collected from samples. To tell the truth, this explanation is not necessarily correct.

Statistics can be roughly classified into two categories: estimation theory and descriptive statistics. The one introduced in the prologue is the former. What, then, is descriptive statistics? It is a kind of a statistics that aims to describe the status of a group simply and clearly by organizing data. Descriptive statistics regards the group as the population.

Perhaps this explanation of descriptive statistics is abstract and difficult to understand. Here is an example to help clarify things. Remember when I figured out the mean and standard deviation of Rui's bowling team? This was not because I was trying to estimate the status of a population from the information collected from Rui's team. I calculated the mean and standard deviation purely because I wanted to describe the status of Rui's team simply. That kind of statistics is descriptive statistics.

EXERCISE AND ANSWER

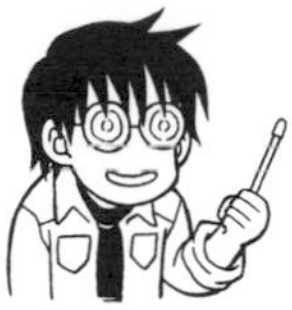

EXERCISE

The table below is a record of a high school girls' 100m track race.

Runner	100m track race (seconds)
Ms. A	16.3
Ms. B	22.4
Ms. C	18.5
Ms. D	18.7
Ms. E	20.1

1. What is the average?
2. What is the median?
3. What is the standard deviation?

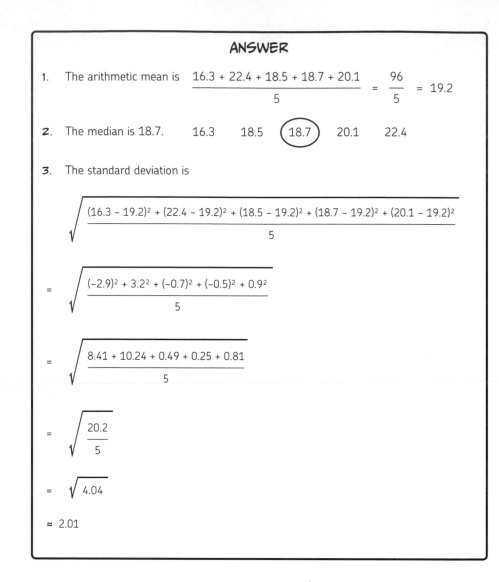

ANSWER

1. The arithmetic mean is $\dfrac{16.3 + 22.4 + 18.5 + 18.7 + 20.1}{5} = \dfrac{96}{5} = 19.2$

2. The median is 18.7.   16.3   18.5   (18.7)   20.1   22.4

3. The standard deviation is

$$\sqrt{\dfrac{(16.3 - 19.2)^2 + (22.4 - 19.2)^2 + (18.5 - 19.2)^2 + (18.7 - 19.2)^2 + (20.1 - 19.2)^2}{5}}$$

$$= \sqrt{\dfrac{(-2.9)^2 + 3.2^2 + (-0.7)^2 + (-0.5)^2 + 0.9^2}{5}}$$

$$= \sqrt{\dfrac{8.41 + 10.24 + 0.49 + 0.25 + 0.81}{5}}$$

$$= \sqrt{\dfrac{20.2}{5}}$$

$$= \sqrt{4.04}$$

$$\approx 2.01$$

SUMMARY

- To visualize the big picture of the data intuitively, create a frequency table or draw a histogram.
- When making a frequency table, the range of class may be determined by the Sturges' Rule.
- To visualize the data mathematically, calculate the arithmetic mean, median, and standard deviation.
- When there is an extremely large or small value in the data set, it is more appropriate to use the median than the arithmetic mean.
- *Standard deviation* is an index to describe "the size of scattering" of the data.

3

GETTING THE BIG PICTURE:
UNDERSTANDING CATEGORICAL DATA

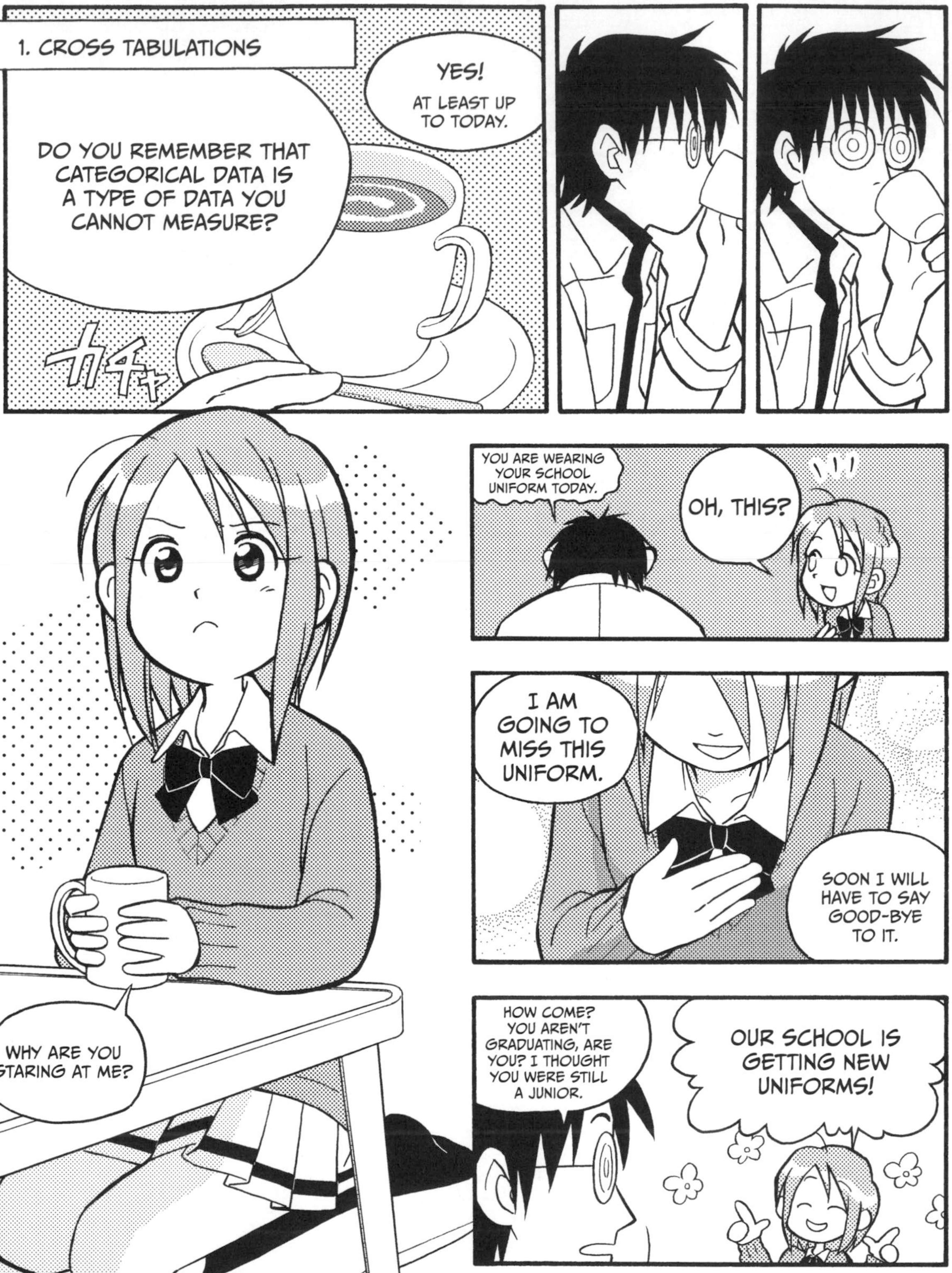
1. CROSS TABULATIONS
DO YOU REMEMBER THAT CATEGORICAL DATA IS A TYPE OF DATA YOU CANNOT MEASURE?
YES! AT LEAST UP TO TODAY.
カチャ
WHY ARE YOU STARING AT ME?
YOU ARE WEARING YOUR SCHOOL UNIFORM TODAY.
OH, THIS?
I AM GOING TO MISS THIS UNIFORM.
SOON I WILL HAVE TO SAY GOOD-BYE TO IT.
HOW COME? YOU AREN'T GRADUATING, ARE YOU? I THOUGHT YOU WERE STILL A JUNIOR.
OUR SCHOOL IS GETTING NEW UNIFORMS!

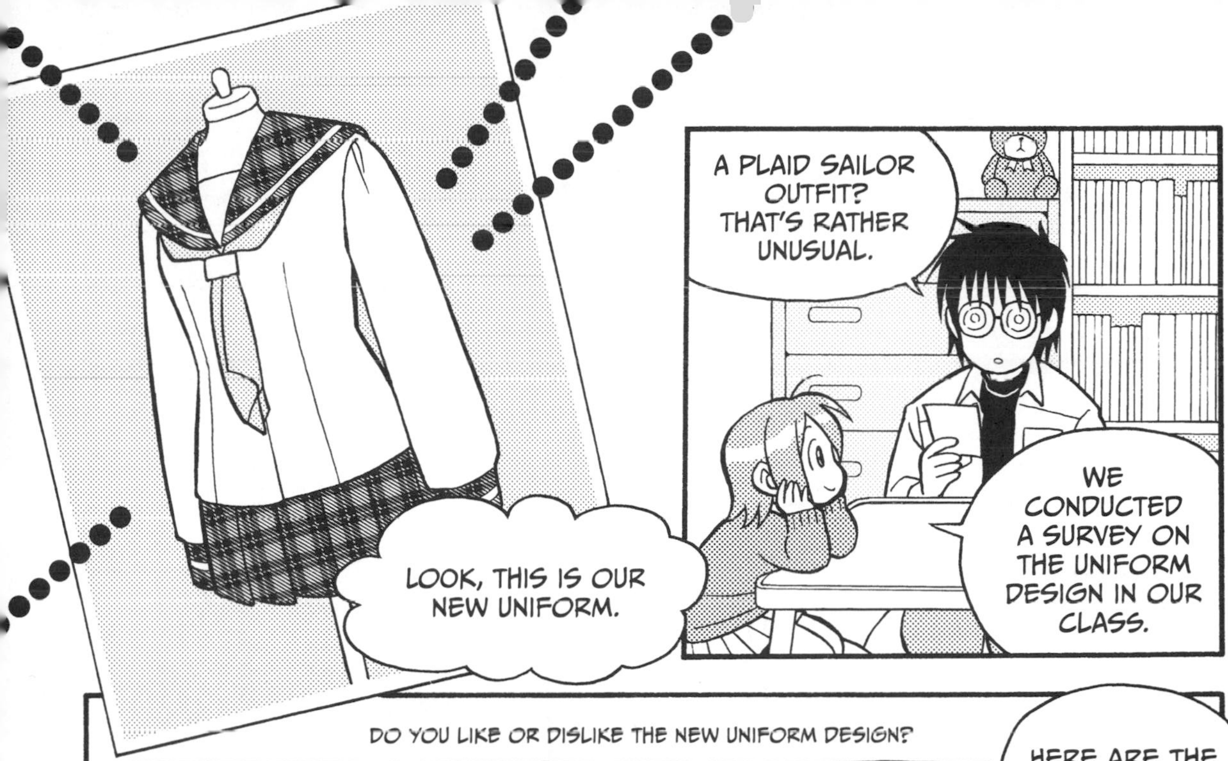
LOOK, THIS IS OUR NEW UNIFORM.
A PLAID SAILOR OUTFIT? THAT'S RATHER UNUSUAL.
WE CONDUCTED A SURVEY ON THE UNIFORM DESIGN IN OUR CLASS.

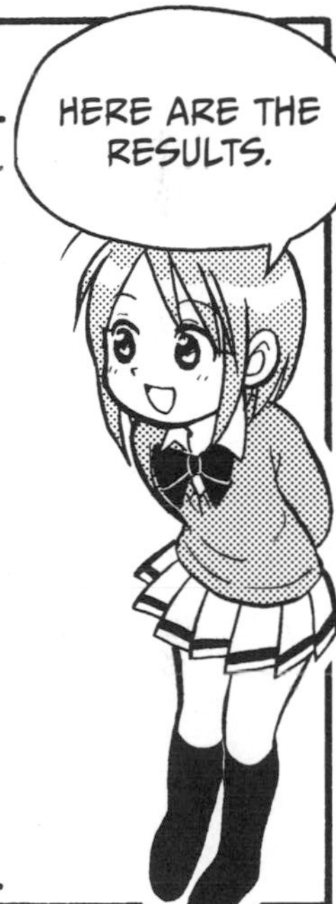
HERE ARE THE RESULTS.

DO YOU LIKE OR DISLIKE THE NEW UNIFORM DESIGN?

	RESPONSE		RESPONSE		RESPONSE
1	LIKE	16	NEITHER	31	NEITHER
2	NEITHER	17	LIKE	32	NEITHER
3	LIKE	18	LIKE	33	LIKE
4	NEITHER	19	LIKE	34	DISLIKE
5	DISLIKE	20	LIKE	35	LIKE
6	LIKE	21	LIKE	36	LIKE
7	LIKE	22	LIKE	37	LIKE
8	LIKE	23	DISLIKE	38	LIKE
9	LIKE	24	NEITHER	39	NEITHER
10	LIKE	25	LIKE	40	LIKE
11	LIKE	26	LIKE		
12	LIKE	27	DISLIKE		
13	NEITHER	28	LIKE		
14	LIKE	29	LIKE		
15	LIKE	30	LIKE		

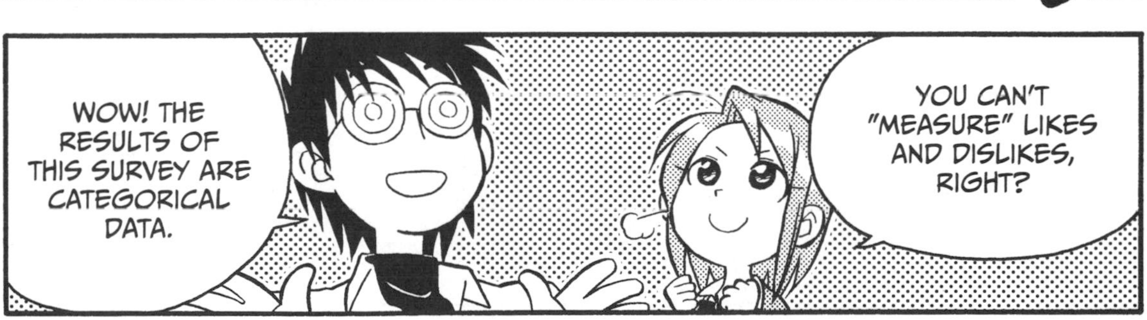
WOW! THE RESULTS OF THIS SURVEY ARE CATEGORICAL DATA.
YOU CAN'T "MEASURE" LIKES AND DISLIKES, RIGHT?

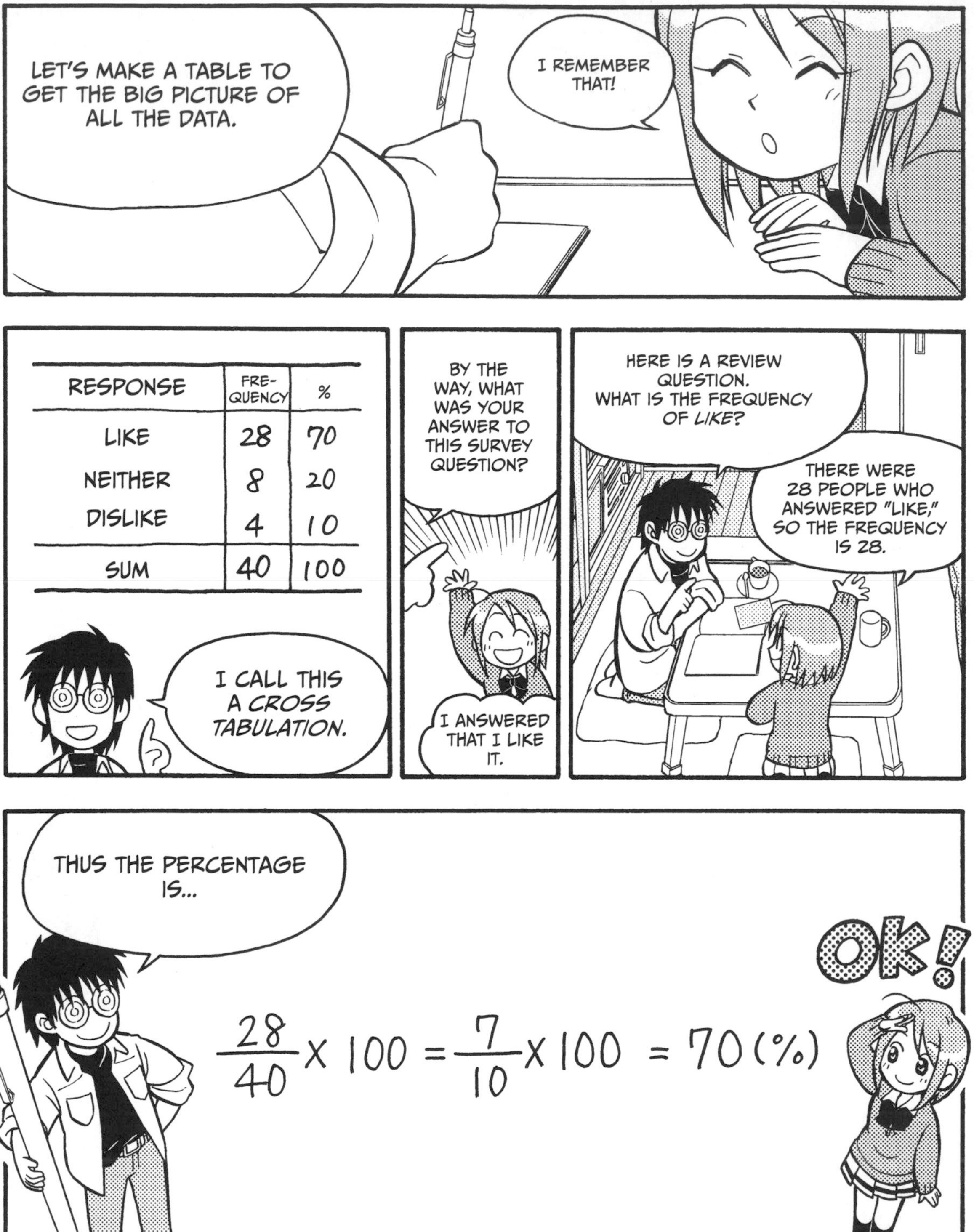
LET'S MAKE A TABLE TO GET THE BIG PICTURE OF ALL THE DATA.
I REMEMBER THAT!
I CALL THIS A *CROSS TABULATION*.
BY THE WAY, WHAT WAS YOUR ANSWER TO THIS SURVEY QUESTION?
I ANSWERED THAT I LIKE IT.
HERE IS A REVIEW QUESTION. WHAT IS THE FREQUENCY OF *LIKE*?
THERE WERE 28 PEOPLE WHO ANSWERED "LIKE," SO THE FREQUENCY IS 28.
THUS THE PERCENTAGE IS...
OK!

RESPONSE	FRE-QUENCY	%
LIKE	28	70
NEITHER	8	20
DISLIKE	4	10
SUM	40	100

$$\frac{28}{40} \times 100 = \frac{7}{10} \times 100 = 70(\%)$$

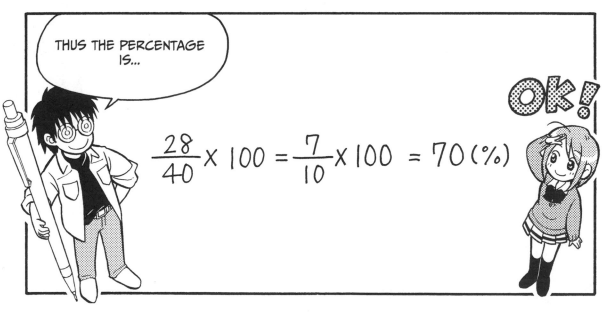

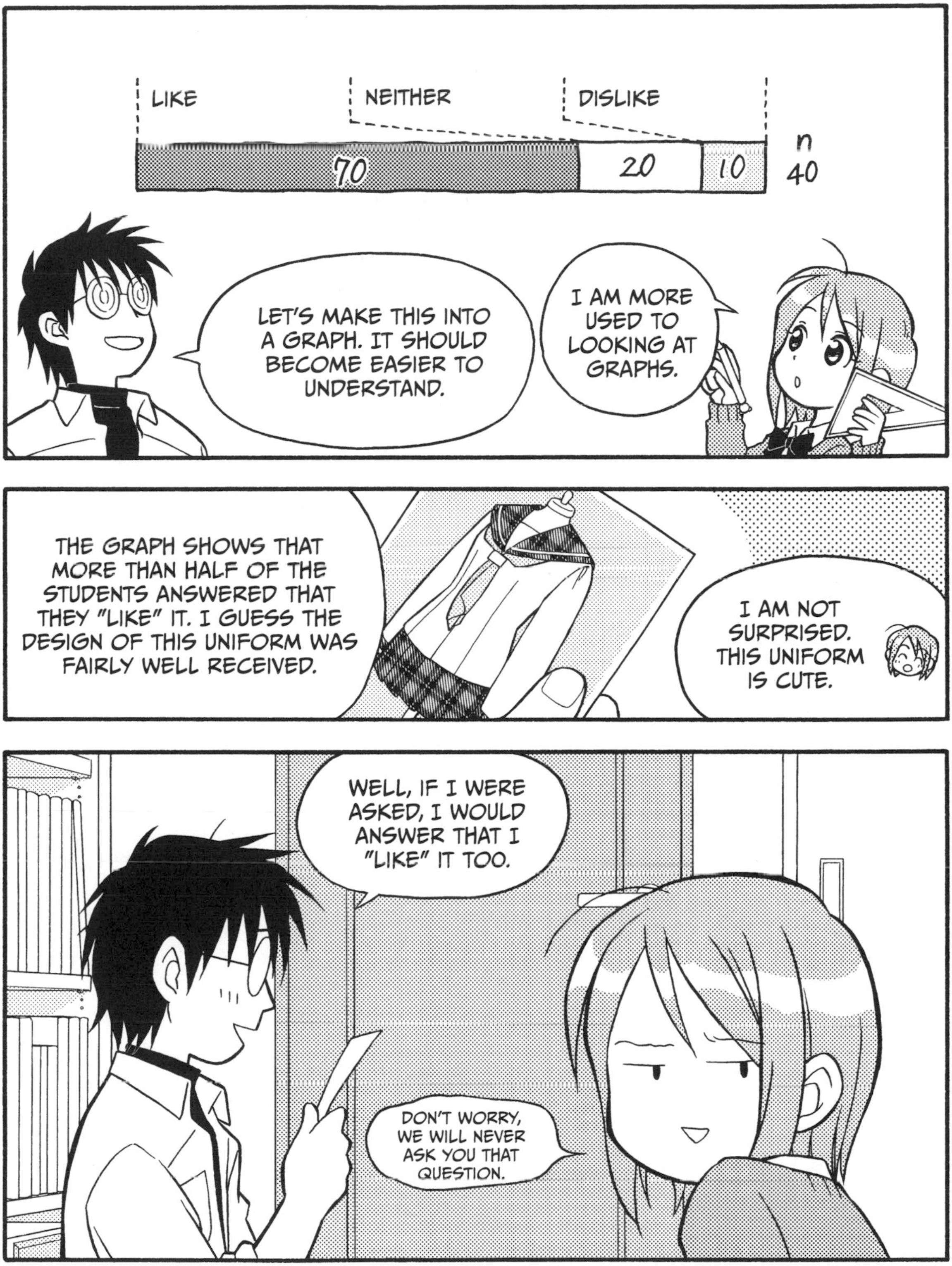
LIKE
NEITHER
DISLIKE
70
20
10
n
40
LET'S MAKE THIS INTO A GRAPH. IT SHOULD BECOME EASIER TO UNDERSTAND.
I AM MORE USED TO LOOKING AT GRAPHS.
THE GRAPH SHOWS THAT MORE THAN HALF OF THE STUDENTS ANSWERED THAT THEY "LIKE" IT. I GUESS THE DESIGN OF THIS UNIFORM WAS FAIRLY WELL RECEIVED.
I AM NOT SURPRISED. THIS UNIFORM IS CUTE.
WELL, IF I WERE ASKED, I WOULD ANSWER THAT I "LIKE" IT TOO.
DON'T WORRY, WE WILL NEVER ASK YOU THAT QUESTION.

EXERCISE AND ANSWER

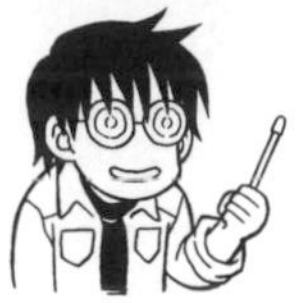

EXERCISE

A newspaper took a survey on political party A, which hopes to win the next election. The results are below.

Respondent	Do you expect party A to win or lose against party B?
1	Lose
2	Lose
3	Lose
4	I don't know
5	Win
6	Lose
7	Win
8	I don't know
9	Lose
10	Lose

Make a cross tabulation from these survey results.

ANSWER

Below is the cross tabulation.

Response	Frequency	%
Win	2	20
I don't know	2	20
Lose	6	60
Sum	10	100

SUMMARY

- One way to see the big picture of all the data is to make a cross tabulation.

4
STANDARD SCORE AND
DEVIATION SCORE

1. NORMALIZATION AND STANDARD SCORE

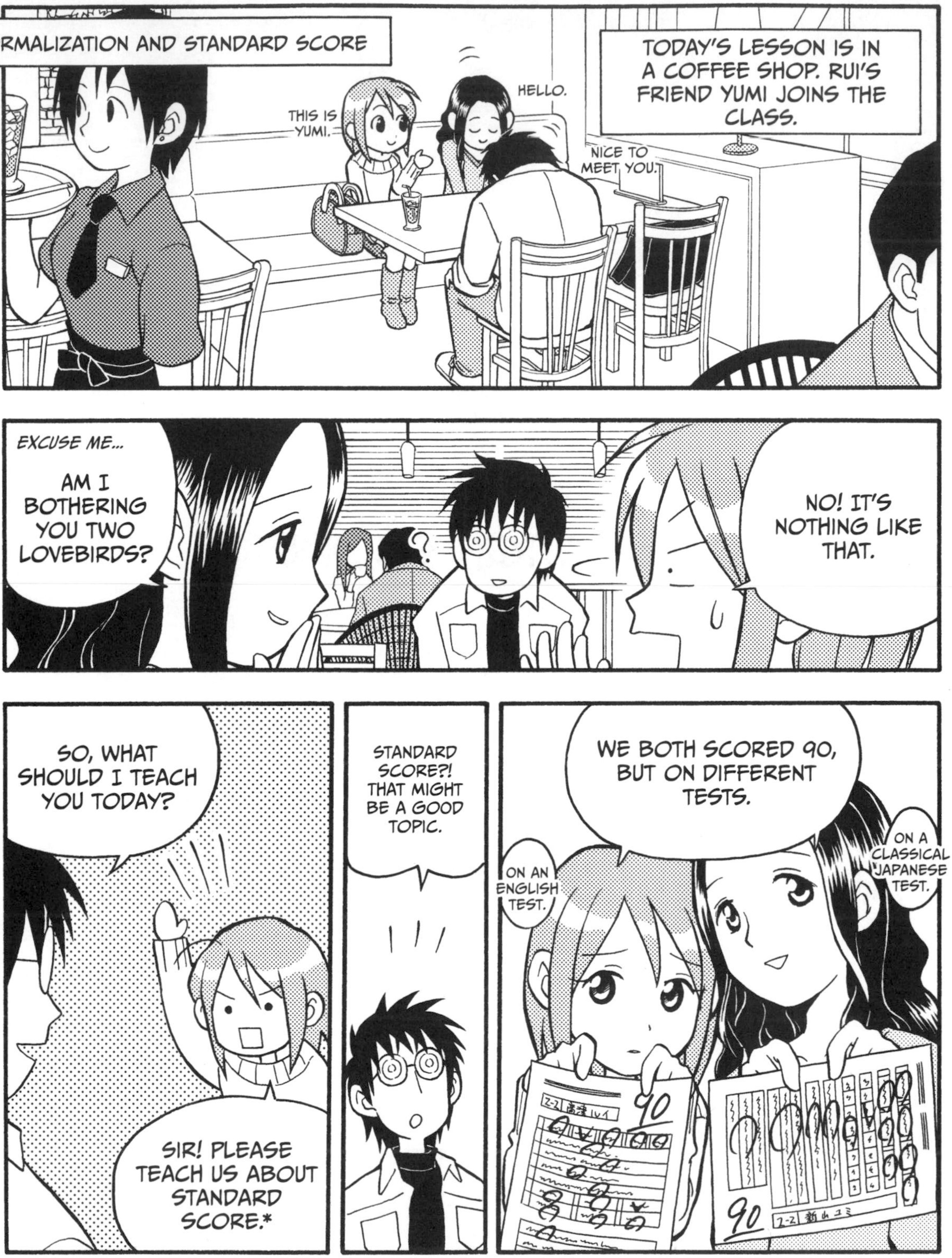
TODAY'S LESSON IS IN A COFFEE SHOP. RUI'S FRIEND YUMI JOINS THE CLASS.
THIS IS YUMI.
HELLO.
NICE TO MEET YOU.
EXCUSE ME...
AM I BOTHERING YOU TWO LOVEBIRDS?
NO! IT'S NOTHING LIKE THAT.
SO, WHAT SHOULD I TEACH YOU TODAY?
SIR! PLEASE TEACH US ABOUT STANDARD SCORE.*
STANDARD SCORE?! THAT MIGHT BE A GOOD TOPIC.
WE BOTH SCORED 90, BUT ON DIFFERENT TESTS.
ON AN ENGLISH TEST.
ON A CLASSICAL JAPANESE TEST.
90
90

* ADJUSTING TEST RESULTS BASED ON STANDARD SCORE IS COMMONLY KNOWN AS *GRADING ON A CURVE*.

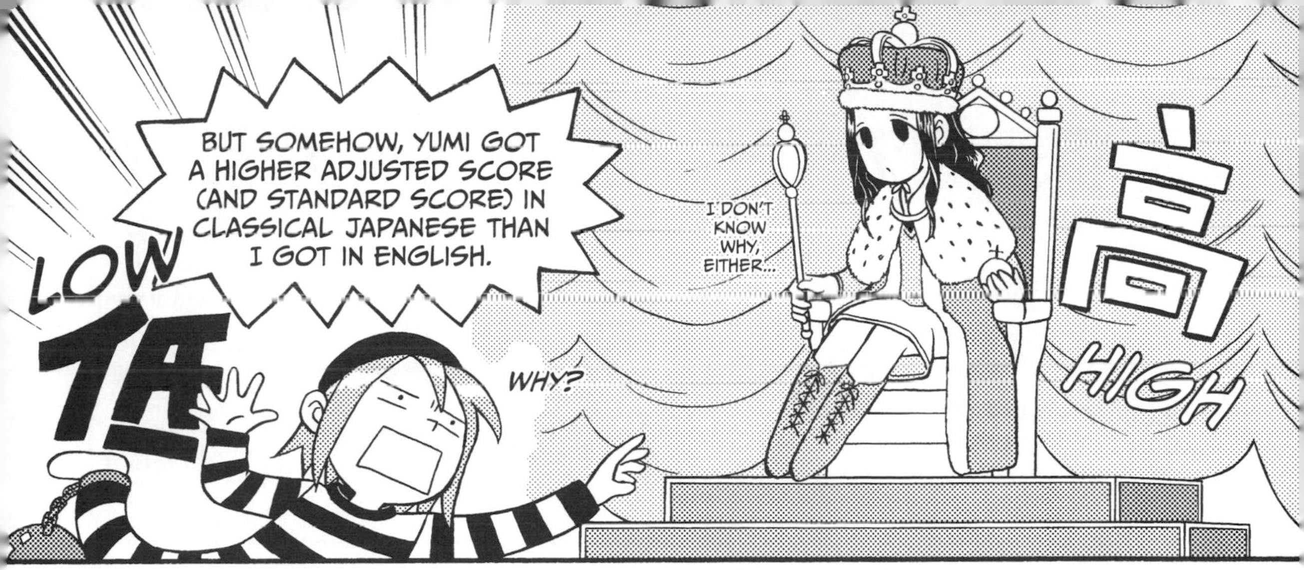
BUT SOMEHOW, YUMI GOT A HIGHER ADJUSTED SCORE (AND STANDARD SCORE) IN CLASSICAL JAPANESE THAN I GOT IN ENGLISH.
LOW
低
WHY?
I DON'T KNOW WHY, EITHER...
高
HIGH

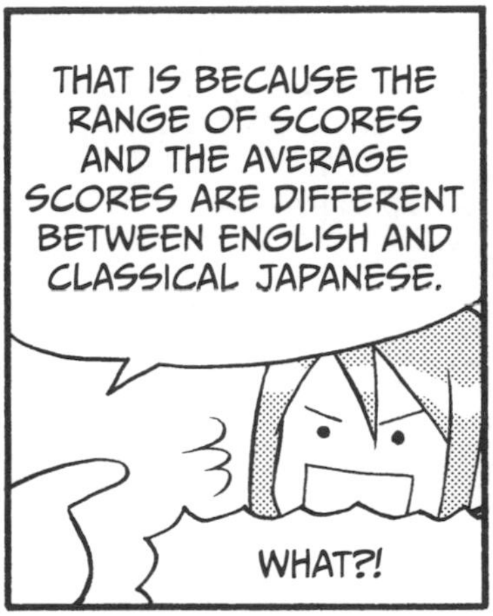
THAT IS BECAUSE THE RANGE OF SCORES AND THE AVERAGE SCORES ARE DIFFERENT BETWEEN ENGLISH AND CLASSICAL JAPANESE.
WHAT?!

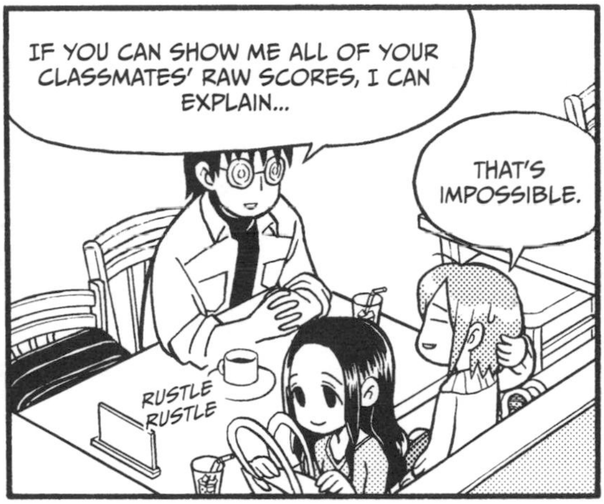
IF YOU CAN SHOW ME ALL OF YOUR CLASSMATES' RAW SCORES, I CAN EXPLAIN...
THAT'S IMPOSSIBLE.
RUSTLE RUSTLE

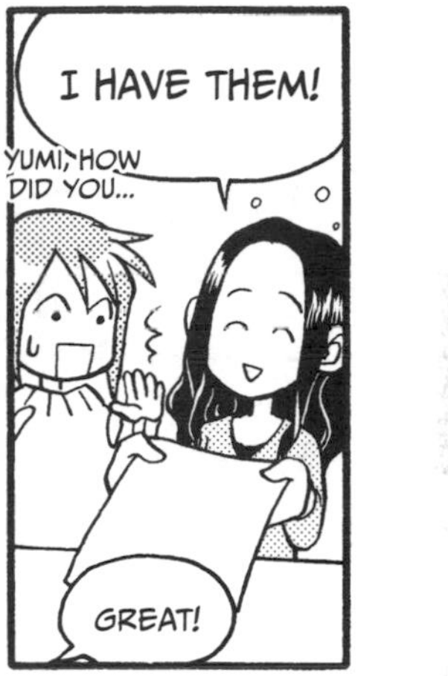
I HAVE THEM!
YUMI, HOW DID YOU...
GREAT!

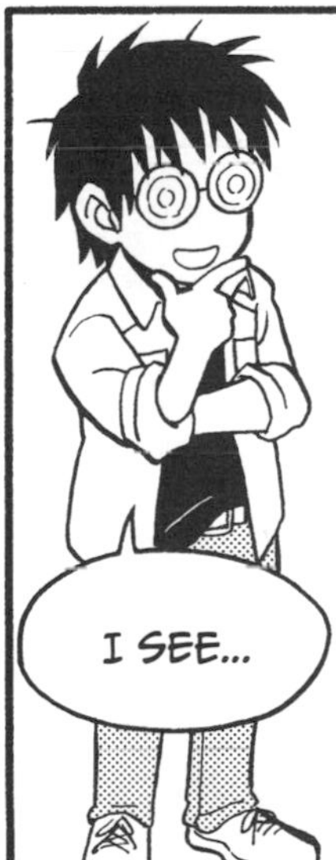
I SEE...

RAW TEST SCORES (OUT OF 100)

STUDENT	ENGLISH	CLASSICAL JAPANESE	STUDENT	ENGLISH	CLASSICAL JAPANESE
RUI	90	71	H	67	85
YUMI	81	90	I	87	93
A	73	79	J	78	89
B	97	70	K	85	78
C	85	67	L	96	74
D	60	66	M	77	65
E	74	60	N	100	78
F	64	83	O	92	53
G	72	57	P	86	80

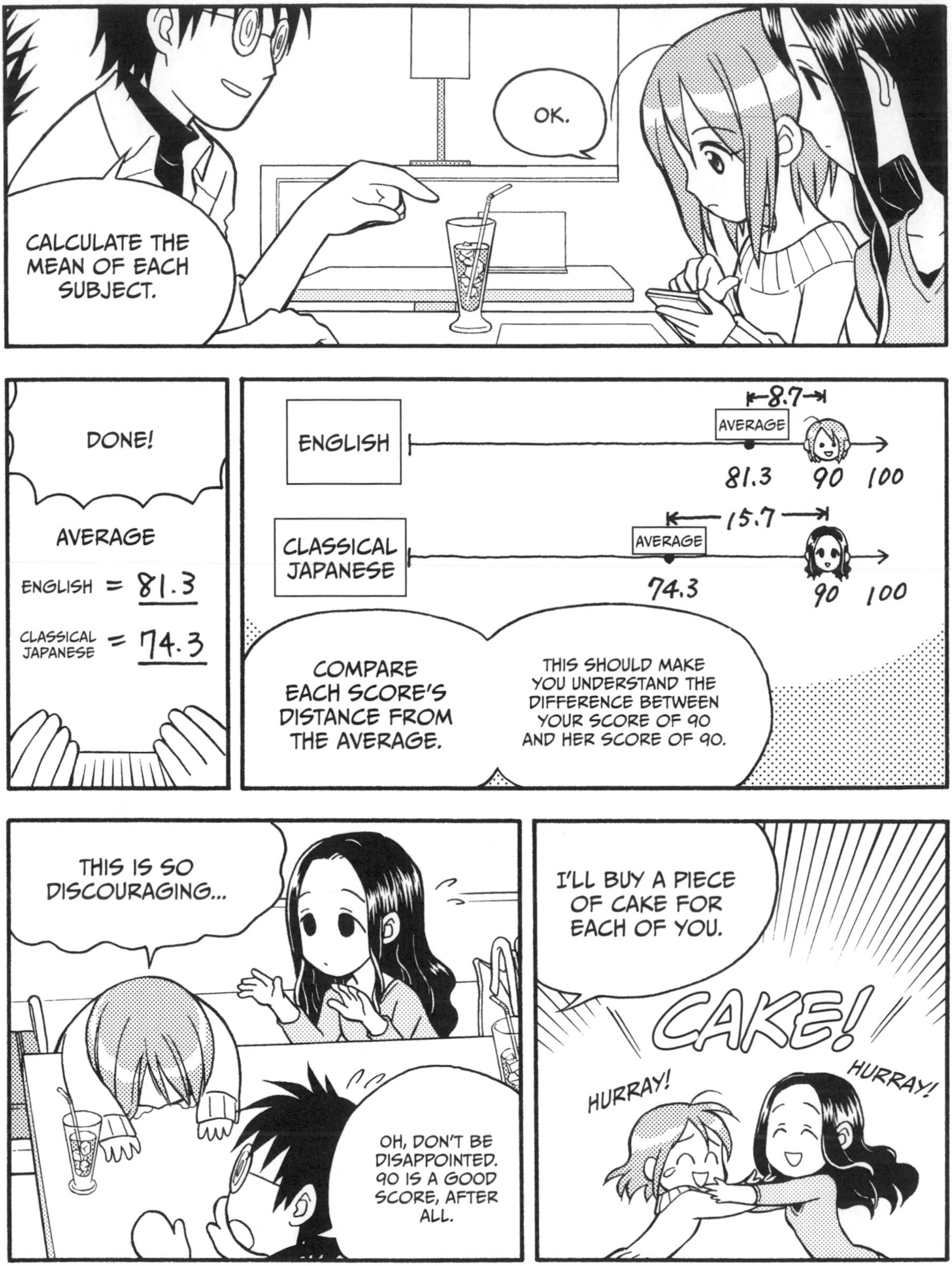

CALCULATE THE MEAN OF EACH SUBJECT.
OK.
DONE!
AVERAGE
ENGLISH = 81.3
CLASSICAL JAPANESE = 74.3
ENGLISH
AVERAGE
8.7
81.3
90
100
CLASSICAL JAPANESE
AVERAGE
15.7
74.3
90
100
COMPARE EACH SCORE'S DISTANCE FROM THE AVERAGE.
THIS SHOULD MAKE YOU UNDERSTAND THE DIFFERENCE BETWEEN YOUR SCORE OF 90 AND HER SCORE OF 90.
THIS IS SO DISCOURAGING...
OH, DON'T BE DISAPPOINTED. 90 IS A GOOD SCORE, AFTER ALL.
I'LL BUY A PIECE OF CAKE FOR EACH OF YOU.
CAKE!
HURRAY!
HURRAY!

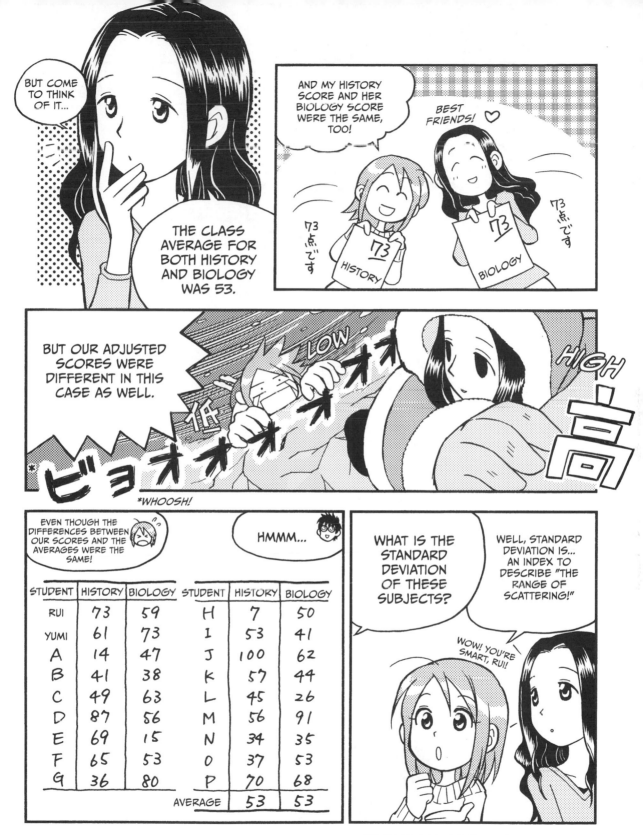

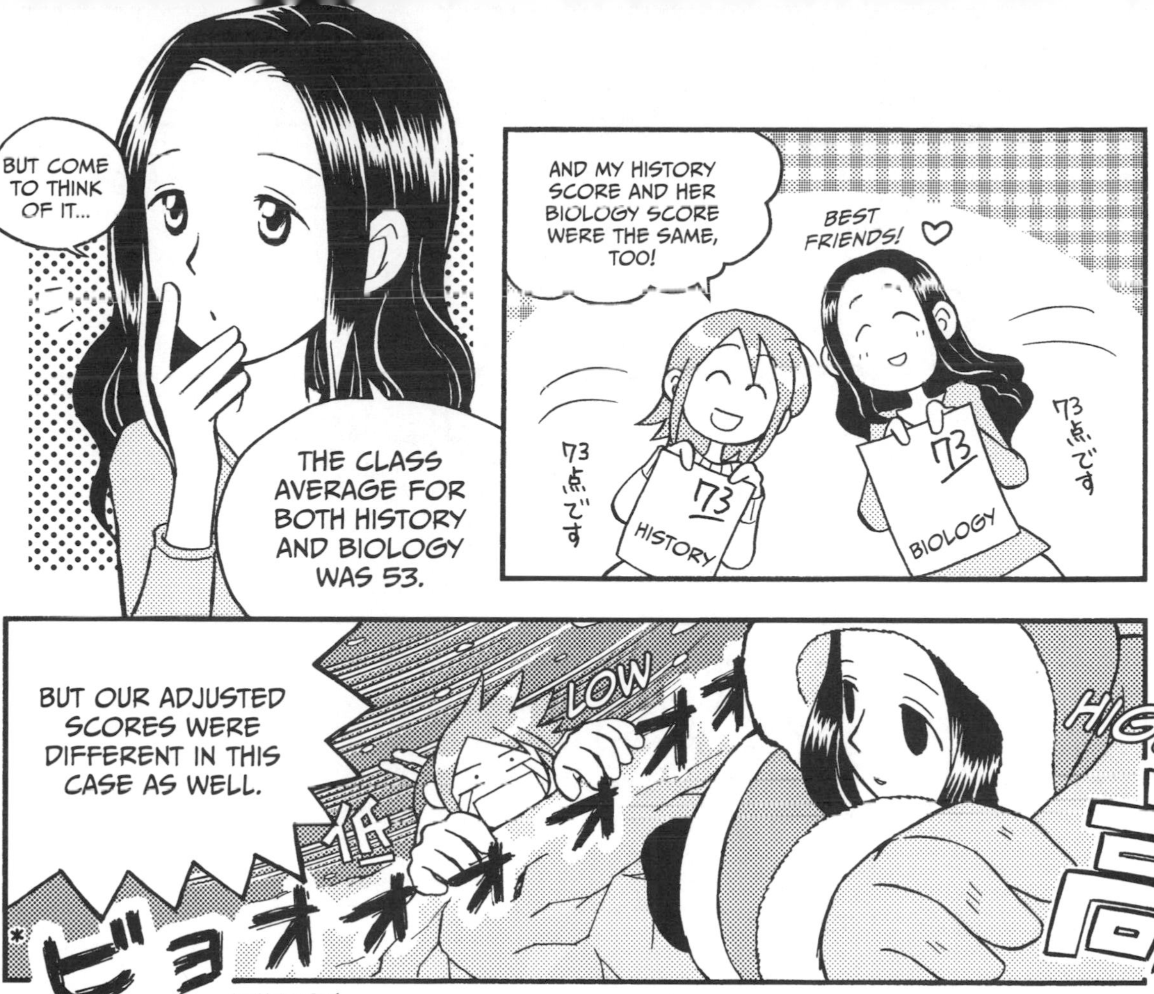
BUT COME TO THINK OF IT...
THE CLASS AVERAGE FOR BOTH HISTORY AND BIOLOGY WAS 53.
AND MY HISTORY SCORE AND HER BIOLOGY SCORE WERE THE SAME, TOO!
BEST FRIENDS! ♡
73点です
73
HISTORY
73
BIOLOGY
73点です
BUT OUR ADJUSTED SCORES WERE DIFFERENT IN THIS CASE AS WELL.
LOW
低
HIGH
高
*ビョオオオ
*WHOOSH!

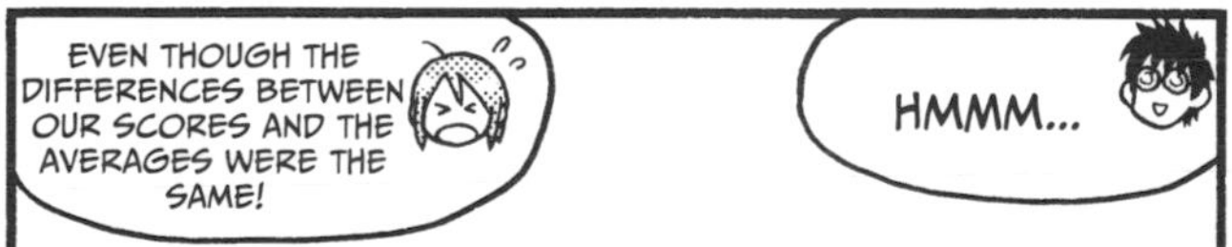
EVEN THOUGH THE DIFFERENCES BETWEEN OUR SCORES AND THE AVERAGES WERE THE SAME!
HMMM...

STUDENT	HISTORY	BIOLOGY	STUDENT	HISTORY	BIOLOGY
RUI	73	59	H	7	50
YUMI	61	73	I	53	41
A	14	47	J	100	62
B	41	38	K	57	44
C	49	63	L	45	26
D	87	56	M	56	91
E	69	15	N	34	35
F	65	53	O	37	53
G	36	80	P	70	68
			AVERAGE	53	53

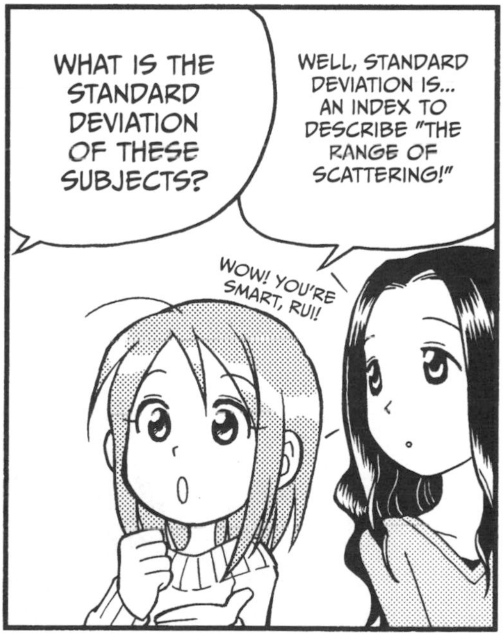
WHAT IS THE STANDARD DEVIATION OF THESE SUBJECTS?
WELL, STANDARD DEVIATION IS... AN INDEX TO DESCRIBE "THE RANGE OF SCATTERING!"
WOW! YOU'RE SMART, RUI!

$$\sqrt{\frac{\text{SUM OF (EACH VALUE − MEAN)}^2}{\text{NUMBER OF VALUES}}}$$

ITS FORMULA IS...

STANDARD DEVIATION

HISTORY = 22.7

BIOLOGY = 18.3

THERE!

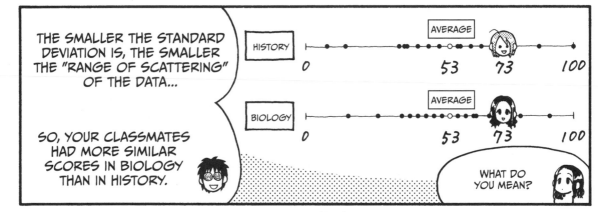

THE SMALLER THE STANDARD DEVIATION IS, THE SMALLER THE "RANGE OF SCATTERING" OF THE DATA...

SO, YOUR CLASSMATES HAD MORE SIMILAR SCORES IN BIOLOGY THAN IN HISTORY.

HISTORY

0   53   73   100

AVERAGE

BIOLOGY

0   53   73   100

AVERAGE

WHAT DO YOU MEAN?

IF I WERE A HIGH SCHOOL JUNIOR APPLYING FOR COLLEGE, I'D STUDY HARD FOR BIOLOGY.

ONE OR TWO POINTS MAY AFFECT YOUR RANK GREATLY.

A HIGH SCHOOL UNIFORM SUITS HIM SO WELL!

TEE-HEE!

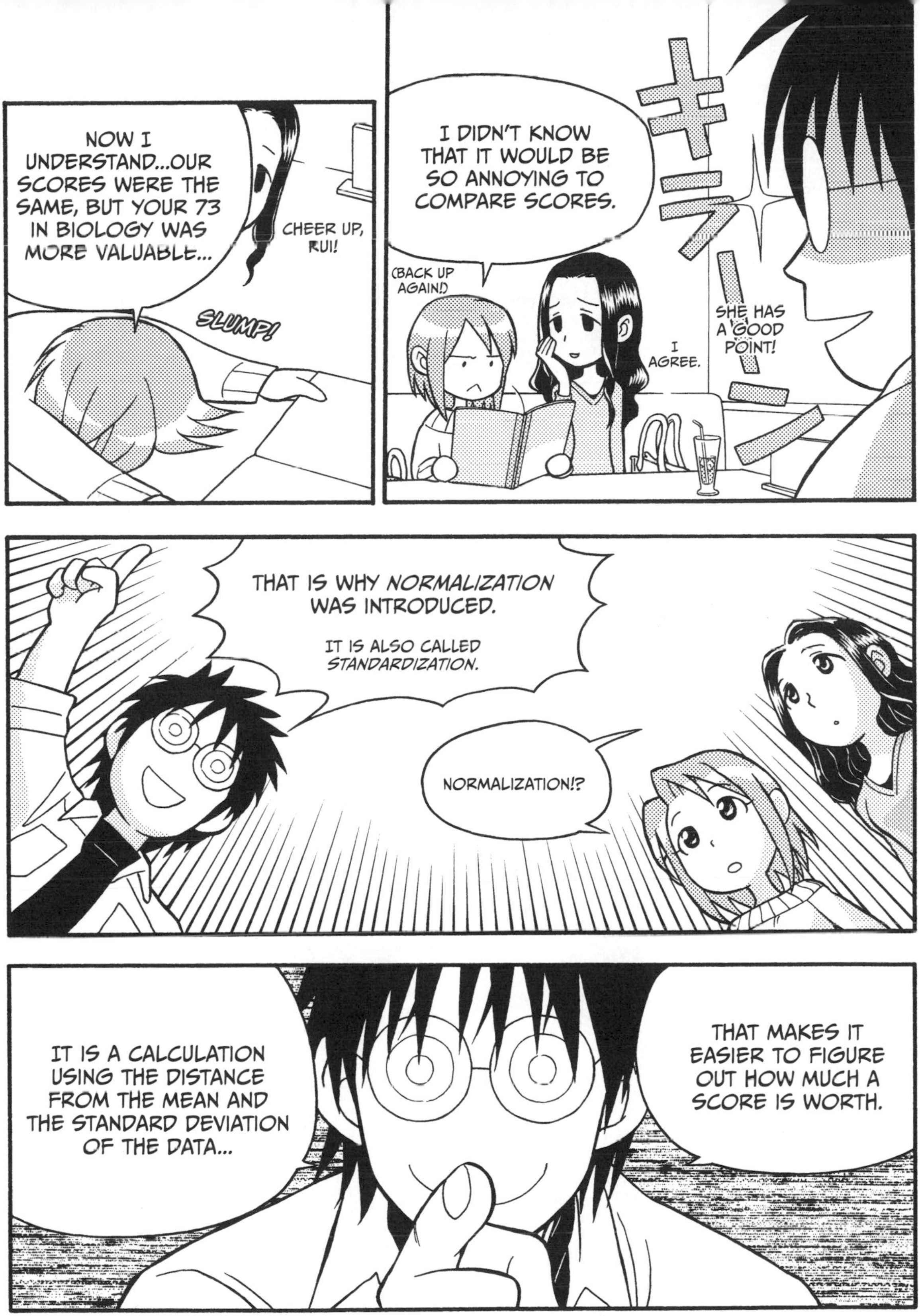
NOW I UNDERSTAND...OUR SCORES WERE THE SAME, BUT YOUR 73 IN BIOLOGY WAS MORE VALUABLE...
CHEER UP, RUI!
SLUMP!
I DIDN'T KNOW THAT IT WOULD BE SO ANNOYING TO COMPARE SCORES.
(BACK UP AGAIN!)
I AGREE.
SHE HAS A GOOD POINT!
THAT IS WHY NORMALIZATION WAS INTRODUCED.
IT IS ALSO CALLED STANDARDIZATION.
NORMALIZATION!?
IT IS A CALCULATION USING THE DISTANCE FROM THE MEAN AND THE STANDARD DEVIATION OF THE DATA...
THAT MAKES IT EASIER TO FIGURE OUT HOW MUCH A SCORE IS WORTH.

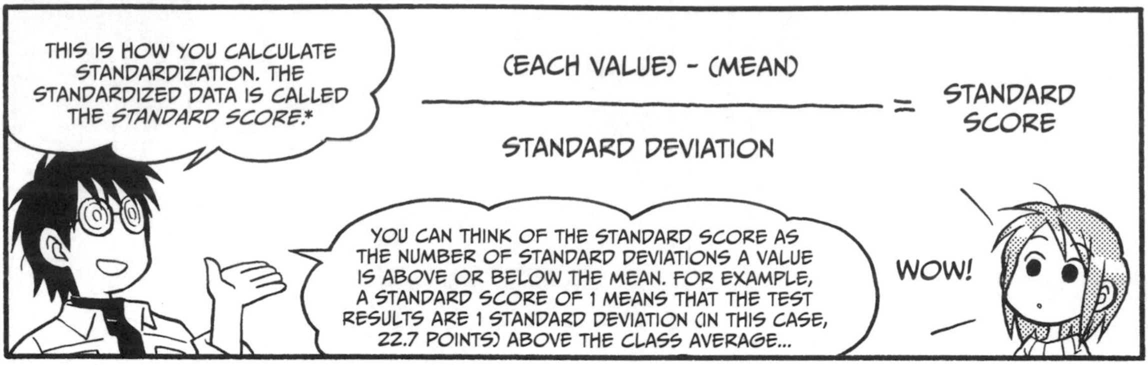
THIS IS HOW YOU CALCULATE STANDARDIZATION. THE STANDARDIZED DATA IS CALLED THE *STANDARD SCORE*.*
(EACH VALUE) - (MEAN) / STANDARD DEVIATION = STANDARD SCORE
YOU CAN THINK OF THE STANDARD SCORE AS THE NUMBER OF STANDARD DEVIATIONS A VALUE IS ABOVE OR BELOW THE MEAN. FOR EXAMPLE, A STANDARD SCORE OF 1 MEANS THAT THE TEST RESULTS ARE 1 STANDARD DEVIATION (IN THIS CASE, 22.7 POINTS) ABOVE THE CLASS AVERAGE...
WOW!

* STANDARD SCORE IS ALSO CALLED Z-SCORE.

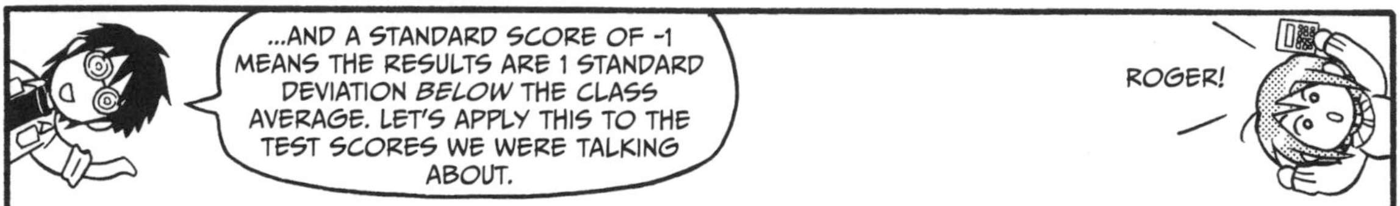
...AND A STANDARD SCORE OF -1 MEANS THE RESULTS ARE 1 STANDARD DEVIATION *BELOW* THE CLASS AVERAGE. LET'S APPLY THIS TO THE TEST SCORES WE WERE TALKING ABOUT.
ROGER!

RESULTS AND STANDARD SCORES OF HISTORY AND BIOLOGY TESTS

STUDENT	HISTORY	BIOLOGY	STANDARD SCORE OF HISTORY	STANDARD SCORE OF BIOLOGY
RUI	73	59	0.88	0.33
YUMI	61	73	0.35	1.09
A	14	47	-1.71	-0.33
B	41	38	-0.53	-0.82
C	49	63	-0.18	0.55
D	87	56	1.49	0.16
E	69	15	0.70	-2.08
F	65	53	0.53	0
G	36	80	-0.75	1.48
H	7	50	-2.02	-0.16
I	53	41	0	-0.66
J	100	62	2.07	0.49
K	57	44	0.18	-0.49
L	45	26	-0.35	-1.48
M	56	91	0.13	2.08
N	34	35	-0.84	-0.98
O	37	53	-0.70	0
P	70	68	0.75	0.82
AVERAGE	53	53	0	0
STANDARD DEVIATION	22.7	18.3	1	1

STANDARD SCORE OF RUI'S HISTORY TEST $\frac{73-53}{22.7} = \frac{20}{22.7} = 0.88$

STANDARD SCORE OF YUMI'S BIOLOGY TEST $\frac{73-53}{18.3} = \frac{20}{18.3} = 1.09$

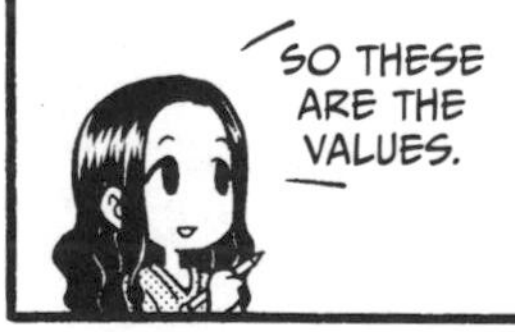
SO THESE ARE THE VALUES.

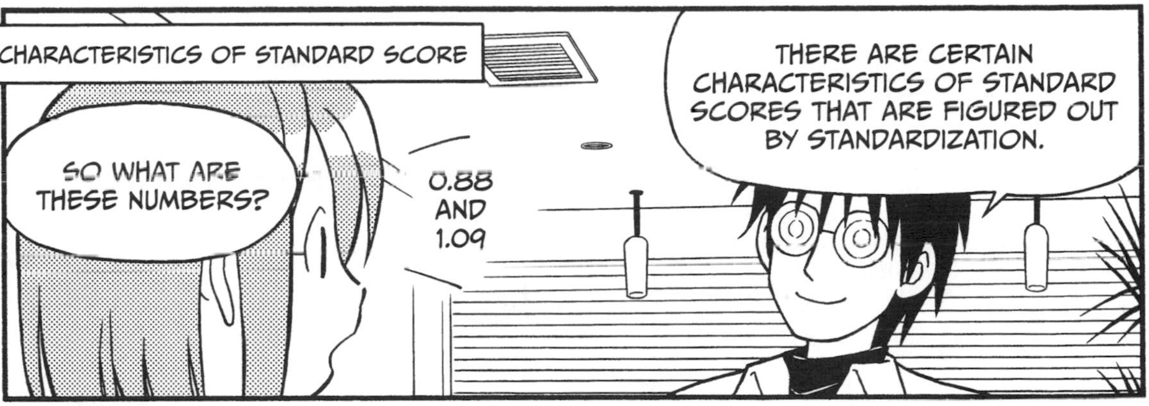
2. CHARACTERISTICS OF STANDARD SCORE
SO WHAT ARE THESE NUMBERS?
0.88 AND 1.09
THERE ARE CERTAIN CHARACTERISTICS OF STANDARD SCORES THAT ARE FIGURED OUT BY STANDARDIZATION.

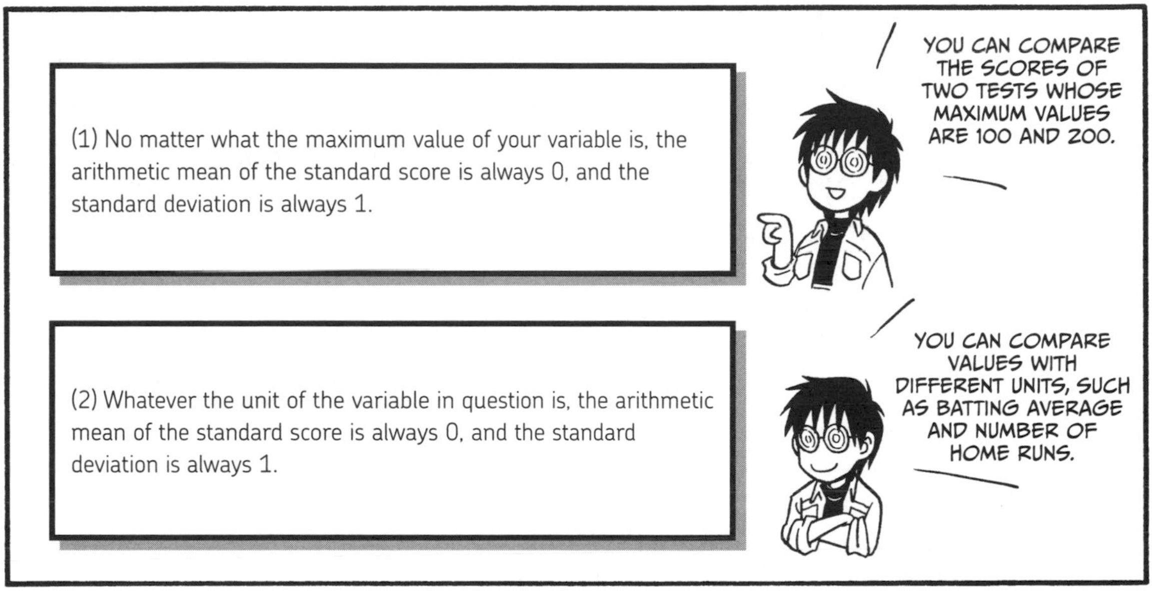
(1) No matter what the maximum value of your variable is, the arithmetic mean of the standard score is always 0, and the standard deviation is always 1.
YOU CAN COMPARE THE SCORES OF TWO TESTS WHOSE MAXIMUM VALUES ARE 100 AND 200.
(2) Whatever the unit of the variable in question is, the arithmetic mean of the standard score is always 0, and the standard deviation is always 1.
YOU CAN COMPARE VALUES WITH DIFFERENT UNITS, SUCH AS BATTING AVERAGE AND NUMBER OF HOME RUNS.

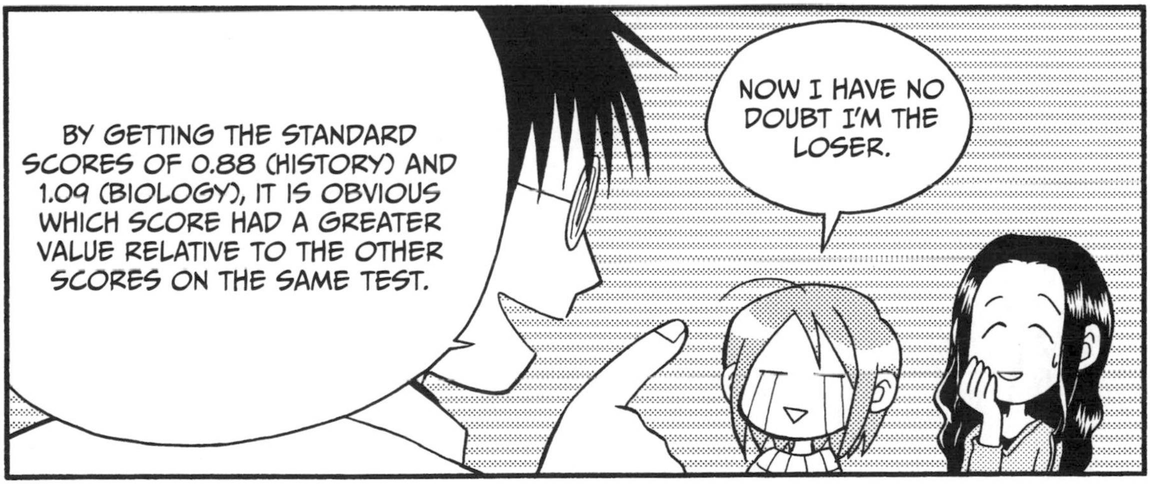
BY GETTING THE STANDARD SCORES OF 0.88 (HISTORY) AND 1.09 (BIOLOGY), IT IS OBVIOUS WHICH SCORE HAD A GREATER VALUE RELATIVE TO THE OTHER SCORES ON THE SAME TEST.
NOW I HAVE NO DOUBT I'M THE LOSER.

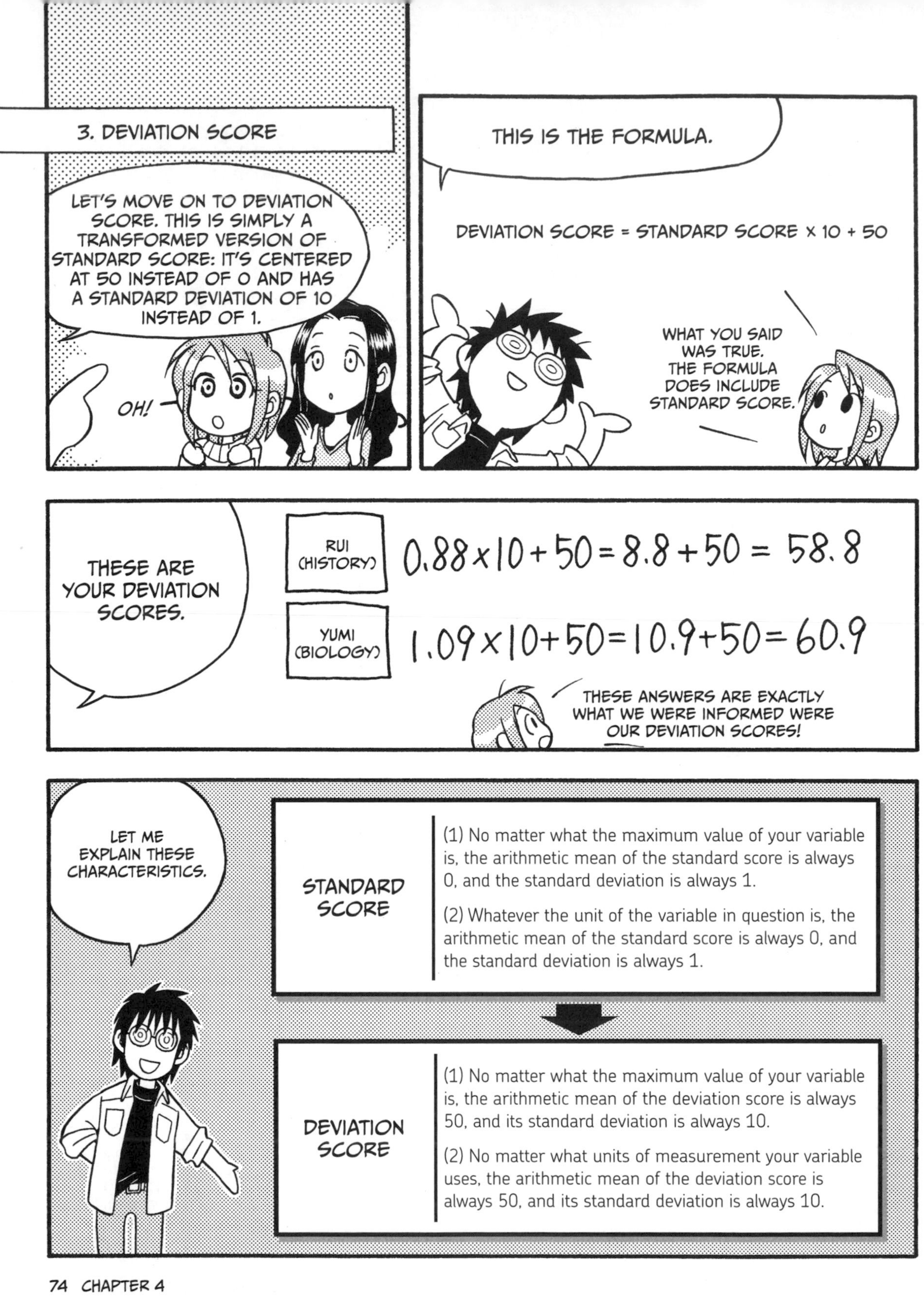
3. DEVIATION SCORE
LET'S MOVE ON TO DEVIATION SCORE. THIS IS SIMPLY A TRANSFORMED VERSION OF STANDARD SCORE: IT'S CENTERED AT 50 INSTEAD OF 0 AND HAS A STANDARD DEVIATION OF 10 INSTEAD OF 1.
OH!
THIS IS THE FORMULA.
DEVIATION SCORE = STANDARD SCORE × 10 + 50
WHAT YOU SAID WAS TRUE. THE FORMULA DOES INCLUDE STANDARD SCORE.
THESE ARE YOUR DEVIATION SCORES.
RUI (HISTORY)
0.88×10+50=8.8+50 = 58.8
YUMI (BIOLOGY)
1.09×10+50=10.9+50=60.9
THESE ANSWERS ARE EXACTLY WHAT WE WERE INFORMED WERE OUR DEVIATION SCORES!
LET ME EXPLAIN THESE CHARACTERISTICS.
STANDARD SCORE
(1) No matter what the maximum value of your variable is, the arithmetic mean of the standard score is always 0, and the standard deviation is always 1.
(2) Whatever the unit of the variable in question is, the arithmetic mean of the standard score is always 0, and the standard deviation is always 1.
DEVIATION SCORE
(1) No matter what the maximum value of your variable is, the arithmetic mean of the deviation score is always 50, and its standard deviation is always 10.
(2) No matter what units of measurement your variable uses, the arithmetic mean of the deviation score is always 50, and its standard deviation is always 10.

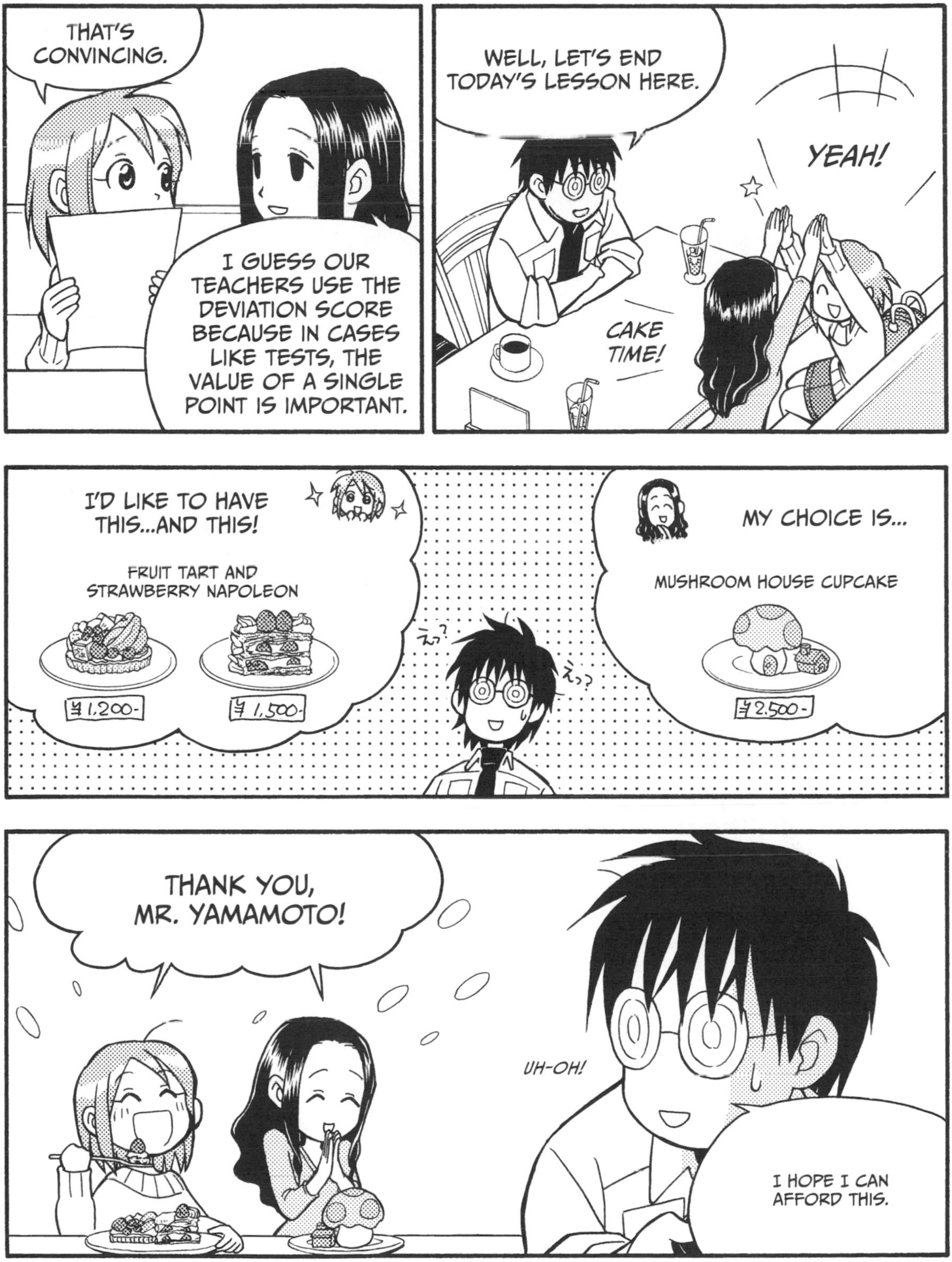

THAT'S CONVINCING.
I GUESS OUR TEACHERS USE THE DEVIATION SCORE BECAUSE IN CASES LIKE TESTS, THE VALUE OF A SINGLE POINT IS IMPORTANT.
WELL, LET'S END TODAY'S LESSON HERE.
YEAH!
CAKE TIME!
I'D LIKE TO HAVE THIS...AND THIS!
FRUIT TART AND STRAWBERRY NAPOLEON
¥1,200-
¥1,500-
えっ?
えっ?
MY CHOICE IS...
MUSHROOM HOUSE CUPCAKE
¥2,500-
THANK YOU, MR. YAMAMOTO!
UH-OH!
I HOPE I CAN AFFORD THIS.

4. INTERPRETATION OF DEVIATION SCORE

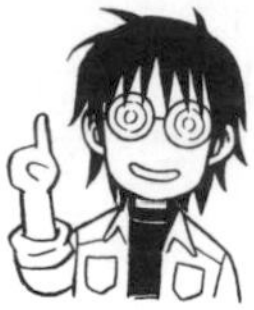

Special caution is necessary when interpreting deviation scores. As explained on page 74, the definition of deviation score is:

$$\text{deviation score} = \text{standard score} \times 10 + 50 = \frac{(\text{each value} - \text{mean})}{\text{standard deviation}} \times 10 + 50$$

As mentioned on page 62, Rui's class has a total of 40 students, and as mentioned on page 40, there are 18 girls in the class. The example of deviation score on page 69 is not for the whole class, but is for the girls only. If the story were about the whole class, the mean and standard deviation would have been different from those for the girls only. Naturally, the deviation scores for Rui and Yumi would have been different as well. In fact, when everybody in the class is taken into consideration, Rui has the higher deviation score. Table 4-1 shows the test results for the whole class. Try calculating the deviation score.

To tell you the answer in advance, the deviation score for Rui's history test is 59.1, and that of Yumi's biology test is 56.7.

Suppose the same test is given to students in classes 1 and 2. The mean and standard deviation of class 1 are calculated individually, and deviation scores are obtained according to those amounts. Similarly, mean, standard deviation, and deviation scores for class 2 are obtained. Student A in class 1 has a deviation score of 57. Student B in class 2 has the same deviation score of 57. Outwardly, students A and B seem to have the same ability. However, the mean and standard deviation used to calculate these two deviation scores differ, because they come from two different classes. Unless the mean and standard deviation of the two classes are equal, you cannot compare the deviation scores of the two students.

Here is another example. Suppose student A takes an entrance exam at a prep school in April and gets a deviation score of 54. After studying hard at a special summer course, student A takes an entrance exam at a different prep school in September. The deviation score is 62. It may seem that student A's proficiency has increased. However, the exam and the students taking it in April are different from the exam and the students taking it in September. Therefore, you cannot compare the deviation scores for these two exams, because the data used to calculate the mean and standard deviation of the April and September exams is different. In exam situations, you can only compare deviation scores for a group of students who all take the same exam. Keep these facts in mind when you interpret deviation scores.

TABLE 4-1: TEST RESULTS OF HISTORY AND BIOLOGY (ALL MEMBERS OF RUI'S CLASS)

Girls	History	Biology	Boys	History	Biology
Rui	73	59	a	54	2
Yumi	61	73	b	93	7
A	14	47	c	91	98
B	41	38	d	37	85
C	49	63	e	44	100
D	87	56	f	16	29
E	69	15	g	12	57
F	65	53	h	44	37
G	36	80	i	4	95
H	7	50	j	17	39
I	53	41	k	66	70
J	100	62	l	53	14
K	57	44	m	14	97
L	45	26	n	73	39
M	56	91	o	6	75
N	34	35	p	22	80
O	37	53	q	69	77
P	70	68	r	95	14
			s	16	24
			t	37	91
			u	14	36
			v	88	76

	History	Biology
Average of the whole class	48.0	54.9
Standard deviation of the whole class	27.5	26.9

EXERCISE AND ANSWER

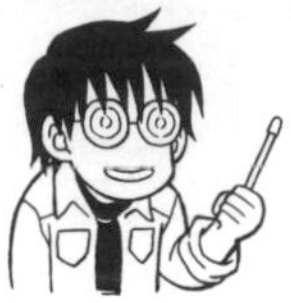

EXERCISE

Below are the results of a high school girls' 100m track race.

Runner	100m track race (seconds)
Ms. A	16.3
Ms. B	22.4
Ms. C	18.5
Ms. D	18.7
Ms. E	20.1
Mean	19.2
Standard deviation	2.01

1. Demonstrate that the mean of the standard scores of the 100m track race is 0.
2. Demonstrate that the standard deviation of the standard score of the 100m track race is 1.

ANSWER

1. Mean of the standard score of the 100m track race

$$= \frac{\left(\frac{16.3 - 19.2}{2.01}\right) + \left(\frac{22.4 - 19.2}{2.01}\right) + \left(\frac{18.5 - 19.2}{2.01}\right) + \left(\frac{18.7 - 19.2}{2.01}\right) + \left(\frac{20.1 - 19.2}{2.01}\right)}{5}$$

$$= \frac{\left\{\frac{(16.3 - 19.2) + (22.4 - 19.2) + (18.5 - 19.2) + (18.7 - 19.2) + (20.1 - 19.2)}{2.01}\right\}}{5}$$

> The numerator has been clarified.

$$= \frac{\left\{\frac{16.3 + 22.4 + 18.5 + 18.7 + 20.1 - 19.2 - 19.2 - 19.2 - 19.2 - 19.2}{2.01}\right\}}{5}$$

> The numerator has been reorganized so that each value and (-19.2) are separate.

$$= \frac{\left\{\frac{96 - 19.2 \times 5}{2.01}\right\}}{5}$$

$$= \frac{\left\{\frac{96 - 96}{2.01}\right\}}{5}$$

$$= \frac{0}{5}$$

$$= 0$$

2. Standard deviation of the standard score of the 100m track race

$$= \sqrt{\frac{\left(\frac{16.3 - 19.2}{2.01} - 0\right)^2 + \left(\frac{22.4 - 19.2}{2.01} - 0\right)^2 + \left(\frac{18.5 - 19.2}{2.01} - 0\right)^2 + \left(\frac{18.7 - 19.2}{2.01} - 0\right)^2 + \left(\frac{20.1 - 19.2}{2.01} - 0\right)^2}{5}}$$

$$= \sqrt{\frac{\left(\frac{16.3 - 19.2}{2.01}\right)^2 + \left(\frac{22.4 - 19.2}{2.01}\right)^2 + \left(\frac{18.5 - 19.2}{2.01}\right)^2 + \left(\frac{18.7 - 19.2}{2.01}\right)^2 + \left(\frac{20.1 - 19.2}{2.01}\right)^2}{5}}$$

$$= \sqrt{\frac{\left\{\frac{(16.3 - 19.2)^2 + (22.4 - 19.2)^2 + (18.5 - 19.2)^2 + (18.7 - 19.2)^2 + (20.1 - 19.2)^2}{2.01^2}\right\}}{5}}$$

> The numerator has been clarified.

$$= \sqrt{\frac{1}{2.01^2} \times \frac{(16.3 - 19.2)^2 + (22.4 - 19.2)^2 + (18.5 - 19.2)^2 + (18.7 - 19.2)^2 + (20.1 - 19.2)^2}{5}}$$

> The numerator has been clarified.

$$= \frac{1}{2.01} \times \sqrt{\frac{(16.3 - 19.2)^2 + (22.4 - 19.2)^2 + (18.5 - 19.2)^2 + (18.7 - 19.2)^2 + (20.1 - 19.2)^2}{5}}$$

$$= \frac{1}{\text{standard deviation of the 100m track race}} \times \text{standard deviation of the 100m track race}$$

> Carefully look at the table on page 78.

$$= 1$$

SUMMARY

- *Standardization* helps you examine the value of a data point relative to the rest of your data by using its distance from the mean and "the size of scattering" of the data.
- Use standardization to:

  - Compare variables with different ranges
  - Compare variables that use different units of measurements

- A data point that has been standardized is called the *standard score* for that observation. Deviation score is an application of standard score.

LET'S OBTAIN THE PROBABILITY

1. PROBABILITY DENSITY FUNCTION

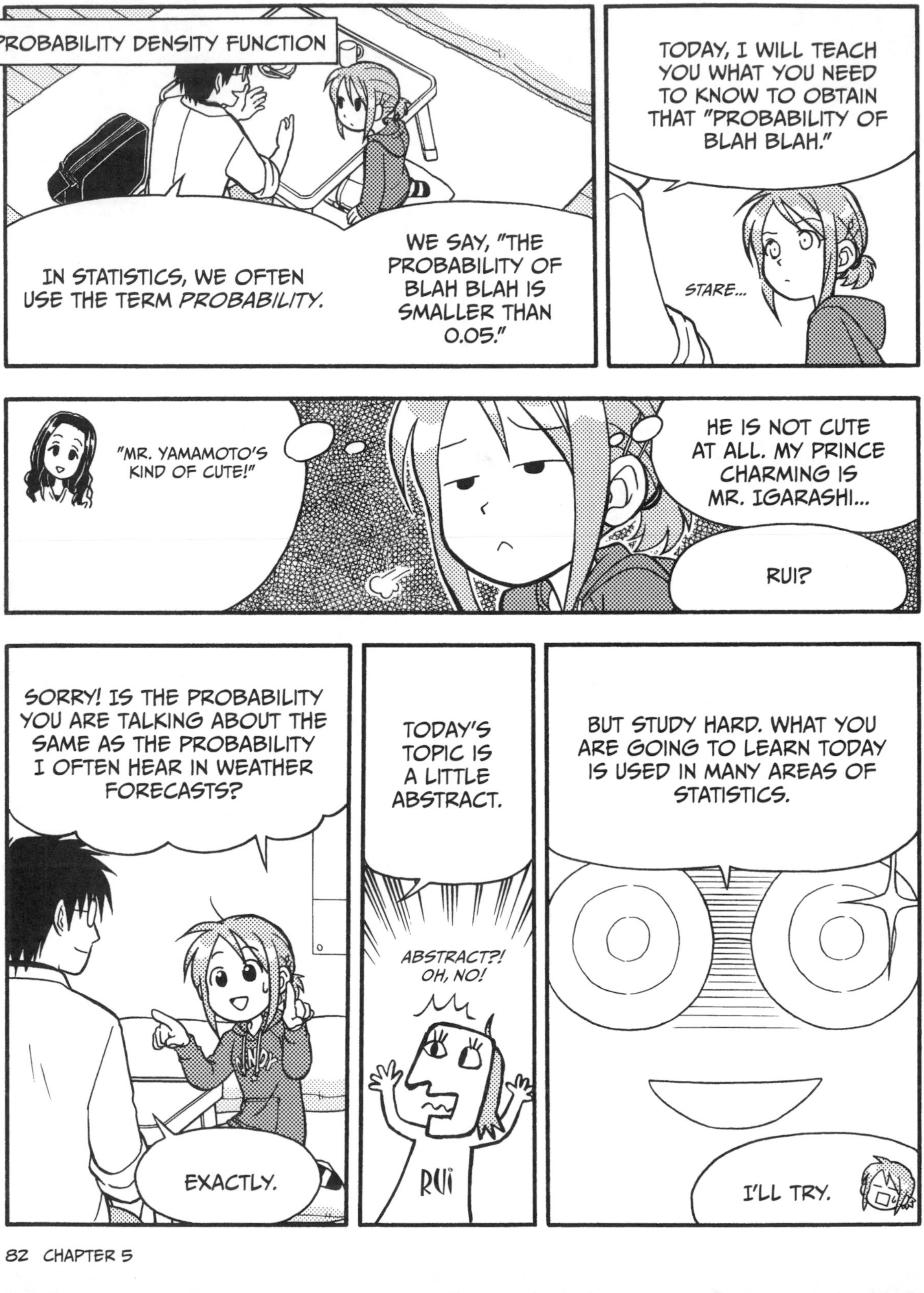
IN STATISTICS, WE OFTEN USE THE TERM *PROBABILITY*.
WE SAY, "THE PROBABILITY OF BLAH BLAH IS SMALLER THAN 0.05."
TODAY, I WILL TEACH YOU WHAT YOU NEED TO KNOW TO OBTAIN THAT "PROBABILITY OF BLAH BLAH."
STARE...
"MR. YAMAMOTO'S KIND OF CUTE!"
HE IS NOT CUTE AT ALL. MY PRINCE CHARMING IS MR. IGARASHI...
RUI?
SORRY! IS THE PROBABILITY YOU ARE TALKING ABOUT THE SAME AS THE PROBABILITY I OFTEN HEAR IN WEATHER FORECASTS?
EXACTLY.
TODAY'S TOPIC IS A LITTLE ABSTRACT.
ABSTRACT?! OH, NO!
RUI
BUT STUDY HARD. WHAT YOU ARE GOING TO LEARN TODAY IS USED IN MANY AREAS OF STATISTICS.
I'LL TRY.

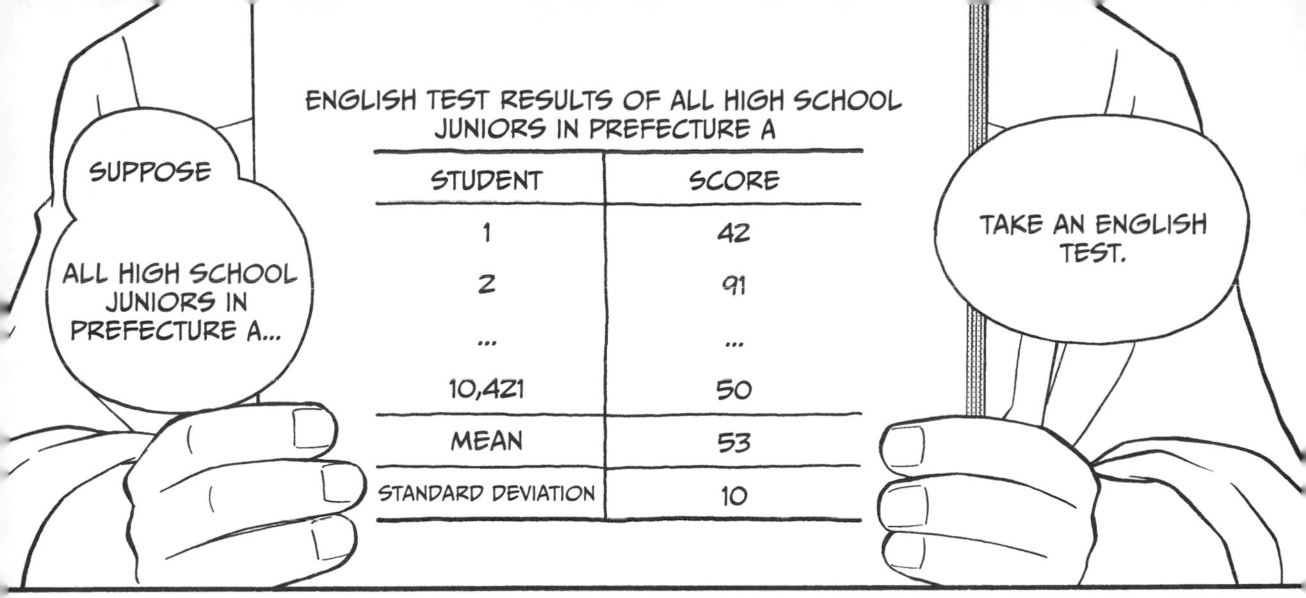
SUPPOSE
ALL HIGH SCHOOL JUNIORS IN PREFECTURE A...
TAKE AN ENGLISH TEST.

ENGLISH TEST RESULTS OF ALL HIGH SCHOOL JUNIORS IN PREFECTURE A

STUDENT	SCORE
1	42
2	91
...	...
10,421	50
MEAN	53
STANDARD DEVIATION	10

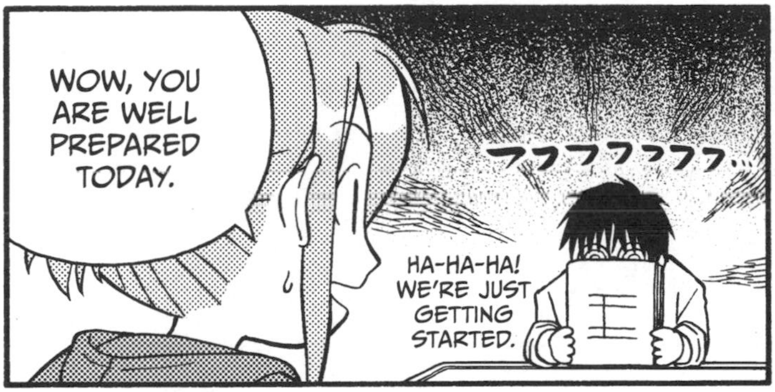
WOW, YOU ARE WELL PREPARED TODAY.
フフフフフフフ...
HA-HA-HA! WE'RE JUST GETTING STARTED.

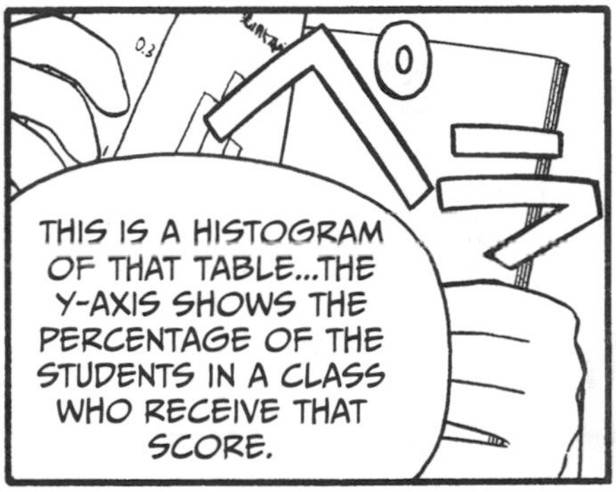
THIS IS A HISTOGRAM OF THAT TABLE...THE Y-AXIS SHOWS THE PERCENTAGE OF THE STUDENTS IN A CLASS WHO RECEIVE THAT SCORE.

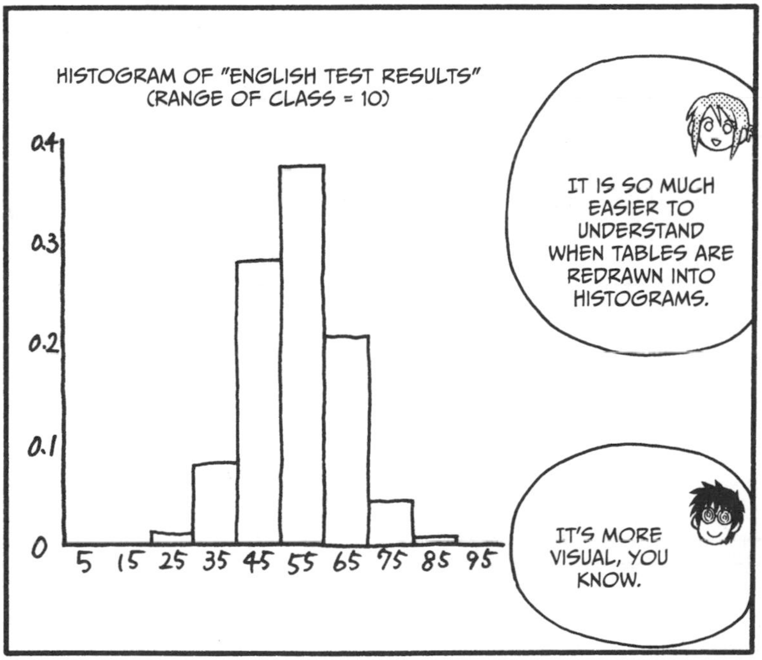
HISTOGRAM OF "ENGLISH TEST RESULTS" (RANGE OF CLASS = 10)
0.4
0.3
0.2
0.1
0
5 15 25 35 45 55 65 75 85 95
IT IS SO MUCH EASIER TO UNDERSTAND WHEN TABLES ARE REDRAWN INTO HISTOGRAMS.
IT'S MORE VISUAL, YOU KNOW.

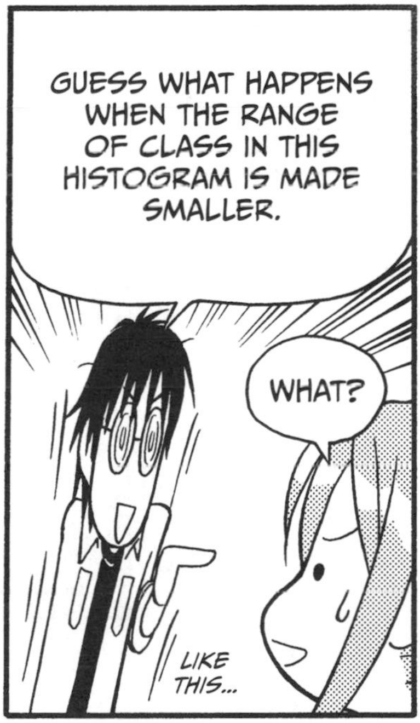
GUESS WHAT HAPPENS WHEN THE RANGE OF CLASS IN THIS HISTOGRAM IS MADE SMALLER.
WHAT?
LIKE THIS...

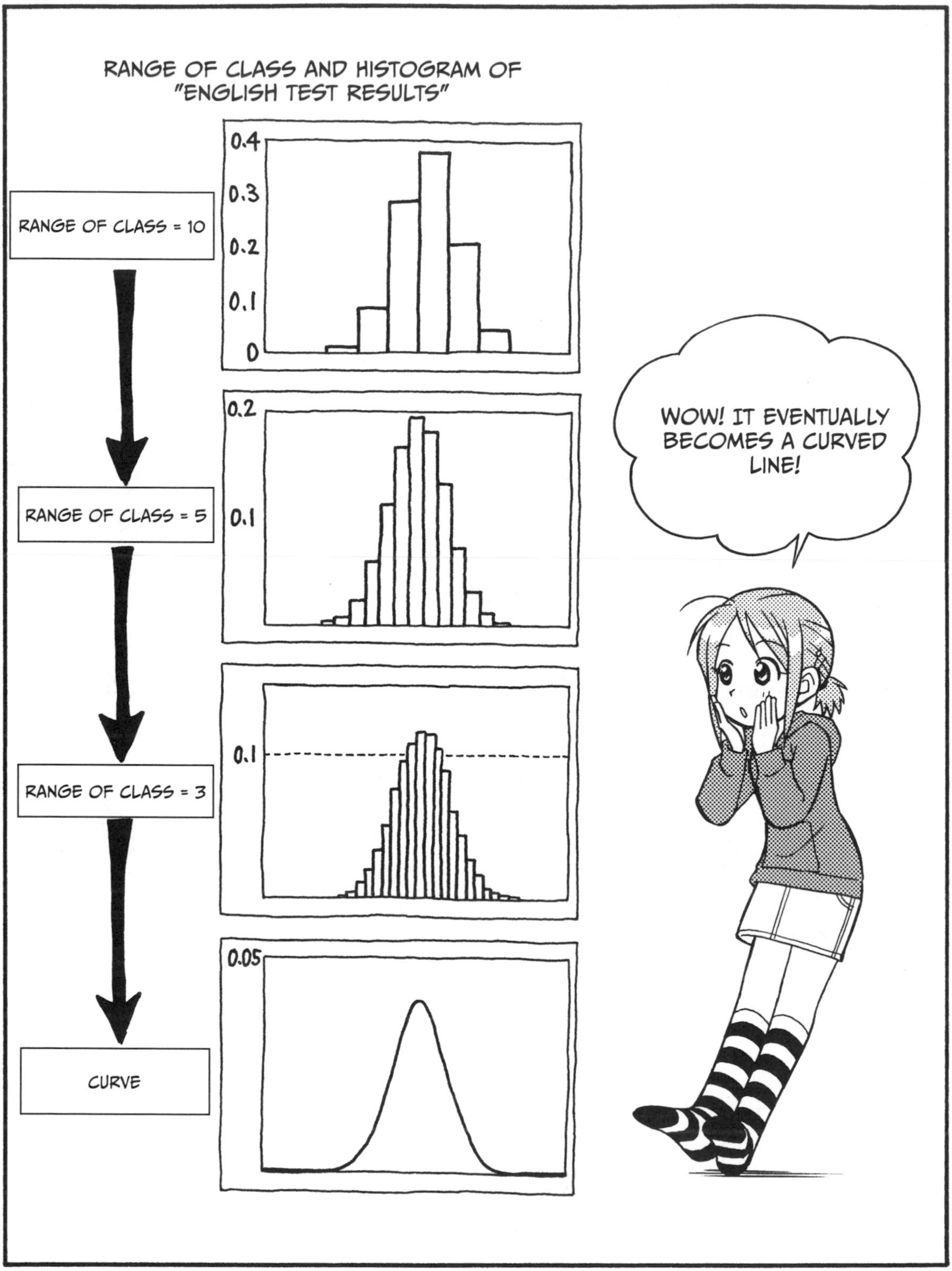

RANGE OF CLASS AND HISTOGRAM OF "ENGLISH TEST RESULTS"
RANGE OF CLASS = 10
0.4
0.3
0.2
0.1
0
RANGE OF CLASS = 5
0.2
0.1
RANGE OF CLASS = 3
0.1
CURVE
0.05
WOW! IT EVENTUALLY BECOMES A CURVED LINE!

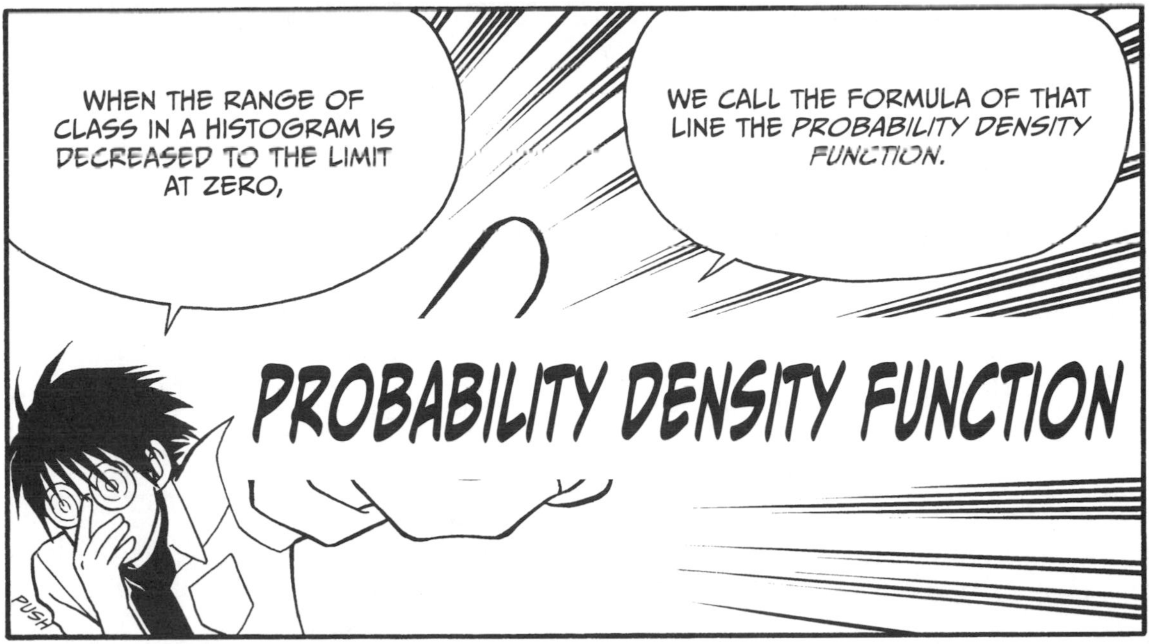

WHEN THE RANGE OF CLASS IN A HISTOGRAM IS DECREASED TO THE LIMIT AT ZERO,
WE CALL THE FORMULA OF THAT LINE THE *PROBABILITY DENSITY FUNCTION*.
PROBABILITY DENSITY FUNCTION
PUSH

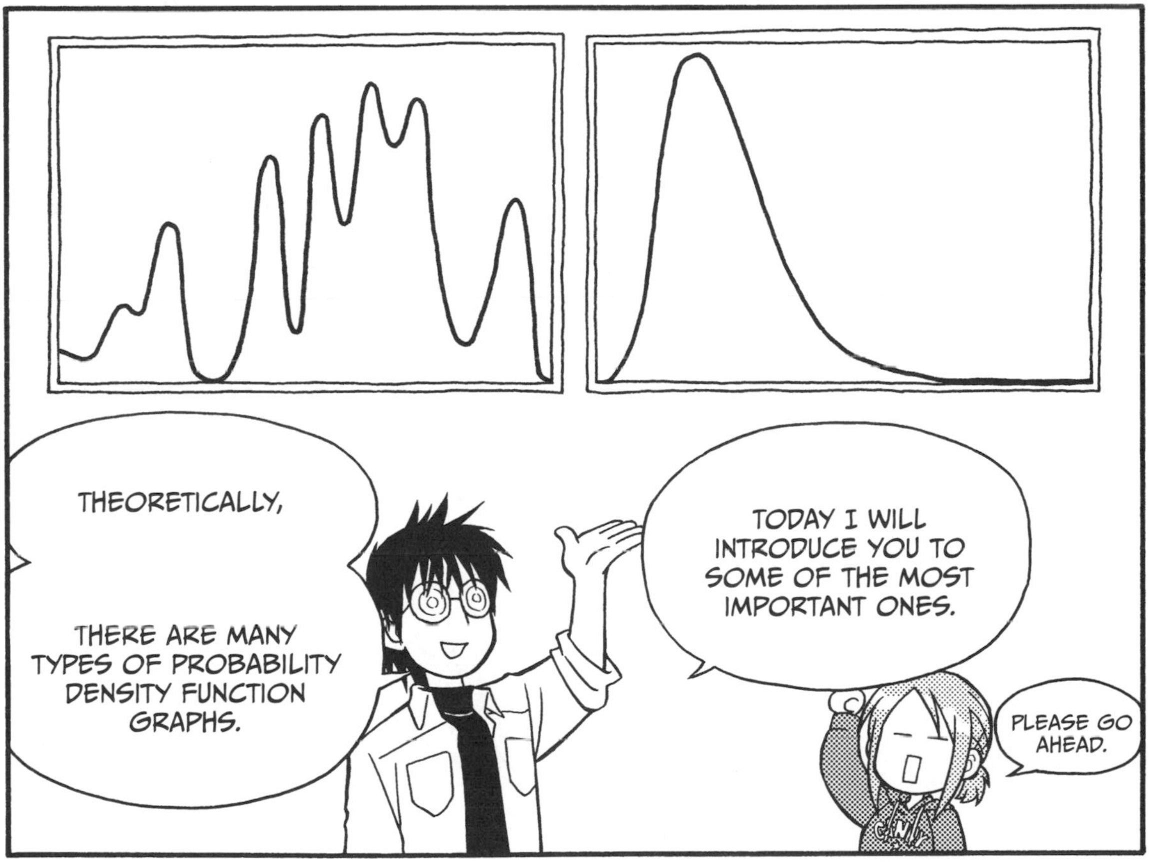

THEORETICALLY,
THERE ARE MANY TYPES OF PROBABILITY DENSITY FUNCTION GRAPHS.
TODAY I WILL INTRODUCE YOU TO SOME OF THE MOST IMPORTANT ONES.
PLEASE GO AHEAD.

2. NORMAL DISTRIBUTION

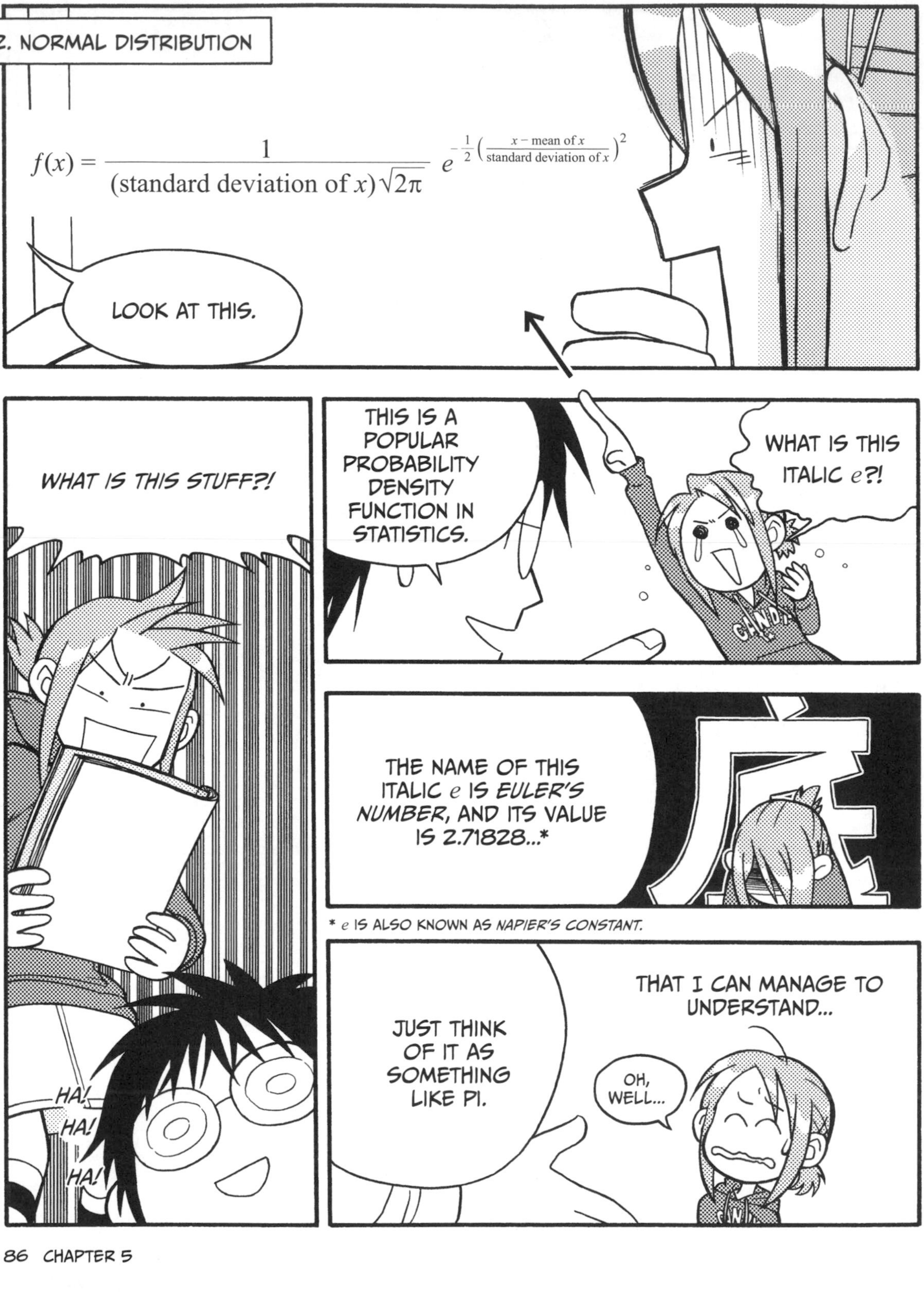
$$f(x) = \frac{1}{(\text{standard deviation of } x)\sqrt{2\pi}}\, e^{-\frac{1}{2}\left(\frac{x - \text{mean of } x}{\text{standard deviation of } x}\right)^2}$$
LOOK AT THIS.
WHAT IS THIS STUFF?!
HA! HA! HA!
THIS IS A POPULAR PROBABILITY DENSITY FUNCTION IN STATISTICS.
WHAT IS THIS ITALIC *e*?!
CANDY
THE NAME OF THIS ITALIC *e* IS *EULER'S NUMBER*, AND ITS VALUE IS 2.71828...*
JUST THINK OF IT AS SOMETHING LIKE PI.
THAT I CAN MANAGE TO UNDERSTAND...
OH, WELL...

* *e* IS ALSO KNOWN AS *NAPIER'S CONSTANT.*

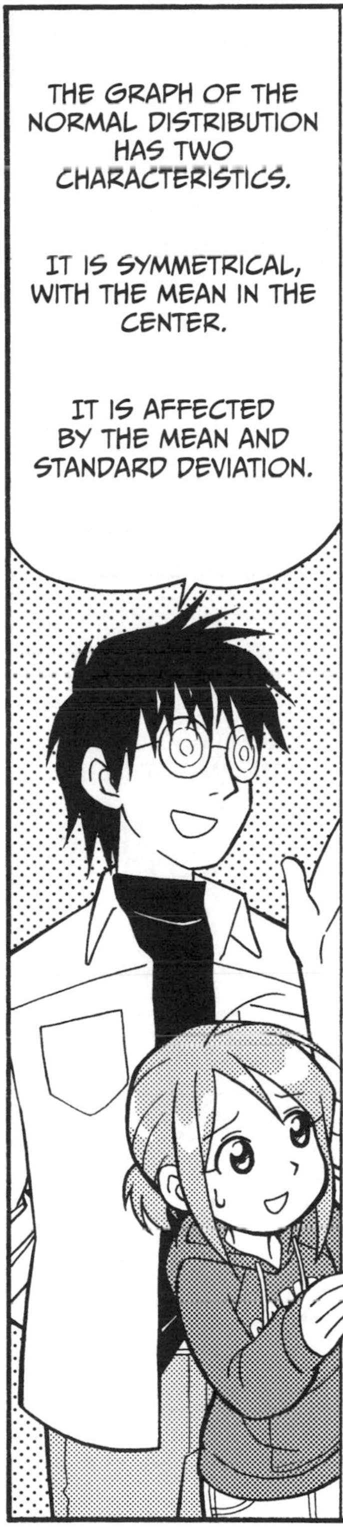
THE GRAPH OF THE NORMAL DISTRIBUTION HAS TWO CHARACTERISTICS.
IT IS SYMMETRICAL, WITH THE MEAN IN THE CENTER.
IT IS AFFECTED BY THE MEAN AND STANDARD DEVIATION.

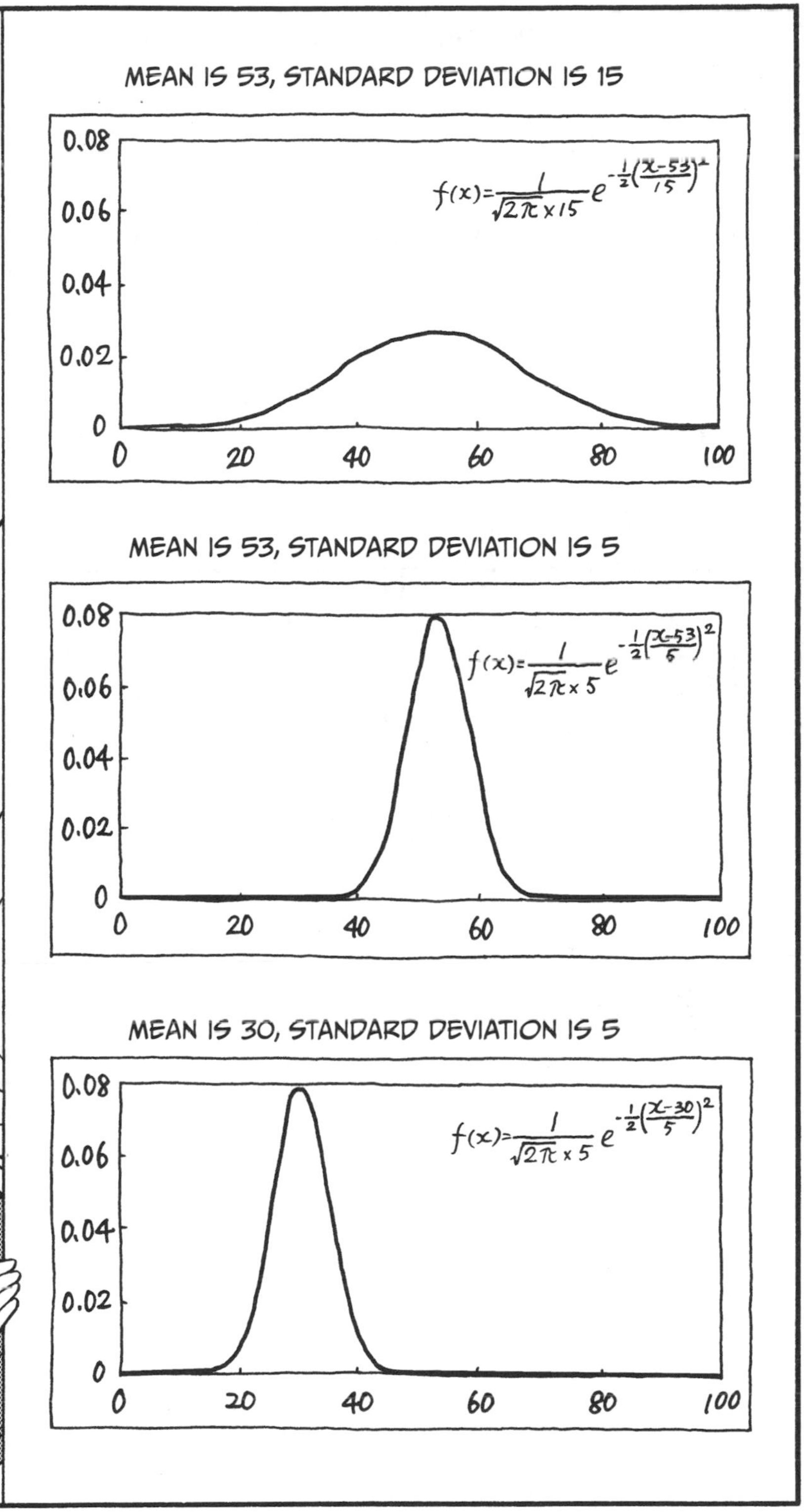
MEAN IS 53, STANDARD DEVIATION IS 15
0.08
0.06
0.04
0.02
0
0 20 40 60 80 100
$f(x)=\frac{1}{\sqrt{2\pi}\times 15}e^{-\frac{1}{2}\left(\frac{x-53}{15}\right)^2}$
MEAN IS 53, STANDARD DEVIATION IS 5
0.08
0.06
0.04
0.02
0
0 20 40 60 80 100
$f(x)=\frac{1}{\sqrt{2\pi}\times 5}e^{-\frac{1}{2}\left(\frac{x-53}{5}\right)^2}$
MEAN IS 30, STANDARD DEVIATION IS 5
0.08
0.06
0.04
0.02
0
0 20 40 60 80 100
$f(x)=\frac{1}{\sqrt{2\pi}\times 5}e^{-\frac{1}{2}\left(\frac{x-30}{5}\right)^2}$

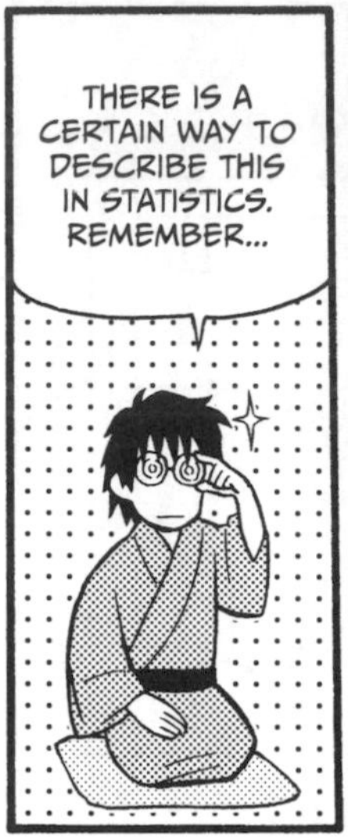
THERE IS A CERTAIN WAY TO DESCRIBE THIS IN STATISTICS. REMEMBER...

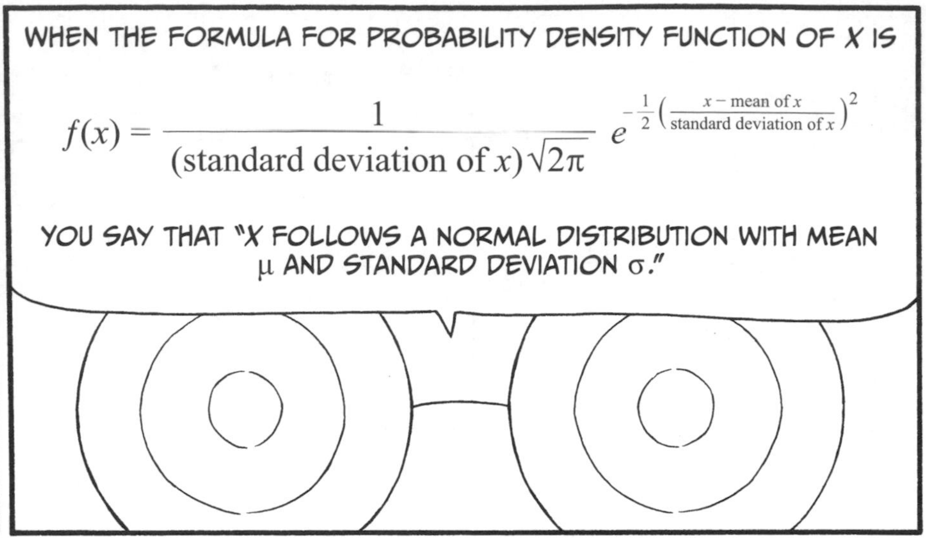
WHEN THE FORMULA FOR PROBABILITY DENSITY FUNCTION OF X IS
$f(x) = \frac{1}{(\text{standard deviation of } x)\sqrt{2\pi}} e^{-\frac{1}{2}\left(\frac{x - \text{mean of } x}{\text{standard deviation of } x}\right)^2}$
YOU SAY THAT "X FOLLOWS A NORMAL DISTRIBUTION WITH MEAN μ AND STANDARD DEVIATION σ."

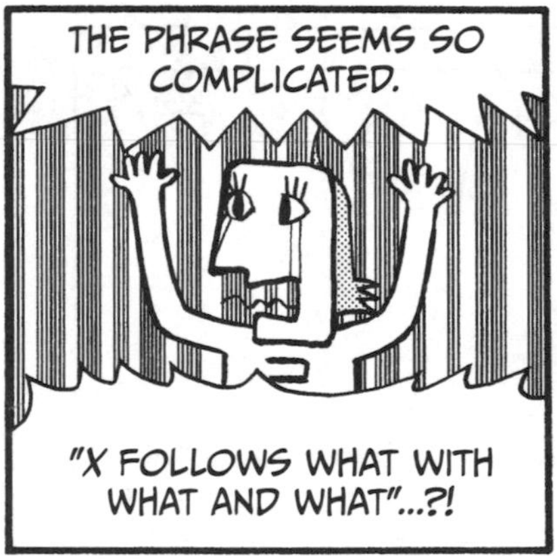
THE PHRASE SEEMS SO COMPLICATED.
"X FOLLOWS WHAT WITH WHAT AND WHAT"...?!

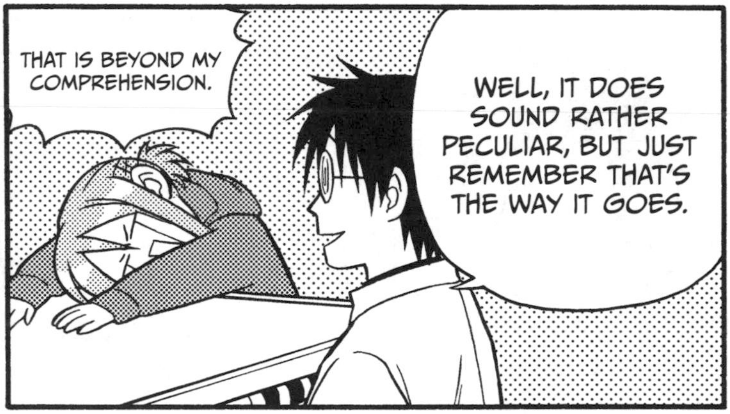
THAT IS BEYOND MY COMPREHENSION.
WELL, IT DOES SOUND RATHER PECULIAR, BUT JUST REMEMBER THAT'S THE WAY IT GOES.

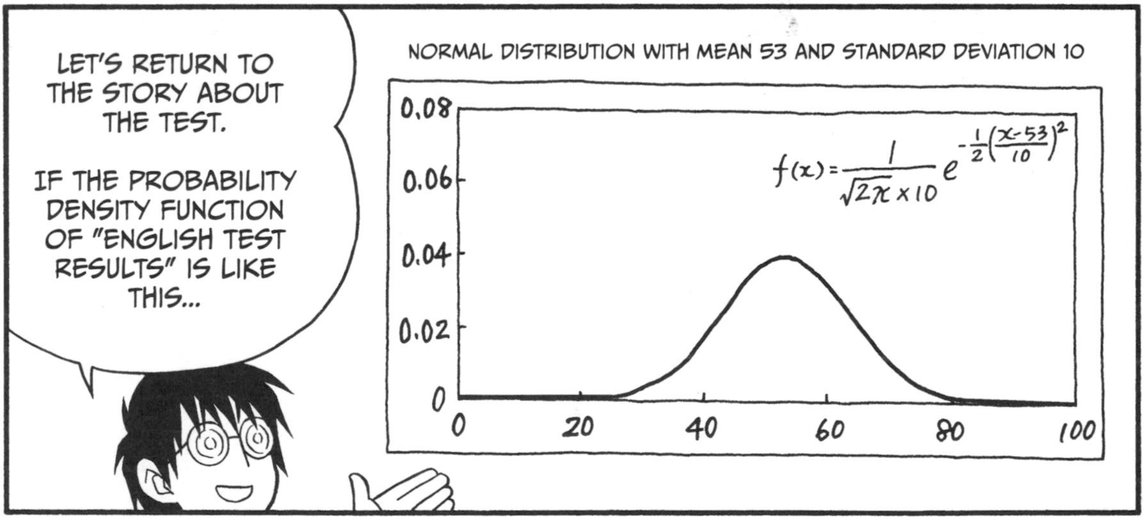
LET'S RETURN TO THE STORY ABOUT THE TEST.
IF THE PROBABILITY DENSITY FUNCTION OF "ENGLISH TEST RESULTS" IS LIKE THIS...
NORMAL DISTRIBUTION WITH MEAN 53 AND STANDARD DEVIATION 10
$f(x) = \frac{1}{\sqrt{2\pi} \times 10} e^{-\frac{1}{2}\left(\frac{x-53}{10}\right)^2}$
0.08
0.06
0.04
0.02
0
0
20
40
60
80
100

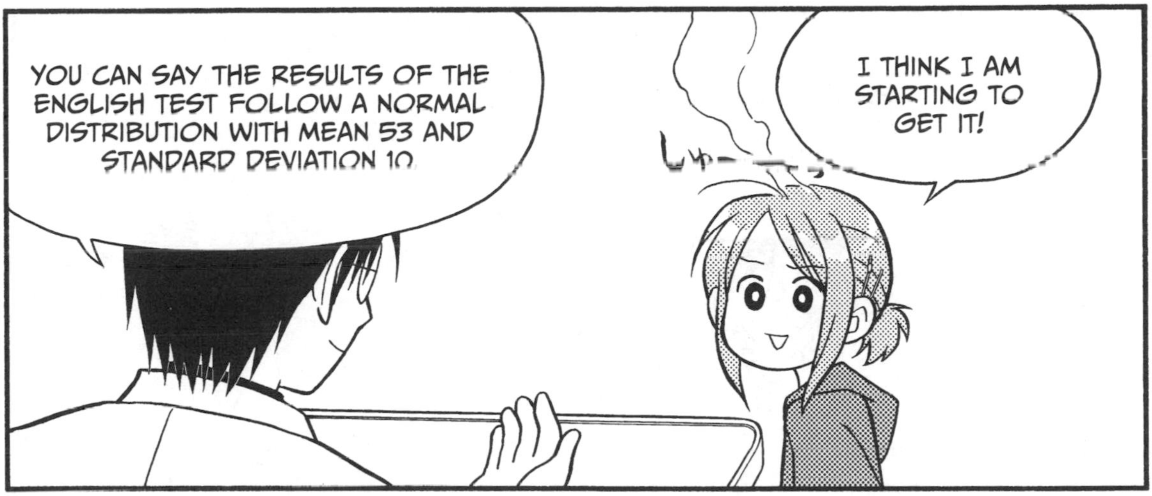
YOU CAN SAY THE RESULTS OF THE ENGLISH TEST FOLLOW A NORMAL DISTRIBUTION WITH MEAN 53 AND STANDARD DEVIATION 10.
I THINK I AM STARTING TO GET IT!

3. STANDARD NORMAL DISTRIBUTION

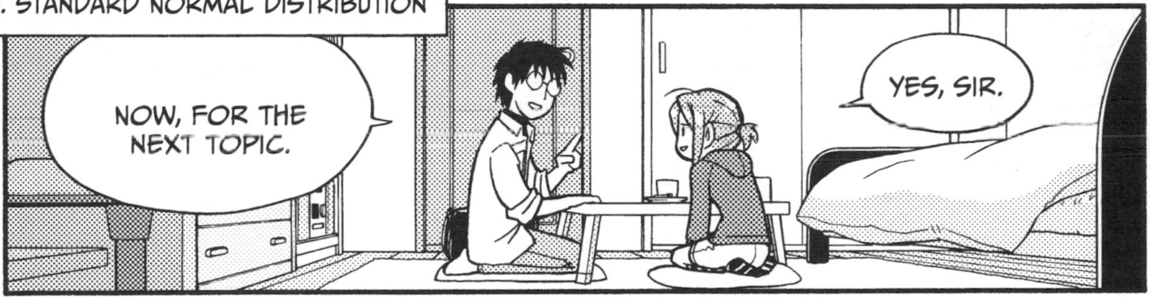
NOW, FOR THE NEXT TOPIC.
YES, SIR.

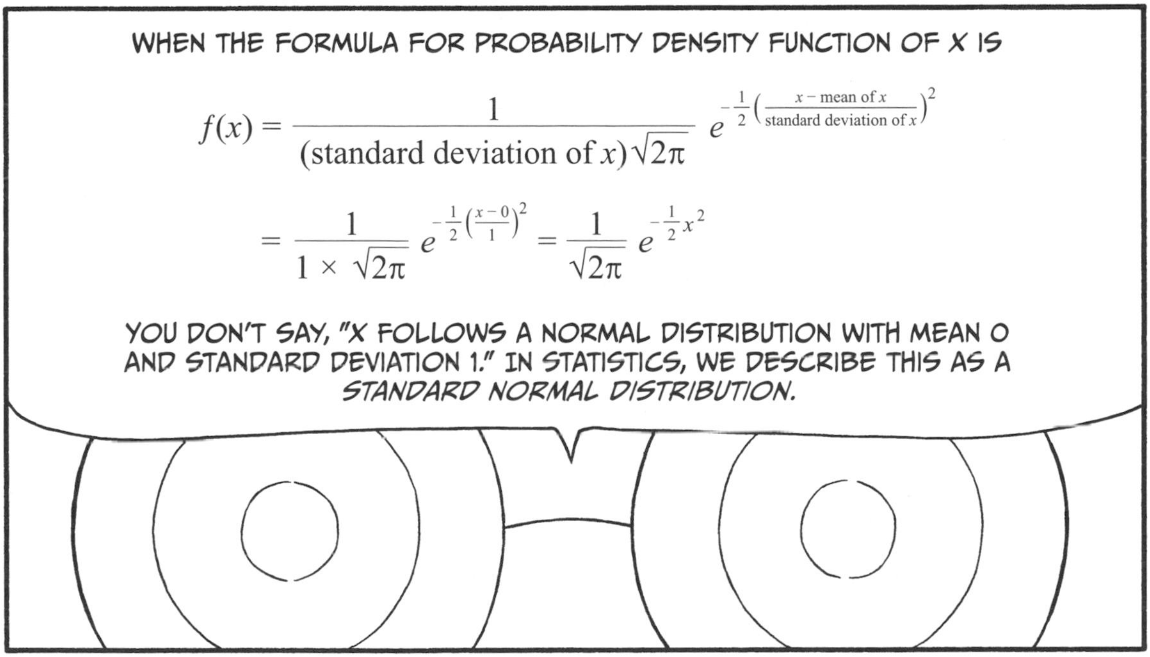
WHEN THE FORMULA FOR PROBABILITY DENSITY FUNCTION OF X IS
$$f(x) = \frac{1}{(\text{standard deviation of } x)\sqrt{2\pi}} e^{-\frac{1}{2}\left(\frac{x - \text{mean of } x}{\text{standard deviation of } x}\right)^2}$$
$$= \frac{1}{1 \times \sqrt{2\pi}} e^{-\frac{1}{2}\left(\frac{x-0}{1}\right)^2} = \frac{1}{\sqrt{2\pi}} e^{-\frac{1}{2}x^2}$$
YOU DON'T SAY, "X FOLLOWS A NORMAL DISTRIBUTION WITH MEAN 0 AND STANDARD DEVIATION 1." IN STATISTICS, WE DESCRIBE THIS AS A *STANDARD NORMAL DISTRIBUTION*.

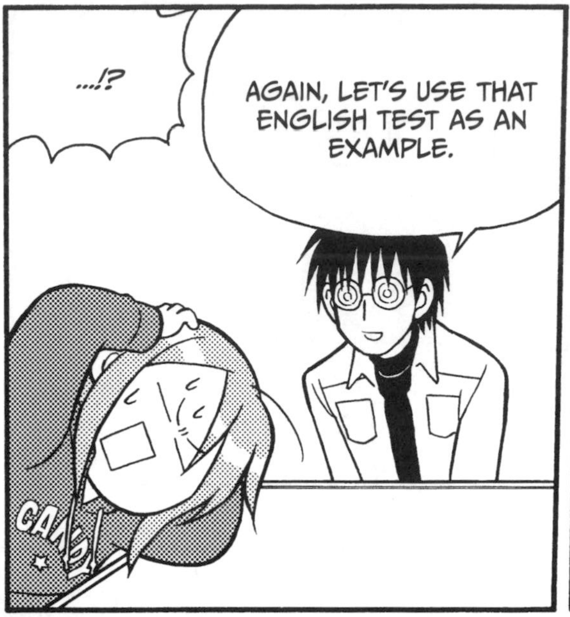
...!?
AGAIN, LET'S USE THAT ENGLISH TEST AS AN EXAMPLE.

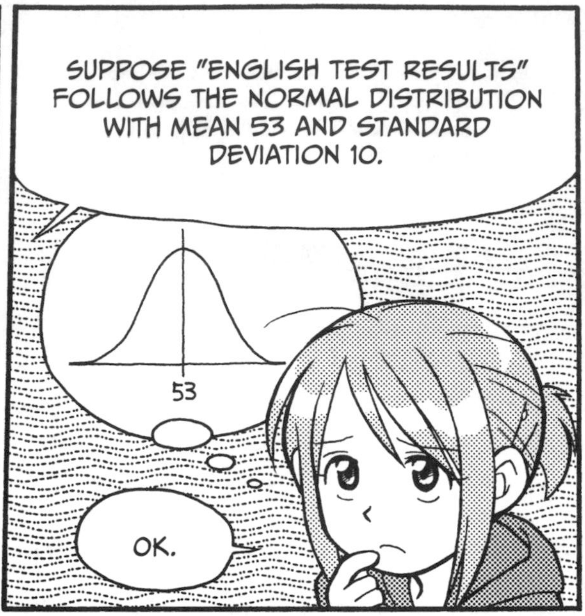
SUPPOSE "ENGLISH TEST RESULTS" FOLLOWS THE NORMAL DISTRIBUTION WITH MEAN 53 AND STANDARD DEVIATION 10.
53
OK.

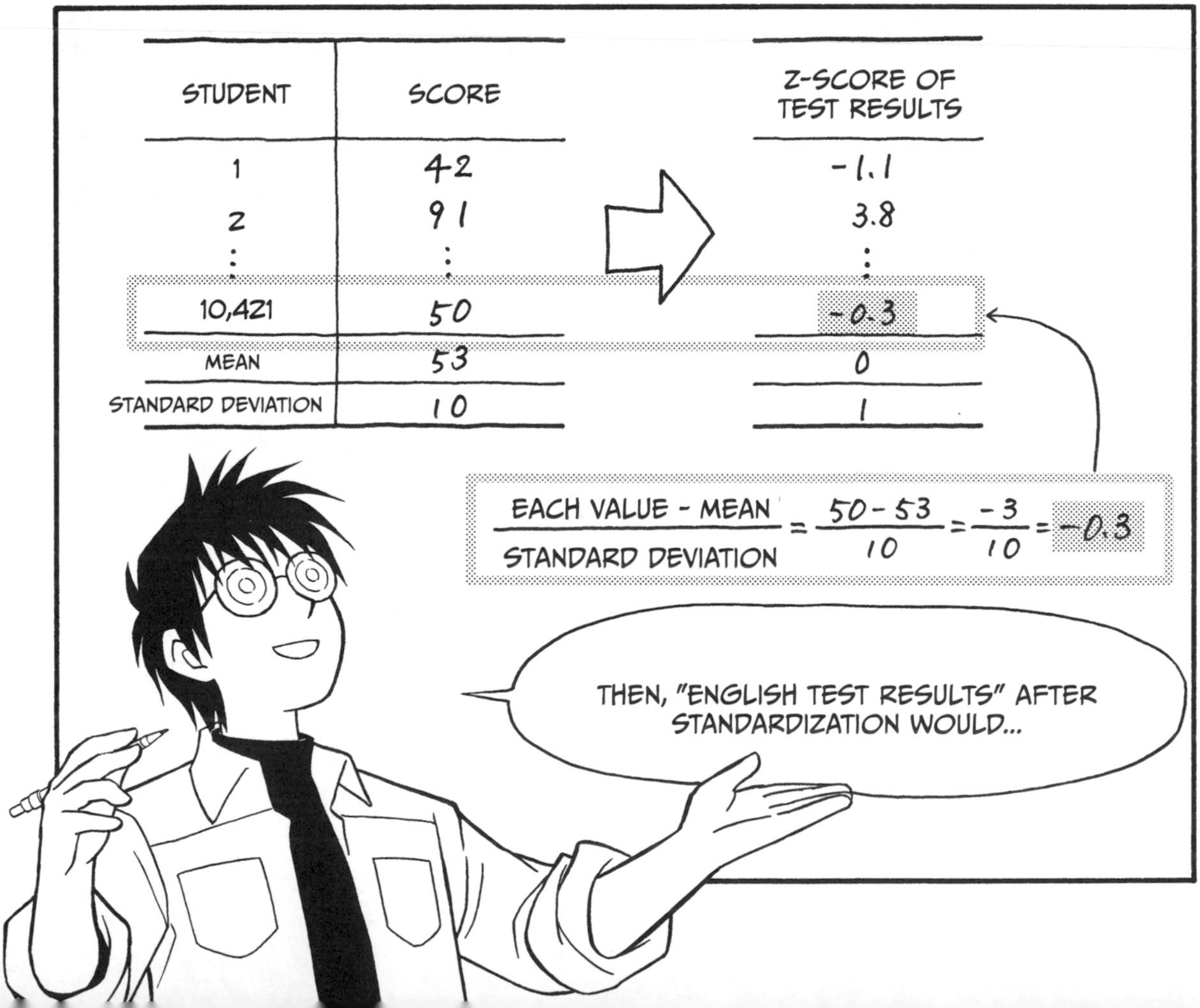
STUDENT | SCORE
1 | 42
2 | 91
⋮ | ⋮
10,421 | 50
MEAN | 53
STANDARD DEVIATION | 10
Z-SCORE OF TEST RESULTS
-1.1
3.8
⋮
-0.3
0
1
EACH VALUE - MEAN / STANDARD DEVIATION = 50 - 53 / 10 = -3 / 10 = -0.3
THEN, "ENGLISH TEST RESULTS" AFTER STANDARDIZATION WOULD...

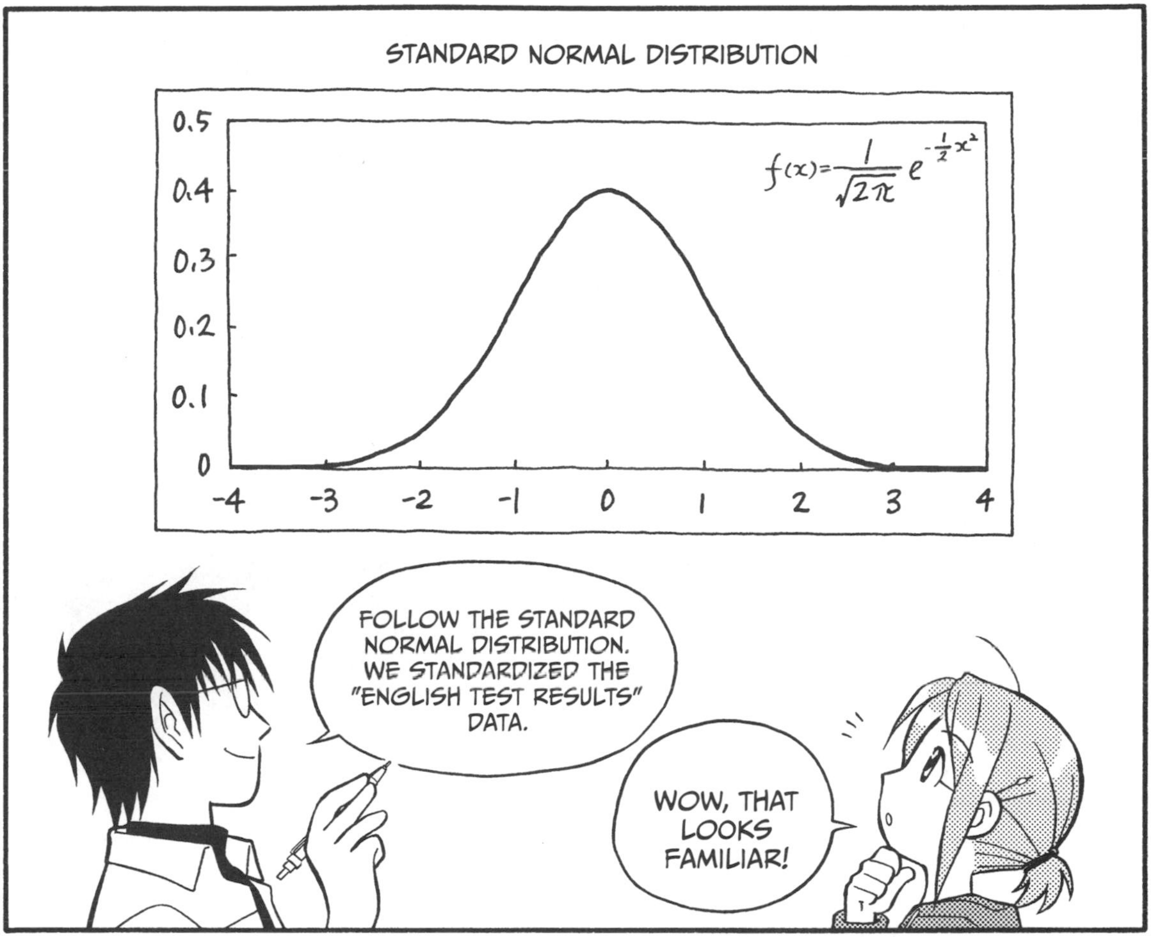
STANDARD NORMAL DISTRIBUTION
0.5
0.4
0.3
0.2
0.1
0
-4 -3 -2 -1 0 1 2 3 4
$f(x)=\frac{1}{\sqrt{2\pi}}e^{-\frac{1}{2}x^2}$
FOLLOW THE STANDARD NORMAL DISTRIBUTION. WE STANDARDIZED THE "ENGLISH TEST RESULTS" DATA.
WOW, THAT LOOKS FAMILIAR!

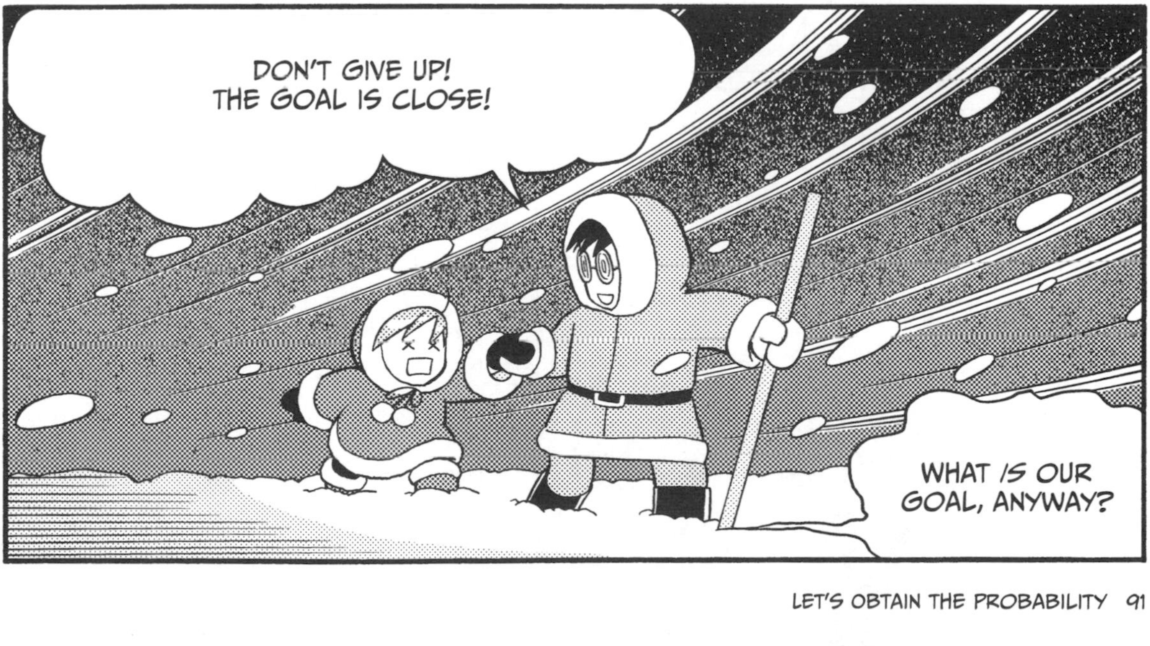
DON'T GIVE UP! THE GOAL IS CLOSE!
WHAT *IS* OUR GOAL, ANYWAY?

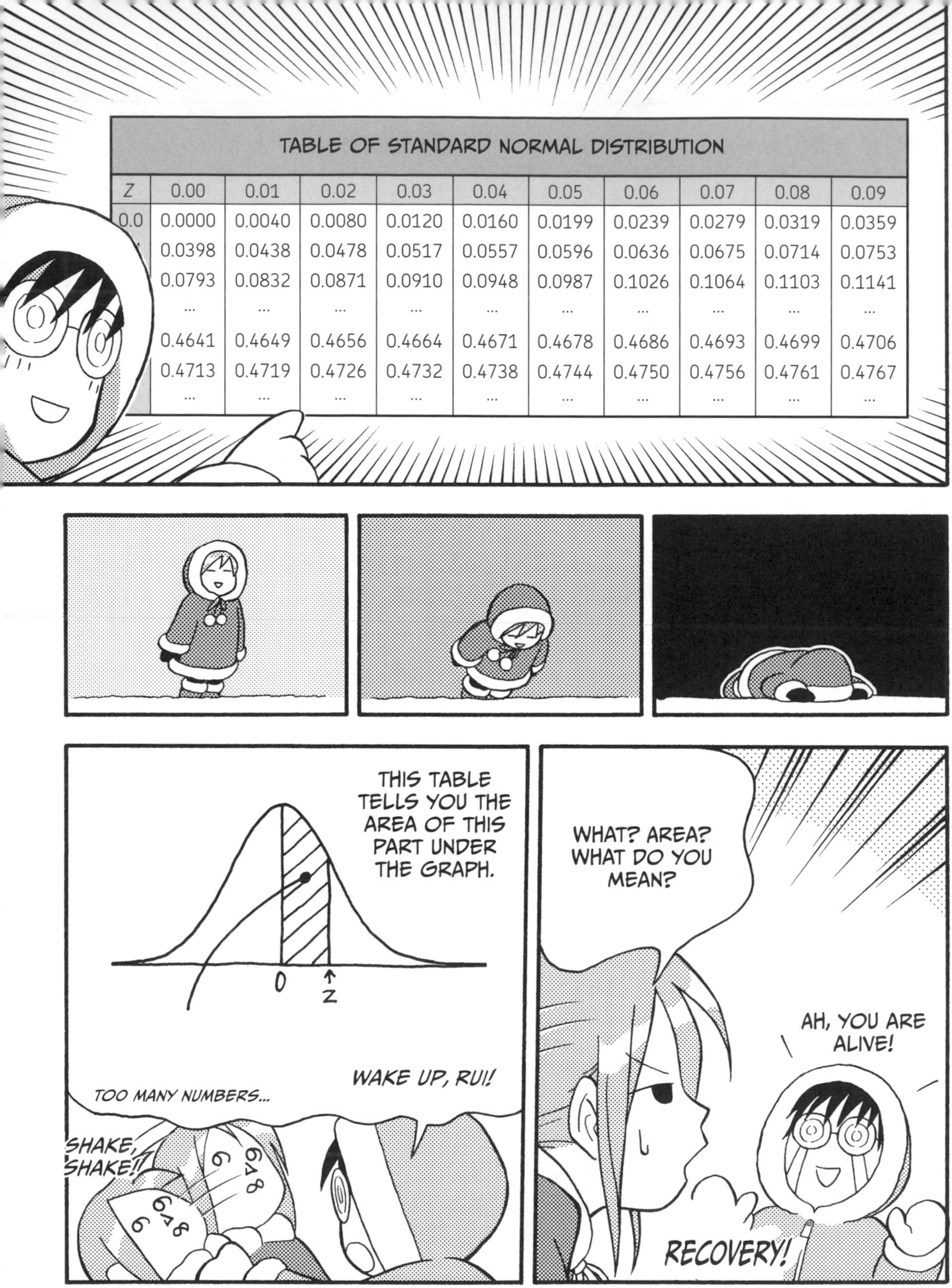

TABLE OF STANDARD NORMAL DISTRIBUTION

Z	0.00	0.01	0.02	0.03	0.04	0.05	0.06	0.07	0.08	0.09
0.0	0.0000	0.0040	0.0080	0.0120	0.0160	0.0199	0.0239	0.0279	0.0319	0.0359
	0.0398	0.0438	0.0478	0.0517	0.0557	0.0596	0.0636	0.0675	0.0714	0.0753
	0.0793	0.0832	0.0871	0.0910	0.0948	0.0987	0.1026	0.1064	0.1103	0.1141
	...	...	...	...	...	...	...	...	...	...
	0.4641	0.4649	0.4656	0.4664	0.4671	0.4678	0.4686	0.4693	0.4699	0.4706
	0.4713	0.4719	0.4726	0.4732	0.4738	0.4744	0.4750	0.4756	0.4761	0.4767
	...	...	...	...	...	...	...	...	...	...

THIS TABLE TELLS YOU THE AREA OF THIS PART UNDER THE GRAPH.
0
Z
WAKE UP, RUI!
TOO MANY NUMBERS...
SHAKE, SHAKE!!
WHAT? AREA? WHAT DO YOU MEAN?
AH, YOU ARE ALIVE!
RECOVERY!

TAKE Z = 1.96, AND THINK ABOUT THIS.

0   1.96

OK.

CONSIDER Z = 1.96

$$Z = 1.9 + 0.06$$

AS Z = 1.9 + 0.06

YOU SEPARATE THE NUMBER BETWEEN THE FIRST DECIMAL AND THE SECOND DECIMAL?

THEN GO BACK TO THE TABLE.

Z	0.00	0.01	0.02	0.03	0.04		0.06	0.07	0.08
0.0	0.0000	0.0040	0.0080	0.0120	0.0160	0.0199	0.0239	0.0279	0.0319
0.1	0.0398	0.0438	0.0478	0.0517	0.0557	0.0596	0.0636	0.0675	0.071
0.2	0.0793	0.0832	0.0871	0.0910	0.0948	0.0987	0.1026	0.1064	0.1141
...	...	...	...	...	...	...	...	...	...
1.8	0.4641	0.4649	0.4656	0.4664	0.4671	0.4678	0.4686	0.4693	0.4699	0.4706
1.9	0.4713	0.4719	0.4726	0.4732	0.4738	0.4744	**0.4750**	0.4756	0.4761	0.4767
...	...	...	...	...	...	...	...	...	...

THE LINE AND ROW FOR 1.9 AND 0.06, RESPECTIVELY, CROSS EACH OTHER AT...0.4750!

0.4750!

0.4750

0   1.96

YES. THAT IS THE AREA WHEN Z = 1.96.

OH, AND I FORGOT TO MENTION—THE AREA BETWEEN THE PROBABILITY DENSITY FUNCTION GRAPH AND THE HORIZONTAL AXIS IS 1, REGARDLESS OF WHETHER IT IS A STANDARD NORMAL DISTRIBUTION OR SOMETHING ELSE.

AREA = 1

AHA!

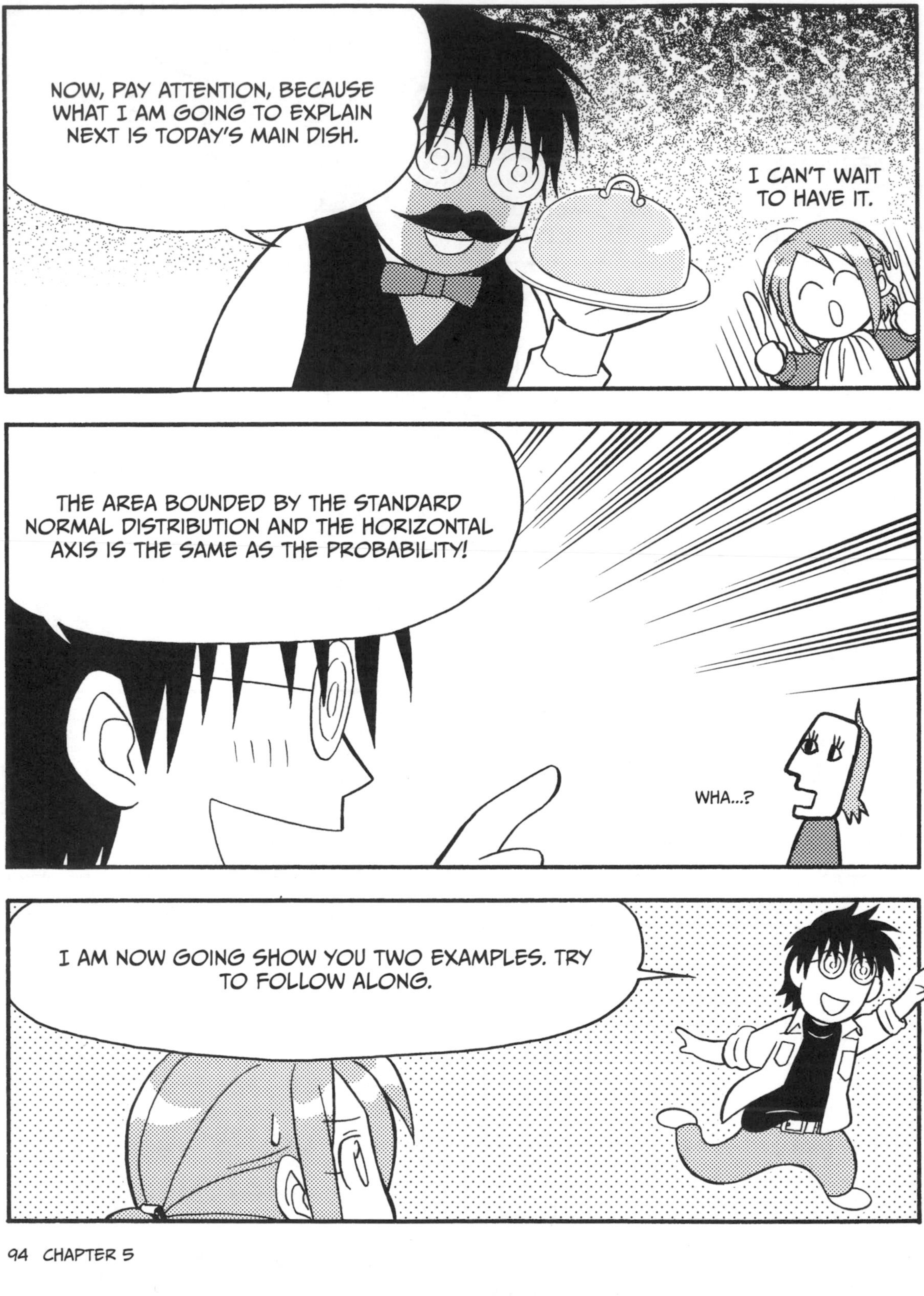
NOW, PAY ATTENTION, BECAUSE WHAT I AM GOING TO EXPLAIN NEXT IS TODAY'S MAIN DISH.
I CAN'T WAIT TO HAVE IT.
THE AREA BOUNDED BY THE STANDARD NORMAL DISTRIBUTION AND THE HORIZONTAL AXIS IS THE SAME AS THE PROBABILITY!
WHA...?
I AM NOW GOING SHOW YOU TWO EXAMPLES. TRY TO FOLLOW ALONG.

EXAMPLE I

All high school freshmen in prefecture B took a math test. After the tests were marked, the test results turned out to follow a normal distribution with a mean of 45 and a standard deviation of 10. Now, think carefully. The five sentences below all have the same meaning.

1. In a normal distribution with an average of 45 and a standard deviation of 10, the shaded area in the chart below is 0.5.

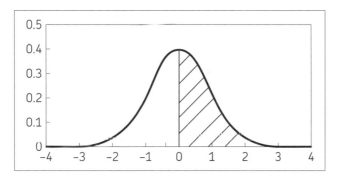

2. The ratio of students who scored 45 points or more is 0.5 (50% of all students tested).

3. When one student is randomly chosen from all students tested, the probability that the student's score is 45 or more is 0.5 (50%).

4. In a normal distribution of standardized "math test results," the ratio of students with a standard score of 0 or more is 0.5 (50% of all students tested).

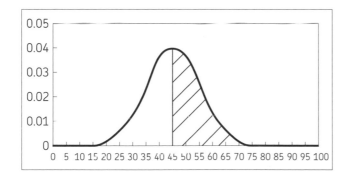

5. When one student's results are randomly chosen from all of those tested in a normal distribution of standardized "math test results," the probability that the selected student's standard score is 0 or more is 0.5 (50%).

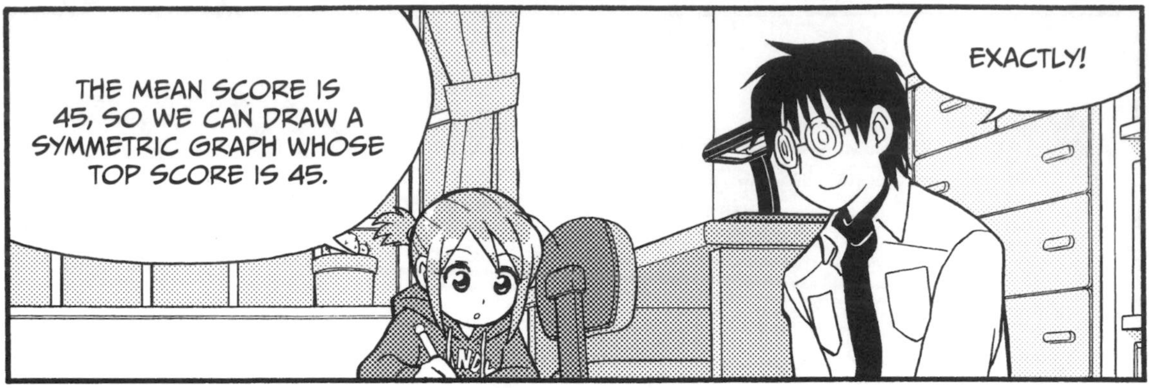
THE MEAN SCORE IS 45, SO WE CAN DRAW A SYMMETRIC GRAPH WHOSE TOP SCORE IS 45.
EXACTLY!

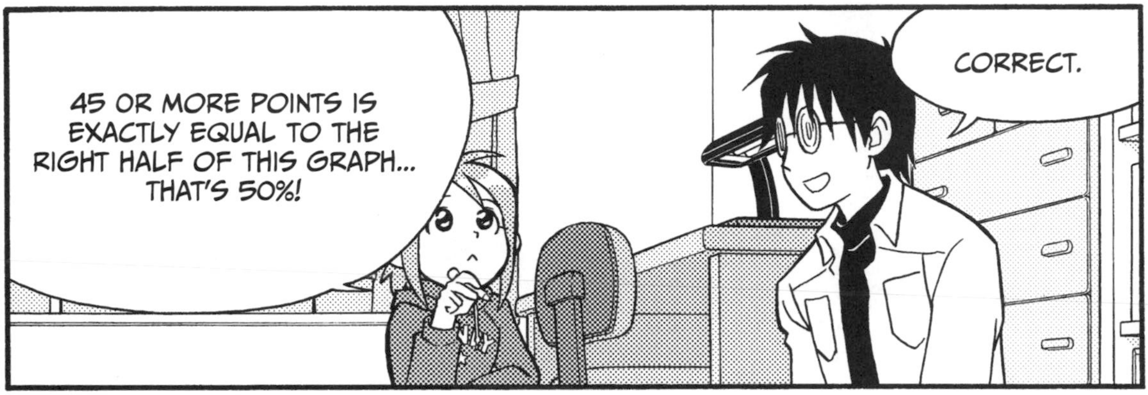
45 OR MORE POINTS IS EXACTLY EQUAL TO THE RIGHT HALF OF THIS GRAPH... THAT'S 50%!
CORRECT.

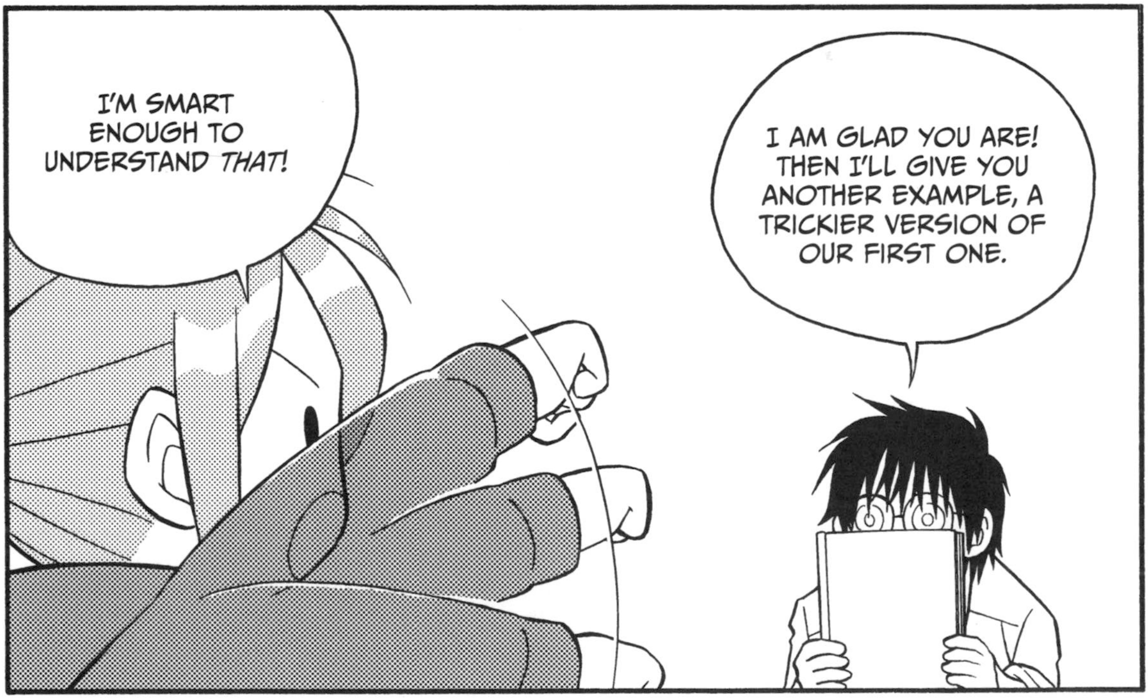
I'M SMART ENOUGH TO UNDERSTAND *THAT*!
I AM GLAD YOU ARE! THEN I'LL GIVE YOU ANOTHER EXAMPLE, A TRICKIER VERSION OF OUR FIRST ONE.

EXAMPLE II

All high school freshmen in prefecture B took a math test. Now, think carefully.
The five sentences below all have the same meaning.

1. In a normal distribution with a mean of 45 and a standard deviation of 10, the shaded area in the chart below is 0.5 – 0.4641 = 0.0359.

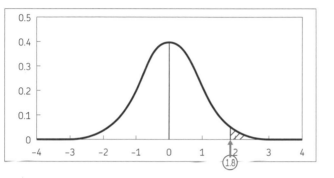

2. The ratio of students who scored 63 points or more is 0.5 – 0.4641 = 0.0359 (3.59% of all students tested).

3. When one student is randomly chosen from all those tested, the probability that the student's score is 63 or more is 0.5 – 0.4641 = 0.0359 (3.59%).

4. In a normal distribution of standardized test results, the ratio of students with standard scores (or z-scores) of 1.8 or more [(each value – average) ÷ standard deviation = (63 – 45) ÷ 10 = 18 ÷ 10 = 1.8] is 3.59% (0.5 – 0.4641 = 0.0359). You can also obtain this value from a table of standard normal distribution.

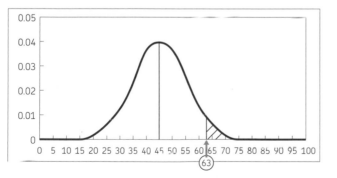

5. When one student is randomly chosen from all those tested in a normal distribution of standardized "math test results," the probability that the student's standard score is 1.8 or more is 0.5 – 0.4641 = 0.0359 (3.59%).

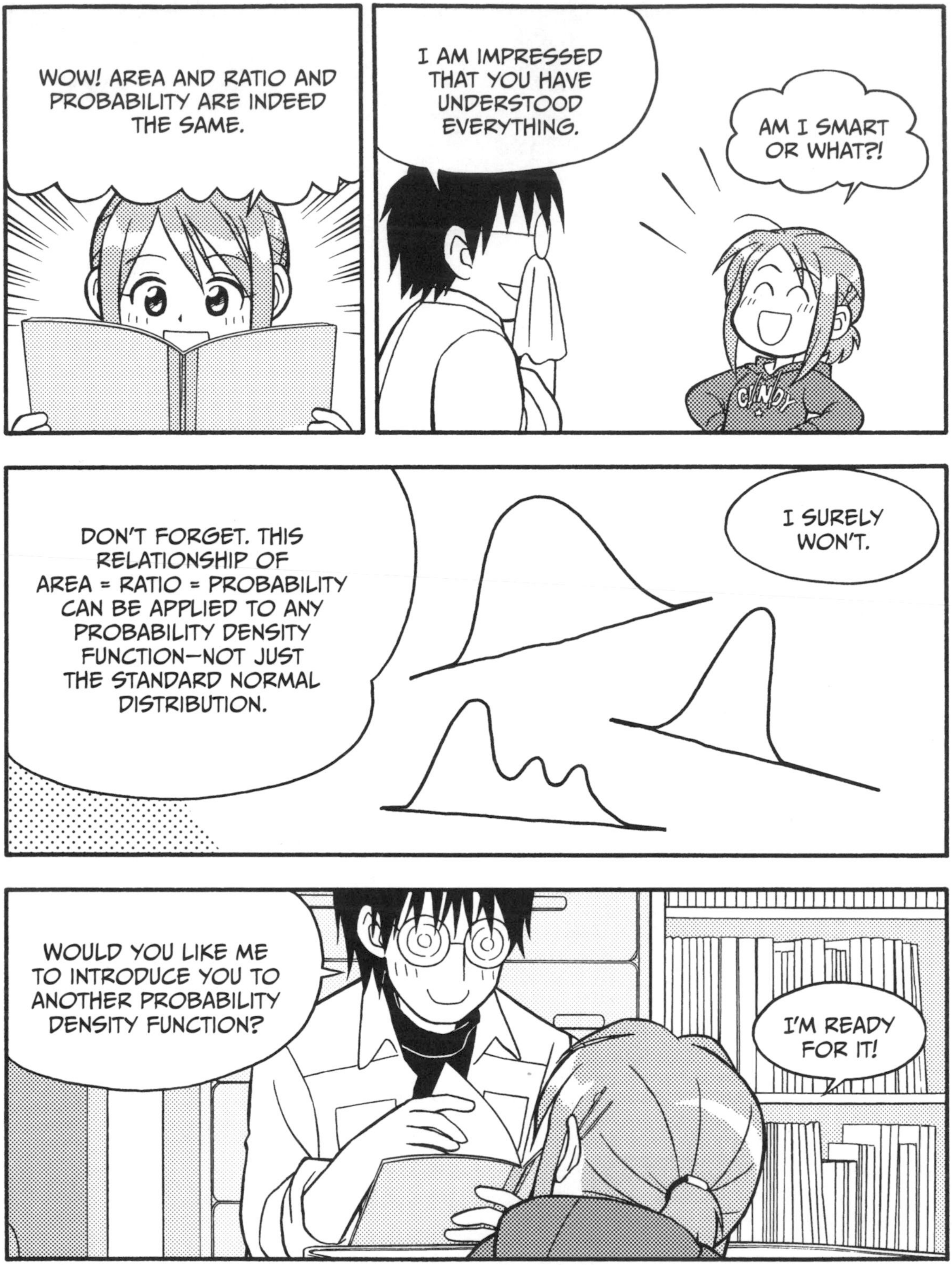
WOW! AREA AND RATIO AND PROBABILITY ARE INDEED THE SAME.
I AM IMPRESSED THAT YOU HAVE UNDERSTOOD EVERYTHING.
AM I SMART OR WHAT?!
CINDY
DON'T FORGET. THIS RELATIONSHIP OF AREA = RATIO = PROBABILITY CAN BE APPLIED TO ANY PROBABILITY DENSITY FUNCTION—NOT JUST THE STANDARD NORMAL DISTRIBUTION.
I SURELY WON'T.
WOULD YOU LIKE ME TO INTRODUCE YOU TO ANOTHER PROBABILITY DENSITY FUNCTION?
I'M READY FOR IT!

4. CHI-SQUARE DISTRIBUTION

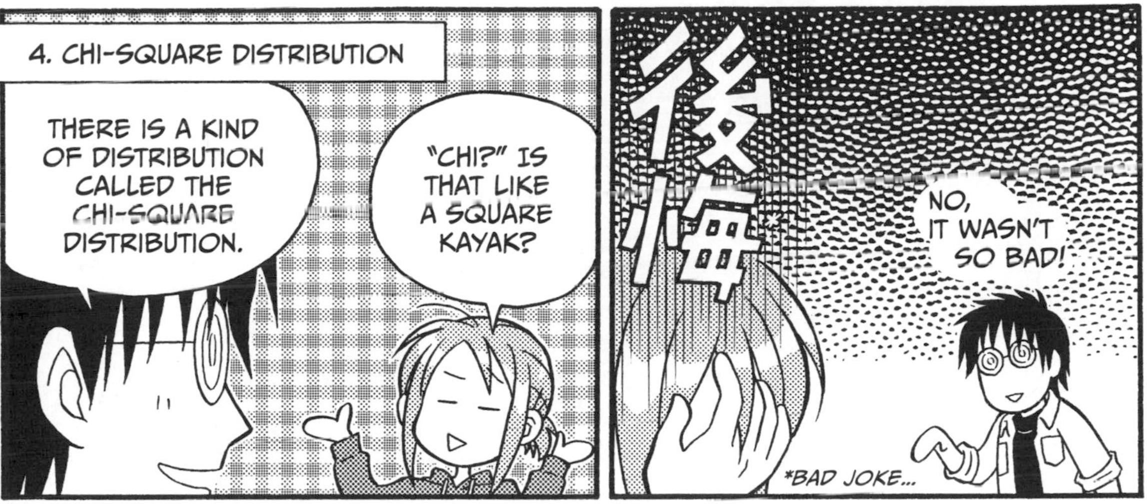
THERE IS A KIND OF DISTRIBUTION CALLED THE CHI-SQUARE DISTRIBUTION.
"CHI?" IS THAT LIKE A SQUARE KAYAK?
後悔*
NO, IT WASN'T SO BAD!
*BAD JOKE...

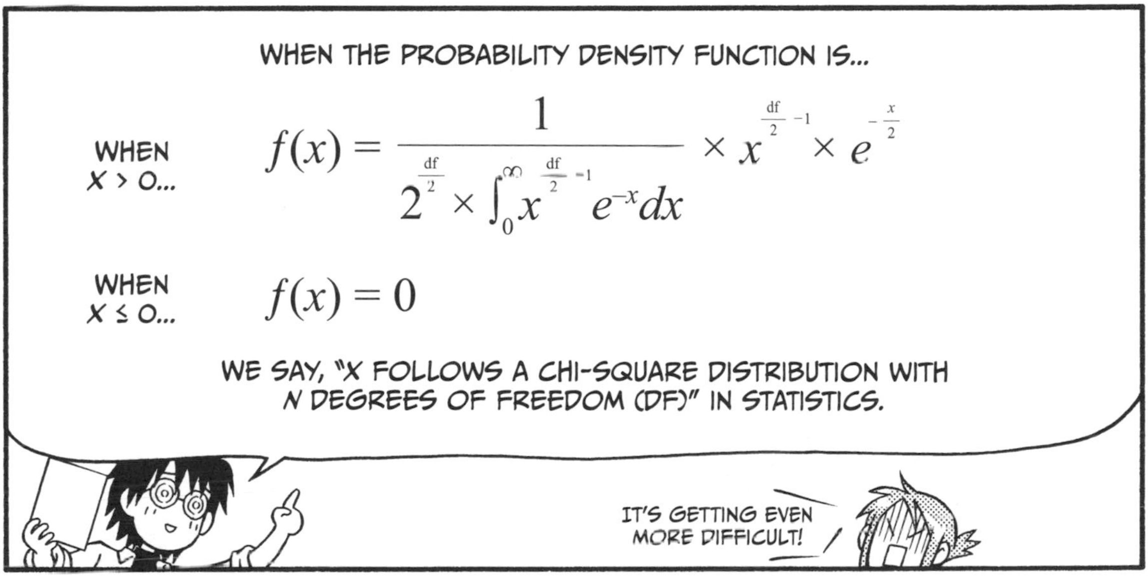
WHEN THE PROBABILITY DENSITY FUNCTION IS...
WHEN $x > 0$...
$f(x) = \frac{1}{2^{\frac{df}{2}} \times \int_0^\infty x^{\frac{df}{2}-1} e^{-x} dx} \times x^{\frac{df}{2}-1} \times e^{-\frac{x}{2}}$
WHEN $x \leq 0$...
$f(x) = 0$
WE SAY, "X FOLLOWS A CHI-SQUARE DISTRIBUTION WITH N DEGREES OF FREEDOM (DF)" IN STATISTICS.
IT'S GETTING EVEN MORE DIFFICULT!

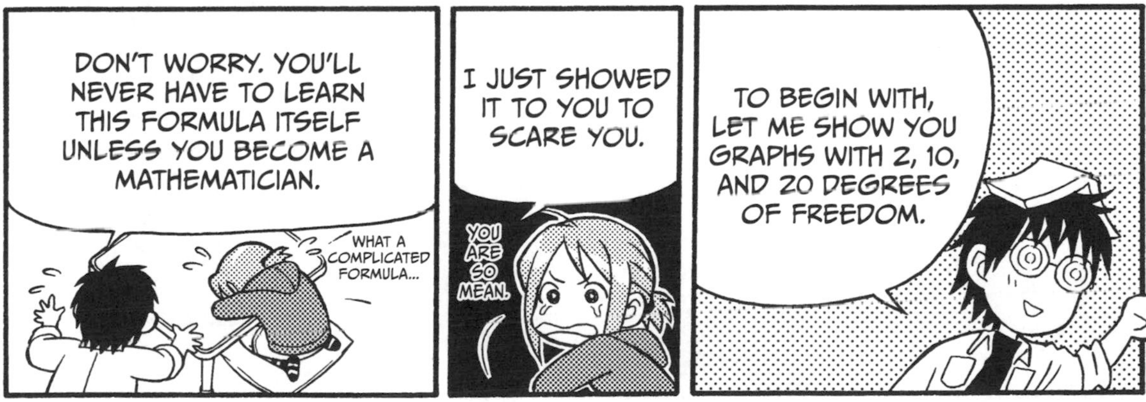
DON'T WORRY. YOU'LL NEVER HAVE TO LEARN THIS FORMULA ITSELF UNLESS YOU BECOME A MATHEMATICIAN.
WHAT A COMPLICATED FORMULA...
I JUST SHOWED IT TO YOU TO SCARE YOU.
YOU ARE SO MEAN.
TO BEGIN WITH, LET ME SHOW YOU GRAPHS WITH 2, 10, AND 20 DEGREES OF FREEDOM.

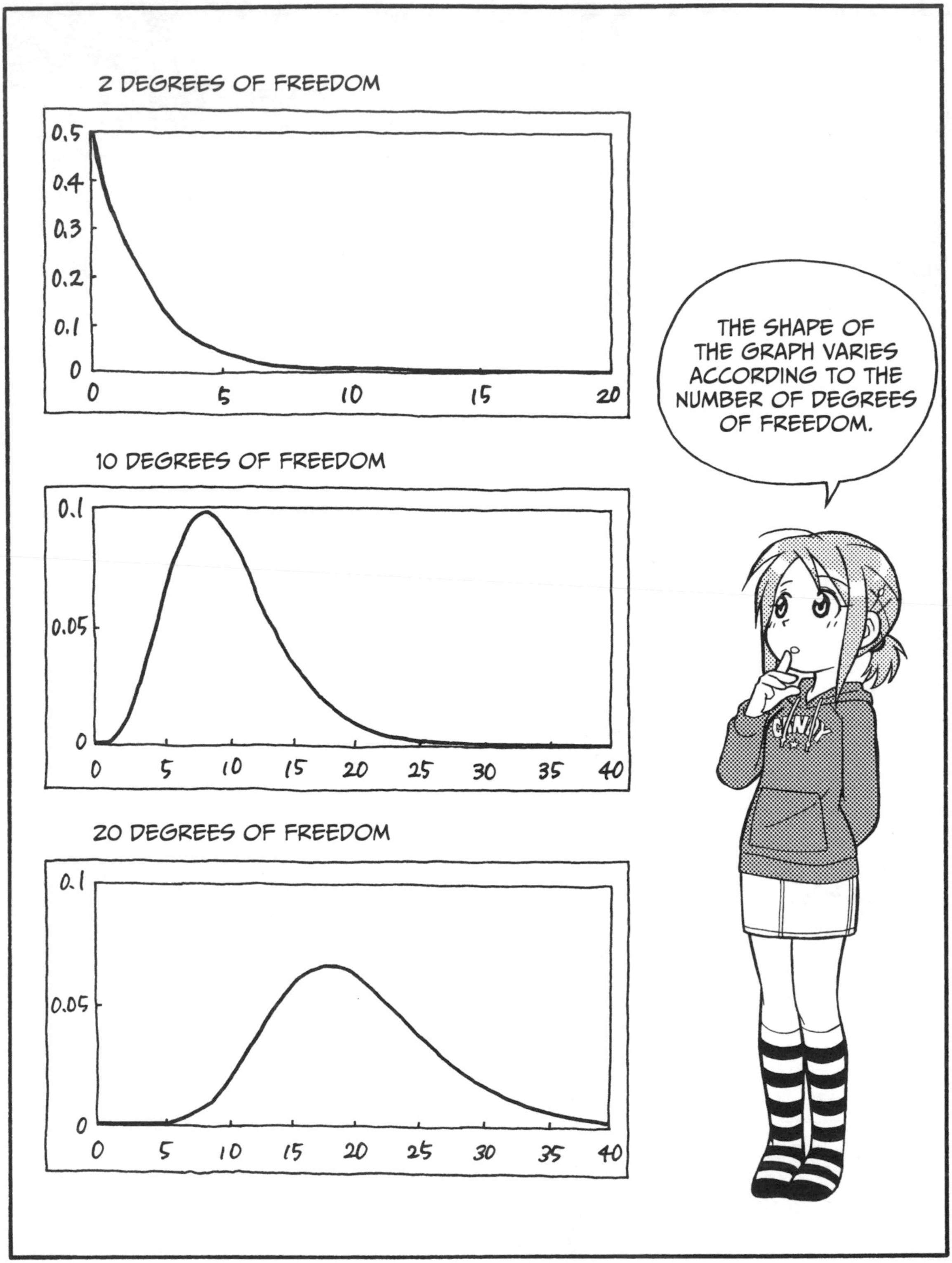
2 DEGREES OF FREEDOM
0.5
0.4
0.3
0.2
0.1
0
0 5 10 15 20
10 DEGREES OF FREEDOM
0.1
0.05
0
0 5 10 15 20 25 30 35 40
20 DEGREES OF FREEDOM
0.1
0.05
0
0 5 10 15 20 25 30 35 40
THE SHAPE OF THE GRAPH VARIES ACCORDING TO THE NUMBER OF DEGREES OF FREEDOM.

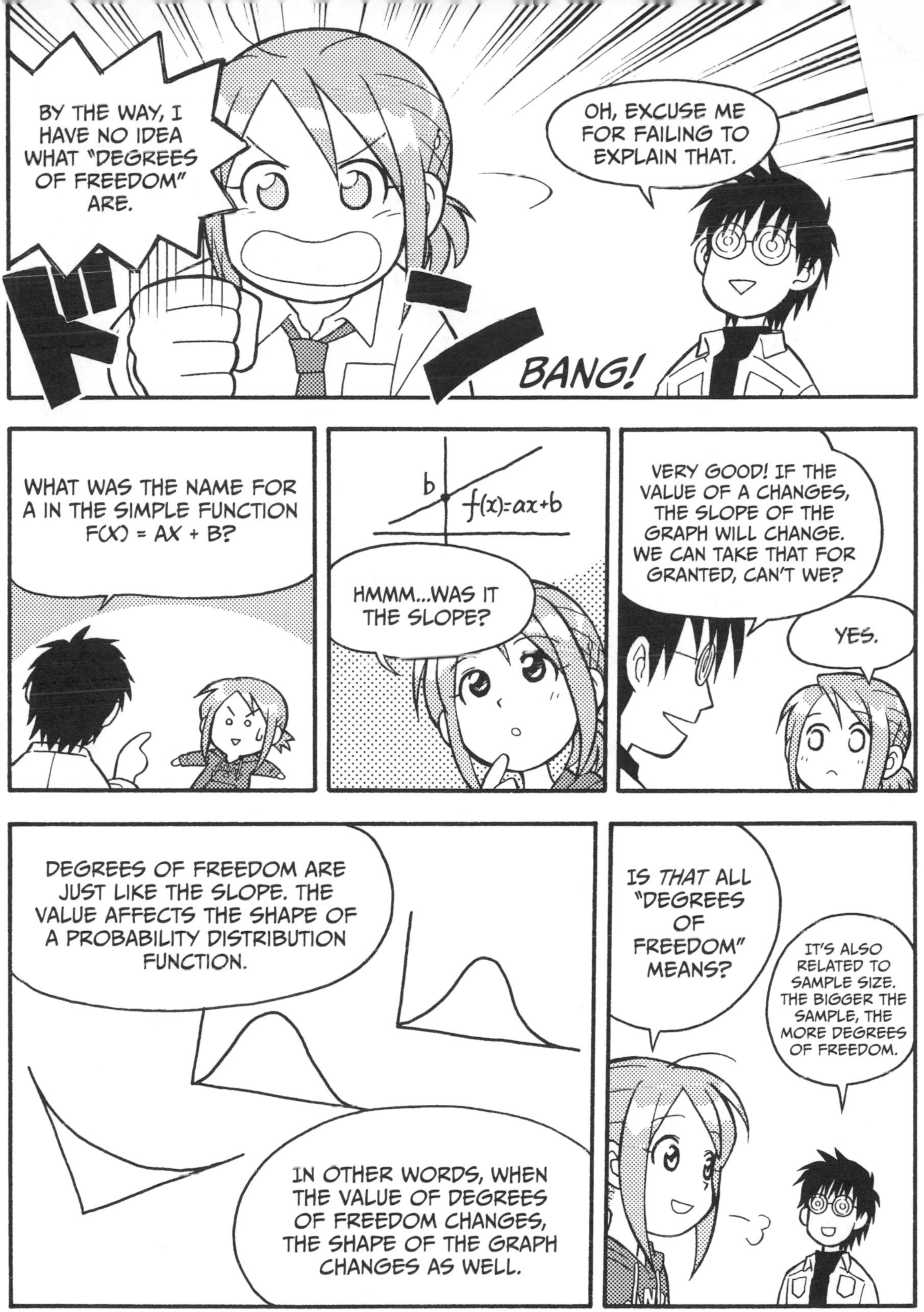
BY THE WAY, I HAVE NO IDEA WHAT "DEGREES OF FREEDOM" ARE.
BANG!
OH, EXCUSE ME FOR FAILING TO EXPLAIN THAT.
WHAT WAS THE NAME FOR A IN THE SIMPLE FUNCTION F(X) = AX + B?
b
f(x)=ax+b
HMMM...WAS IT THE SLOPE?
VERY GOOD! IF THE VALUE OF A CHANGES, THE SLOPE OF THE GRAPH WILL CHANGE. WE CAN TAKE THAT FOR GRANTED, CAN'T WE?
YES.
DEGREES OF FREEDOM ARE JUST LIKE THE SLOPE. THE VALUE AFFECTS THE SHAPE OF A PROBABILITY DISTRIBUTION FUNCTION.
IN OTHER WORDS, WHEN THE VALUE OF DEGREES OF FREEDOM CHANGES, THE SHAPE OF THE GRAPH CHANGES AS WELL.
IS *THAT* ALL "DEGREES OF FREEDOM" MEANS?
IT'S ALSO RELATED TO SAMPLE SIZE. THE BIGGER THE SAMPLE, THE MORE DEGREES OF FREEDOM.

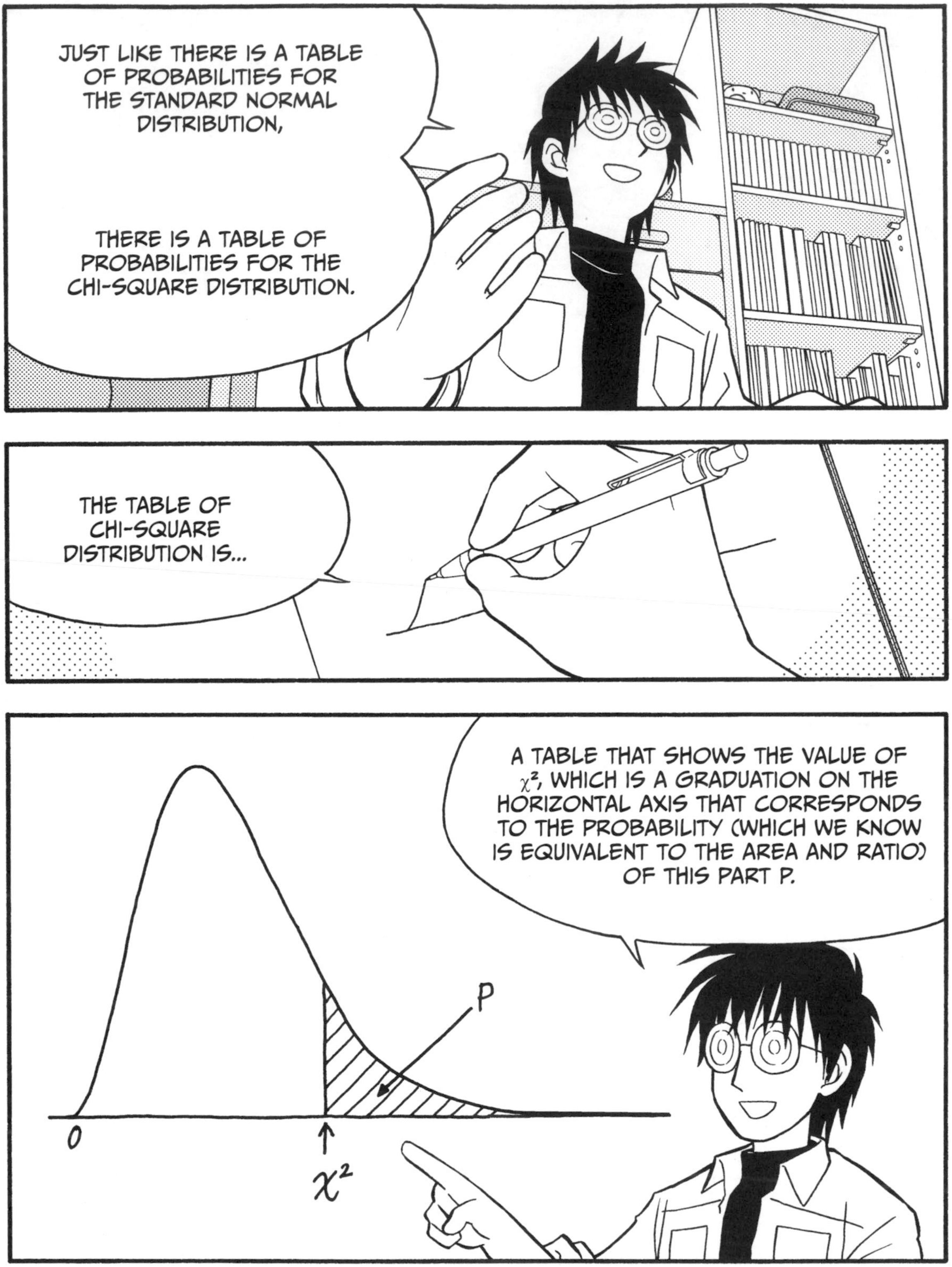
JUST LIKE THERE IS A TABLE OF PROBABILITIES FOR THE STANDARD NORMAL DISTRIBUTION,
THERE IS A TABLE OF PROBABILITIES FOR THE CHI-SQUARE DISTRIBUTION.
THE TABLE OF CHI-SQUARE DISTRIBUTION IS...
A TABLE THAT SHOWS THE VALUE OF χ^2, WHICH IS A GRADUATION ON THE HORIZONTAL AXIS THAT CORRESPONDS TO THE PROBABILITY (WHICH WE KNOW IS EQUIVALENT TO THE AREA AND RATIO) OF THIS PART P.
P
0
χ^2

WHAT IS THIS SYMBOL?

χ^2

THAT IS CHI-SQUARE. CHI IS A GREEK LETTER THAT LOOKS A BIT LIKE OUR LETTER X.

OH, SO THIS IS WHAT WE WERE TALKING ABOUT.

HERE IS THE TABLE.

DEGREES OF FREEDOM

TABLE OF CHI-SQUARE DISTRIBUTION

P	0.995	0.99	0.975	0.95	0.05	0.025	0.01	0.005
1	0.000039	0.0002	0.0010	0.0039	3.8415	5.0239	6.6349	7.8794
2	0.0100	0.0201	0.0506	0.1026	5.9915	7.3778	9.2104	10.5965
3	0.0717	0.1148	0.2158	0.3518	7.8147	9.3484	11.3449	12.8381
4	0.2070	0.2971	0.4844	0.7107	9.4877	11.1433	13.2767	14.8602
5	0.4118	0.5543	0.8312	1.1455	11.0705	12.8325	15.0863	16.7496
6	0.6757	0.8721	1.2373	1.6354	12.5916	14.4494	16.8119	18.5475
7	0.9893	1.2390	1.6899	2.1673	14.0671	16.0128	18.4753	20.2777
8	1.3444	1.6465	2.1797	2.7326	15.5073	17.5345	20.0902	21.9549
9	1.7349	2.0879	2.7004	3.3251	16.9190	19.0228	21.6660	23.5893
10	2.1558	2.5582	3.2470	3.9403	18.3070	20.4832	23.2093	25.1881
...	...	...	...	...	...	...	...	...

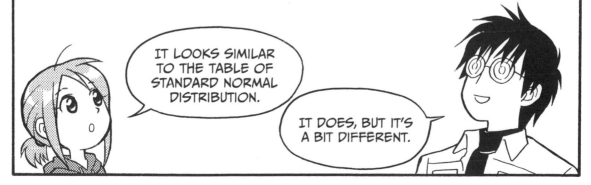

IT LOOKS SIMILAR TO THE TABLE OF STANDARD NORMAL DISTRIBUTION.

IT DOES, BUT IT'S A BIT DIFFERENT.

WITH THE TABLE OF STANDARD NORMAL DISTRIBUTION, YOU PROVIDE THE X-COORDINATE, AND IT TELLS YOU THE ASSOCIATED PROBABILITY.

IT PROVIDED THE PROBABILITY (= AREA = RATIO)

WITH THE TABLE OF CHI-SQUARE DISTRIBUTION, YOU PROVIDE THE PROBABILITY, AND IT TELLS YOU THE ASSOCIATED X-COORDINATE.

THIS VALUE!

χ^2

I AM GETTING CONFUSED...!

CALM DOWN!

* TABLE OF CHI-SQUARE DISTRIBUTION

LET'S FIND OUT WHAT THE VALUE OF χ^2 IS IN A CASE IN WHICH THE DEGREE OF FREEDOM IS 1 AND P IS 0.05.

THE LINE FOR 1 AND THE ROW FOR 0.05 CROSS AT...

	0.99	0.975	0.95	0.05	0.025	0.01
	0.0002	0.0010	0.0039	3.8415	5.0239	6.6349
	201	0.0506	0.1026		7.3778	9.2104
	148	0.2158	0.3518	915	9.3484	11.3449
	2971	0.4844	0.71	147	11.1433	
	0.9893	0.8721	0.8312	9.4877	12.83	
	1.344	1.2390	1.2373	1.14	14.4	
		1.6899	1.63	16.0		
			2.16	17.53		
			2.7	19.0228		

3.8415.

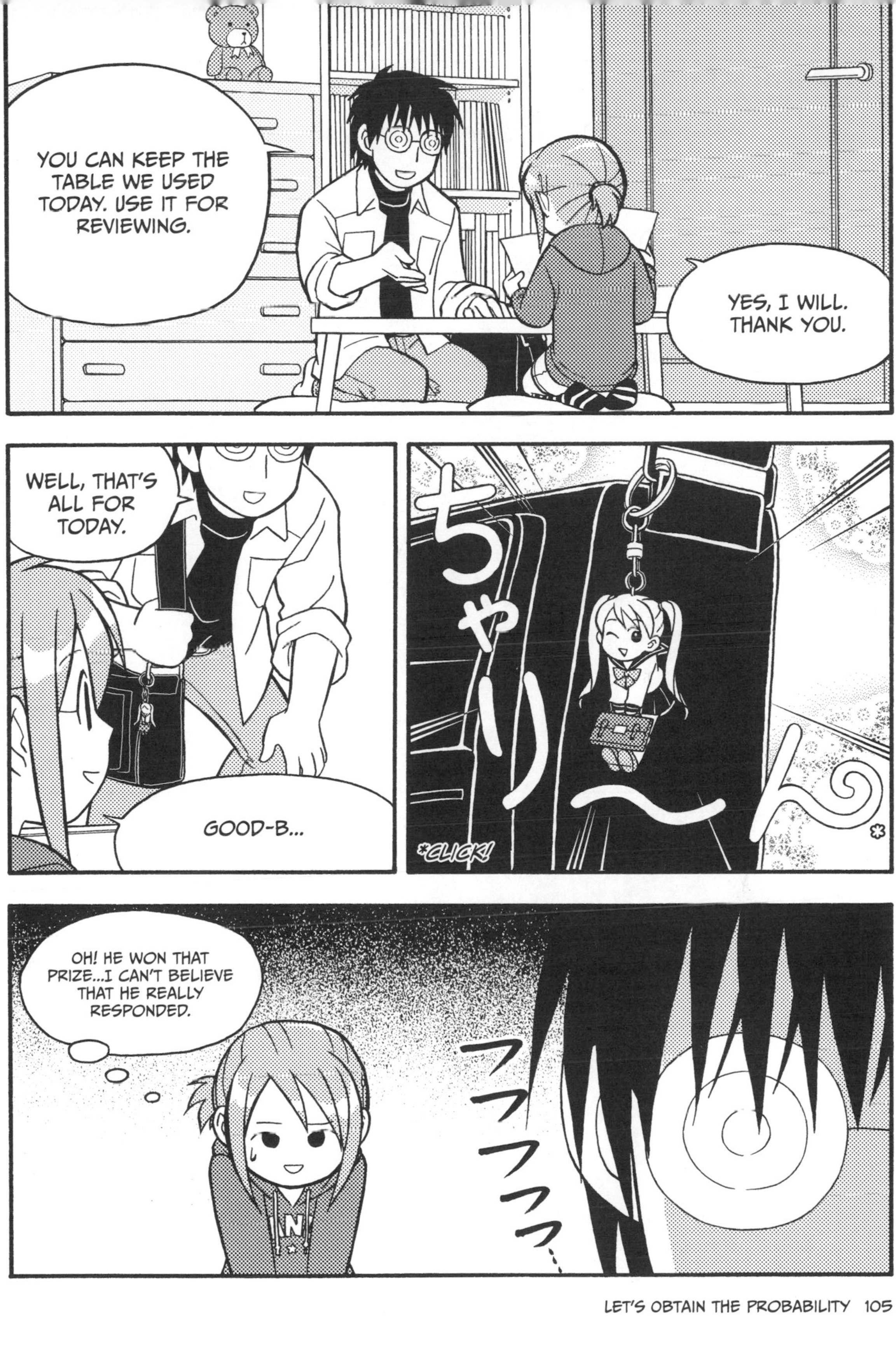
YOU CAN KEEP THE TABLE WE USED TODAY. USE IT FOR REVIEWING.
YES, I WILL. THANK YOU.
WELL, THAT'S ALL FOR TODAY.
GOOD-B...
ちゃり～ん
*CLICK!
OH! HE WON THAT PRIZE...I CAN'T BELIEVE THAT HE REALLY RESPONDED.
フフフフフ...

5. T DISTRIBUTION

The probability density function below is a popular topic in statistics.

$$f(x) = \frac{\int_0^\infty x^{\frac{\text{df}+1}{2}-1} e^{-x}dx}{\sqrt{\text{df} \times \pi} \times \int_0^\infty x^{\frac{\text{df}}{2}-1} e^{-x}dx} \times \left(1 + \frac{x^2}{\text{df}}\right)^{-\frac{\text{df}+1}{2}}$$

When the probability density function for *x* looks like this, we say, "*x* follows a t distribution with *n* degrees of freedom."

Here is a case with 5 degrees of freedom:

0.6
0.5
0.4
0.3
0.2
0.1
0
-6
-4
-2
0
2
4
6

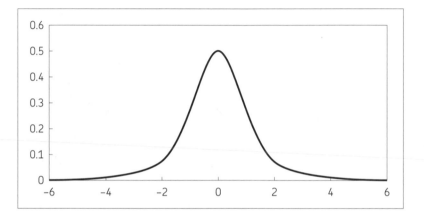

6. F DISTRIBUTION

The probability density function below is a popular topic in statistics.

when $x > 0$:

$$f(x) = \frac{\left(\int_0^\infty x^{\frac{\left(\text{first df} + \text{second df}\right)}{2}-1} e^{-x}dx\right) \times (\text{first df})^{\frac{\text{first df}}{2}} \times (\text{second df})^{\frac{\text{second df}}{2}}}{\left(\int_0^\infty x^{\frac{\text{first df}}{2}-1} e^{-x}dx\right) \times \left(\int_0^\infty x^{\frac{\text{second df}}{2}-1} e^{-x}dx\right)} \times \frac{x^{\frac{\text{first df}}{2}-1}}{(\text{first df} \times x + \text{second df})^{\frac{\left(\text{first df} + \text{second df}\right)}{2}}}$$

when $x \leq 0$: $f(x) = 0$

When the probability density function for *x* looks like this, we say, "*x* follows an F distribution with the first degree of freedom *m* and the second degree of freedom *n*."

Here is a case in which the first degree of freedom is 10 and the second degree of freedom is 5:

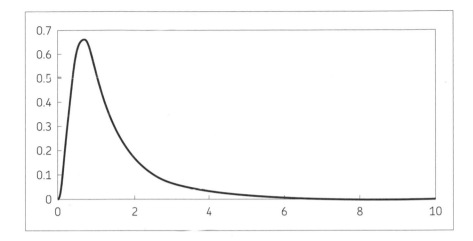

7. DISTRIBUTIONS AND EXCEL

Until the rise of personal computers (roughly speaking, around the beginning of the 1990s), it was difficult for an individual to calculate the probability without tables of standard normal distribution or chi-square distribution. However, these tables of distribution are not used much anymore—you can use Excel functions to find the same values as the ones provided by the tables. This enables individuals to calculate even more types of values than the ones found in the tables of distribution. Table 5-1 summarizes Excel functions related to various distributions. (Refer to the appendix on page 191 for more information on making calculations with Excel.)

TABLE 5-1: EXCEL FUNCTIONS RELATED TO VARIOUS DISTRIBUTIONS

Distribution	Functions	Feature of the function
normal[*]	NORMDIST	Calculates the probability that corresponds to a point on the horizontal axis.
normal	NORMINV	Calculates a point on the horizontal axis that corresponds to the probability.
standard normal	NORMSDIST	Calculates the probability that corresponds to a point on the horizontal axis.
standard normal	NORMSINV	Calculates a point on the horizontal axis that corresponds to the probability.
chi-square	CHIDIST	Calculates the probability that corresponds to a point on the horizontal axis.
chi-square	CHIINV	Calculates a point on the horizontal axis that corresponds to the probability.
t	TDIST	Calculates the probability that corresponds to a point on the horizontal axis.
t	TINV	Calculates a point on the horizontal axis that corresponds to the probability.
F	FDIST	Calculates the probability that corresponds to a point on the horizontal axis.
F	FINV	Calculates a point on the horizontal axis that corresponds to the probability.

[*] The probability density function for normal distribution is affected by the mean and standard deviation. Thus, it is impossible to make a "table of normal distribution," and no such thing exists in this world. However, by using Excel, you can conveniently calculate the values and make a table relevant to the normal distribution.

EXERCISE

1. Calculate the probability (the shaded area in the graph below) using the table of standard normal distribution on page 93.

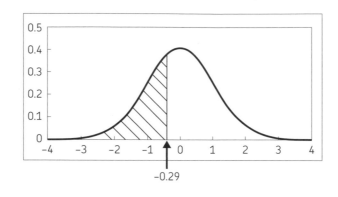

2. Calculate the value of χ^2 when there are 2 degrees of freedom and P is 0.05 using the table of chi-square distribution on page 103.

ANSWER

1. Because the standard normal distribution is symmetrical, the probability in question is equal to the probability shown in the graph below.

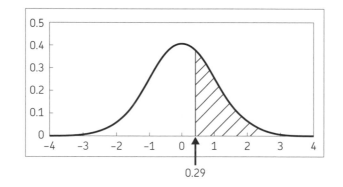

   The probability when $z = 0.29 = 0.2 + 0.09$ is 0.1141 according to the table of standard normal distribution. Therefore, the probability to be obtained is $0.5 - 0.1141 = 0.3859$.

2. The value of χ^2 to be obtained is 5.9915 according to the table of chi-square distribution.

SUMMARY

- Some of the most common probability density functions are:

    - Normal distribution
    - Standard normal distribution
    - Chi-square distribution
    - t distribution
    - F distribution

- The area between the probability density function and the horizontal axis is 1. This area is equivalent to a ratio or a probability.
- By using an Excel function or a table of probabilities for the appropriate distribution, you can calculate:

    - The probability that corresponds to a point on the horizontal axis
    - The point on the horizontal axis that corresponds to the probability

LET'S LOOK AT THE RELATIONSHIP
BETWEEN TWO VARIABLES

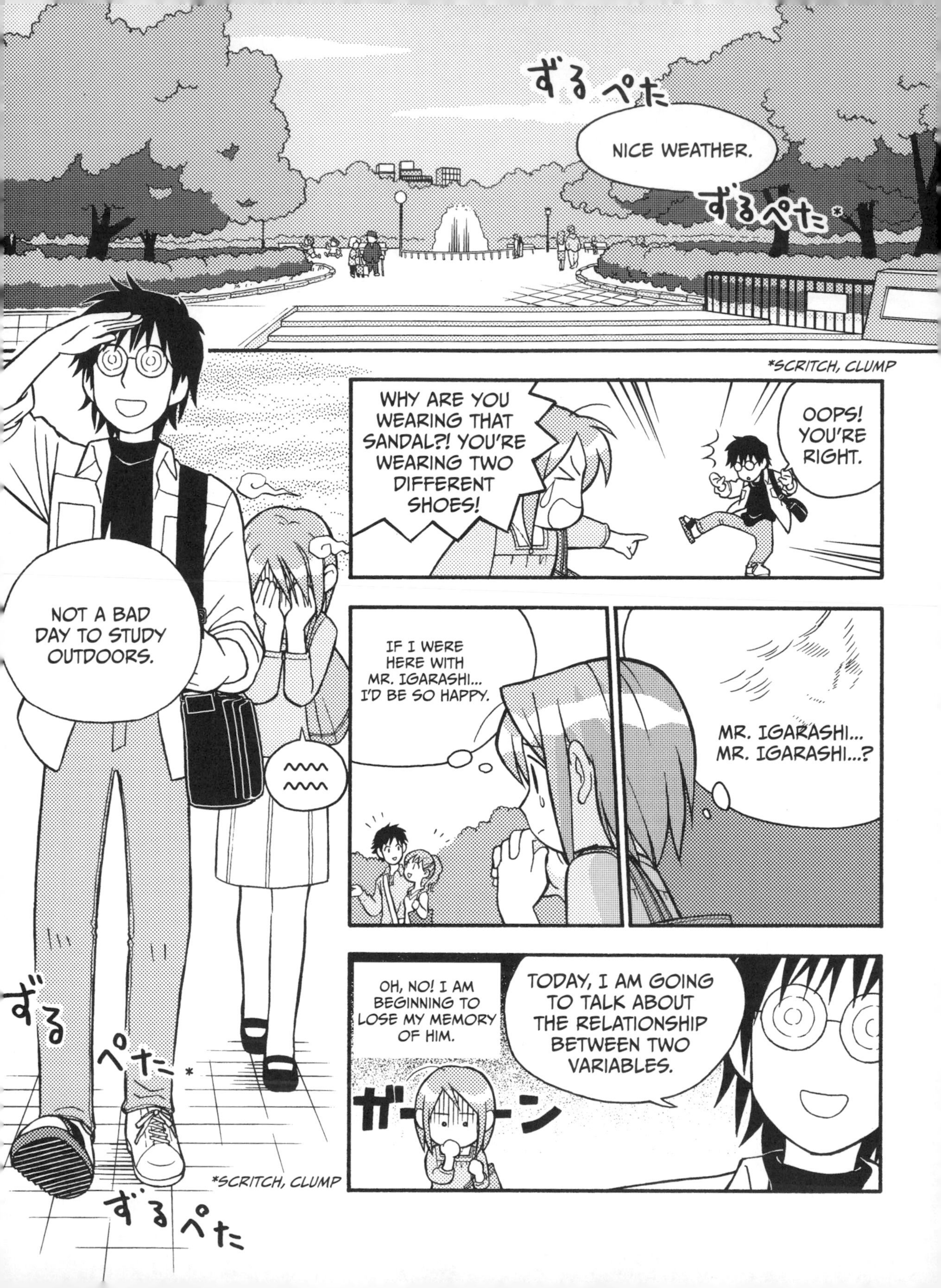
ずるぺた
NICE WEATHER.
ずるぺた*
*SCRITCH, CLUMP
NOT A BAD DAY TO STUDY OUTDOORS.
ずるぺた*
*SCRITCH, CLUMP
ずるぺた
WHY ARE YOU WEARING THAT SANDAL?! YOU'RE WEARING TWO DIFFERENT SHOES!
OOPS! YOU'RE RIGHT.
IF I WERE HERE WITH MR. IGARASHI... I'D BE SO HAPPY.
MR. IGARASHI... MR. IGARASHI...?
OH, NO! I AM BEGINNING TO LOSE MY MEMORY OF HIM.
TODAY, I AM GOING TO TALK ABOUT THE RELATIONSHIP BETWEEN TWO VARIABLES.
ガーン

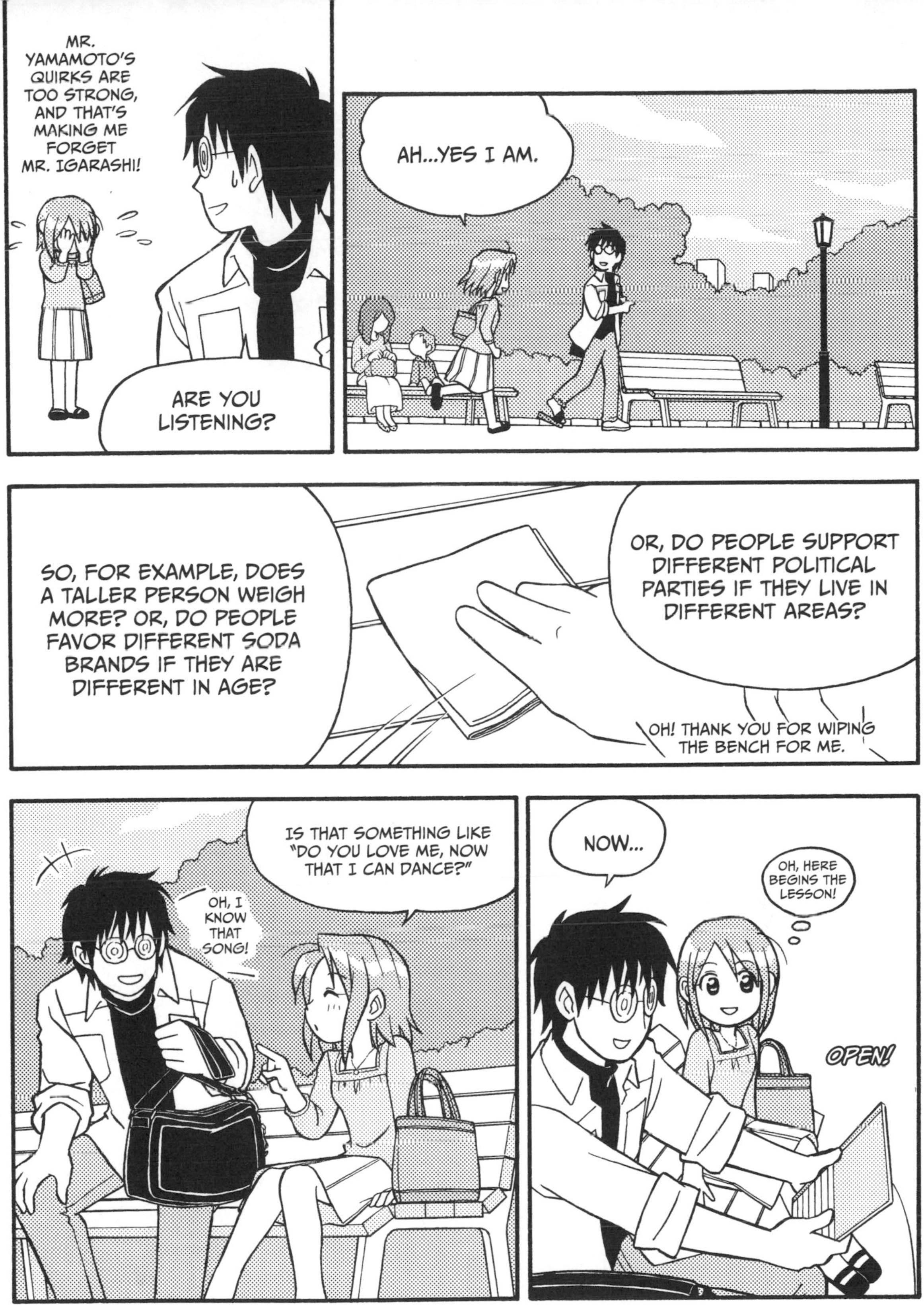
MR. YAMAMOTO'S QUIRKS ARE TOO STRONG, AND THAT'S MAKING ME FORGET MR. IGARASHI!
ARE YOU LISTENING?
AH...YES I AM.
SO, FOR EXAMPLE, DOES A TALLER PERSON WEIGH MORE? OR, DO PEOPLE FAVOR DIFFERENT SODA BRANDS IF THEY ARE DIFFERENT IN AGE?
OR, DO PEOPLE SUPPORT DIFFERENT POLITICAL PARTIES IF THEY LIVE IN DIFFERENT AREAS?
OH! THANK YOU FOR WIPING THE BENCH FOR ME.
IS THAT SOMETHING LIKE "DO YOU LOVE ME, NOW THAT I CAN DANCE?"
OH, I KNOW THAT SONG!
NOW...
OH, HERE BEGINS THE LESSON!
OPEN!

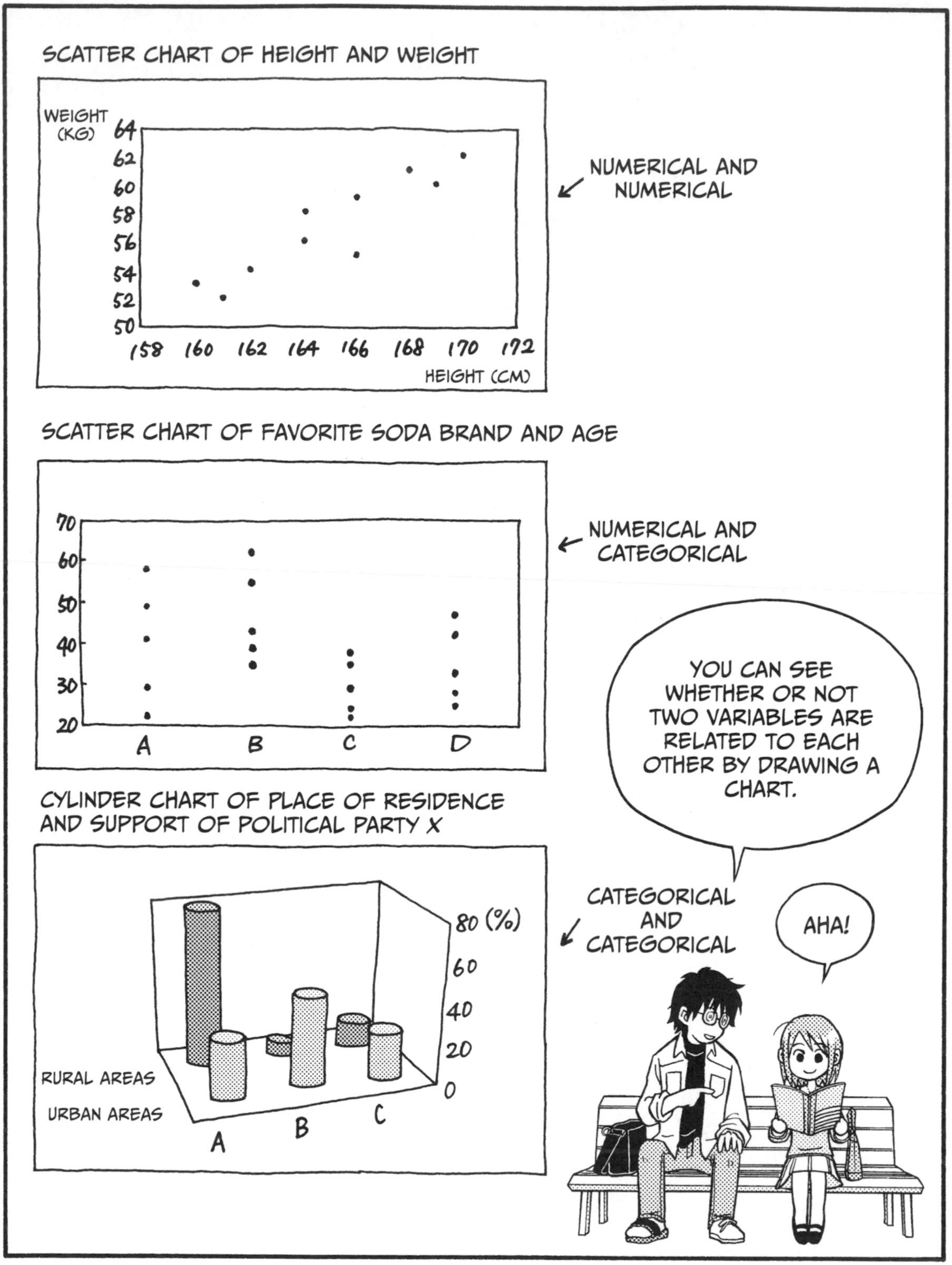
SCATTER CHART OF HEIGHT AND WEIGHT
WEIGHT (KG)
64
62
60
58
56
54
52
50
158 160 162 164 166 168 170 172
HEIGHT (CM)
NUMERICAL AND NUMERICAL
SCATTER CHART OF FAVORITE SODA BRAND AND AGE
70
60
50
40
30
20
A B C D
NUMERICAL AND CATEGORICAL
YOU CAN SEE WHETHER OR NOT TWO VARIABLES ARE RELATED TO EACH OTHER BY DRAWING A CHART.
CYLINDER CHART OF PLACE OF RESIDENCE AND SUPPORT OF POLITICAL PARTY X
80 (%)
60
40
20
0
RURAL AREAS
URBAN AREAS
A B C
CATEGORICAL AND CATEGORICAL
AHA!

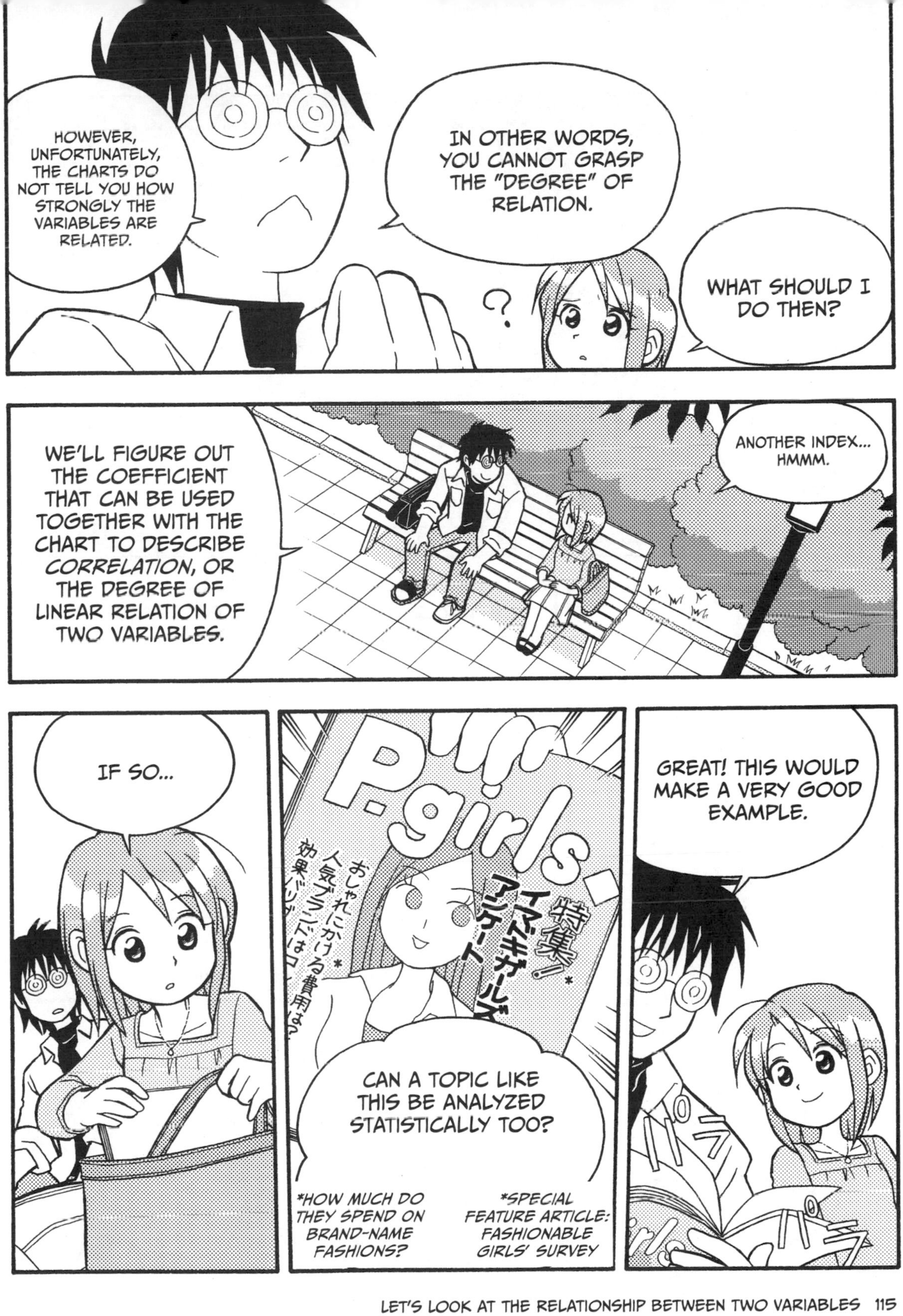
HOWEVER, UNFORTUNATELY, THE CHARTS DO NOT TELL YOU HOW STRONGLY THE VARIABLES ARE RELATED.
IN OTHER WORDS, YOU CANNOT GRASP THE "DEGREE" OF RELATION.
?
WHAT SHOULD I DO THEN?
WE'LL FIGURE OUT THE COEFFICIENT THAT CAN BE USED TOGETHER WITH THE CHART TO DESCRIBE *CORRELATION*, OR THE DEGREE OF LINEAR RELATION OF TWO VARIABLES.
ANOTHER INDEX... HMMM.
IF SO...
P.girls
特集！イマドキガールズアンケート*
おしゃれにかける費用は*
CAN A TOPIC LIKE THIS BE ANALYZED STATISTICALLY TOO?
*HOW MUCH DO THEY SPEND ON BRAND-NAME FASHIONS?
*SPECIAL FEATURE ARTICLE: FASHIONABLE GIRLS' SURVEY
GREAT! THIS WOULD MAKE A VERY GOOD EXAMPLE.

1. CORRELATION COEFFICIENT

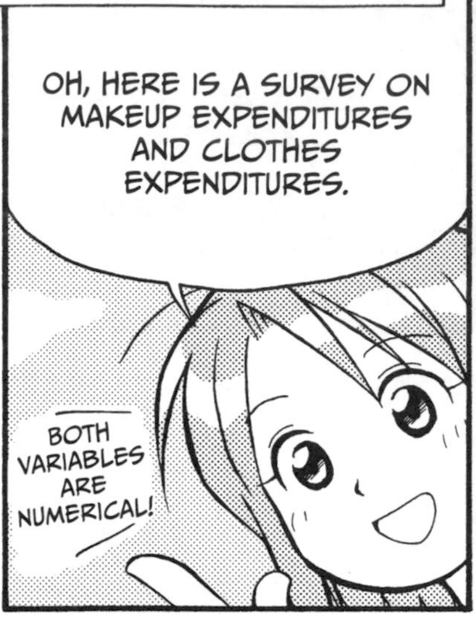

OH, HERE IS A SURVEY ON MAKEUP EXPENDITURES AND CLOTHES EXPENDITURES.
BOTH VARIABLES ARE NUMERICAL!

Street survey!

Ten ladies in their 20s answered
Monthly Expenditures on Makeup and Clothes

Respondent	Amount spent on makeup (¥)	Amount spent on clothes (¥)
Ms. A	3,000	7,000
Ms. B	5,000	8,000
Ms. C	12,000	25,000
Ms. D	2,000	5,000
Ms. E	7,000	12,000
Ms. F	15,000	30,000
Ms. G	5,000	10,000
Ms. H	6,000	15,000
Ms. I	8,000	20,000
Ms. J	10,000	18,000

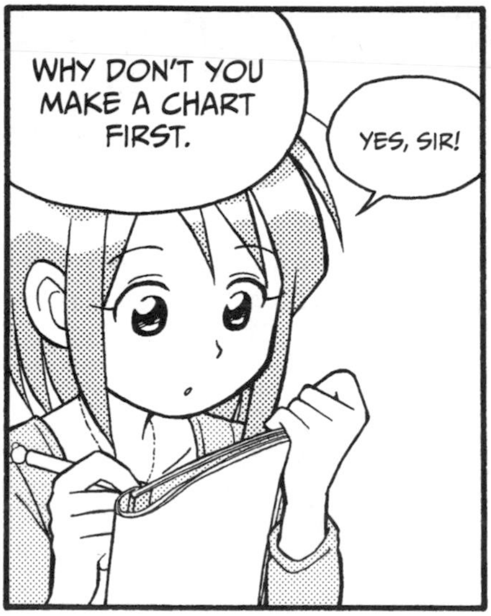

WHY DON'T YOU MAKE A CHART FIRST.
YES, SIR!

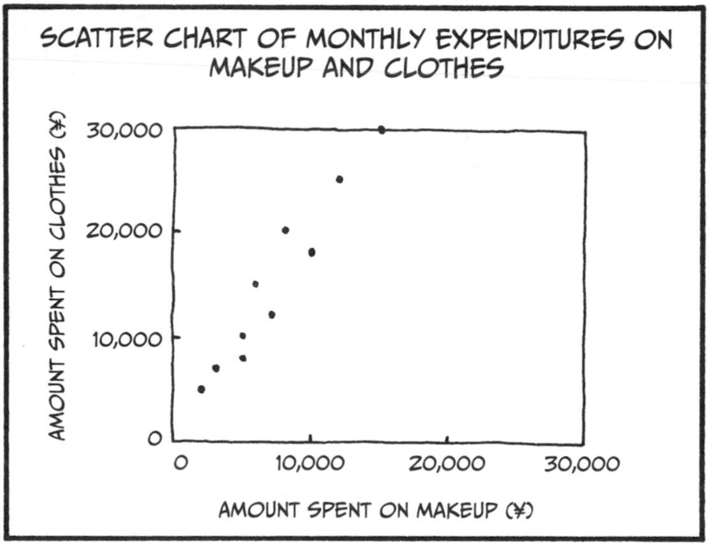

SCATTER CHART OF MONTHLY EXPENDITURES ON MAKEUP AND CLOTHES
AMOUNT SPENT ON CLOTHES (¥)
30,000
20,000
10,000
0
0
10,000
20,000
30,000
AMOUNT SPENT ON MAKEUP (¥)

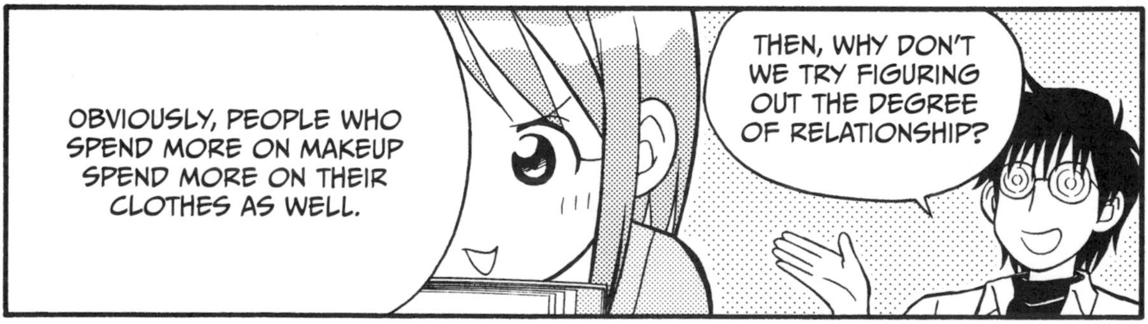

OBVIOUSLY, PEOPLE WHO SPEND MORE ON MAKEUP SPEND MORE ON THEIR CLOTHES AS WELL.
THEN, WHY DON'T WE TRY FIGURING OUT THE DEGREE OF RELATIONSHIP?

Data types	Index	Value range	Formula
Numerical and numerical	Correlation coefficient	–1 – 1	$$\frac{\sum(x-\bar{x})(y-\bar{y})}{\sqrt{\sum(x-\bar{x})^2 \times \sum(y-\bar{y})^2}} = \frac{Sxy}{\sqrt{Sxx \times Syy}}$$
Numerical and categorical	Correlation ratio*	0 – 1	$$\frac{\text{interclass variance}}{\text{intraclass variance} + \text{interclass variance}}$$
Categorical and categorical	Cramer's coefficient*	0 – 1	$$\sqrt{\frac{\chi_0^2}{\text{the total number of values} \times (\min\{\text{the number of lines in the cross tabulation, the number of rows in the cross tabulation}\} - 1)}}$$

* See page 121, "Correlation Ratio," and page 127, "Cramer's Coefficient."

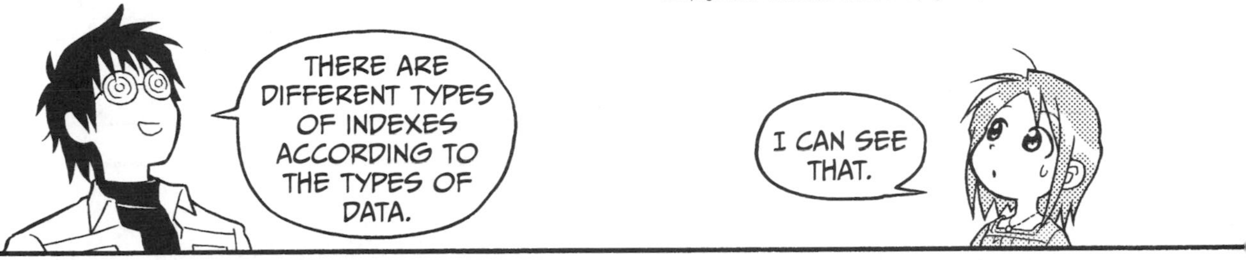

THERE ARE DIFFERENT TYPES OF INDEXES ACCORDING TO THE TYPES OF DATA.

I CAN SEE THAT.

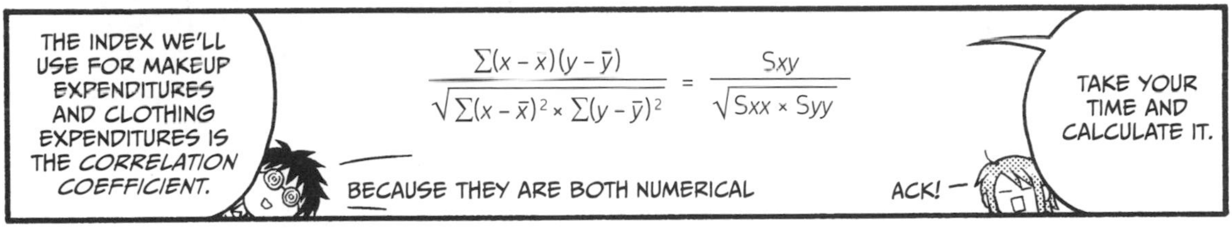

THE INDEX WE'LL USE FOR MAKEUP EXPENDITURES AND CLOTHING EXPENDITURES IS THE *CORRELATION COEFFICIENT*.

$$\frac{\sum(x-x)(y-\bar{y})}{\sqrt{\sum(x-\bar{x})^2 \times \sum(y-\bar{y})^2}} = \frac{Sxy}{\sqrt{Sxx \times Syy}}$$

BECAUSE THEY ARE BOTH NUMERICAL

TAKE YOUR TIME AND CALCULATE IT.

ACK! —

HERE WE GO!

THIS FREAKS ME OUT!

THE PROCESS FOR CALCULATING THE CORRELATION COEFFICIENT FOR MONTHLY EXPENDITURES ON MAKEUP AND CLOTHES

	Amount spent on makeup (¥)	Amount spent on clothes (¥)					
	x	y	$x-\bar{x}$	$y-\bar{y}$	$(x-\bar{x})^2$	$(y-\bar{y})^2$	$(x-\bar{x})(y-\bar{y})$
Ms. A	3,000	7,000	–4,300	–8,000	18,490,000	64,000,000	34,400,000
Ms. B	5,000	8,000	–2,300	–7,000	5,290,000	49,000,000	16,100,000
Ms. C	12,000	25,000	4,700	10,000	22,090,000	100,000,000	47,000,000
Ms. D	2,000	5,000	–5,300	–10,000	28,090,000	100,000,000	53,000,000
Ms. E	7,000	12,000	–300	–3,000	90,000	9,000,000	900,000
Ms. F	15,000	30,000	7,700	15,000	59,290,000	225,000,000	115,500,000
Ms. G	5,000	10,000	–2,300	–5,000	5,290,000	25,000,000	11,500,000
Ms. H	6,000	15,000	1,300	0	1,690,000	0	0
Ms. I	8,000	20,000	700	5,000	490,000	25,000,000	3,500,000
Ms. J	10,000	18,000	2,700	3,000	7,290,000	9,000,000	8,100,000
Sum	73,000	150,000	0	0	148,100,000	606,000,000	290,000,000
Mean	7,300	15,000					

$\bar{x}$ $\bar{y}$ Sxx   Syy   Sxy

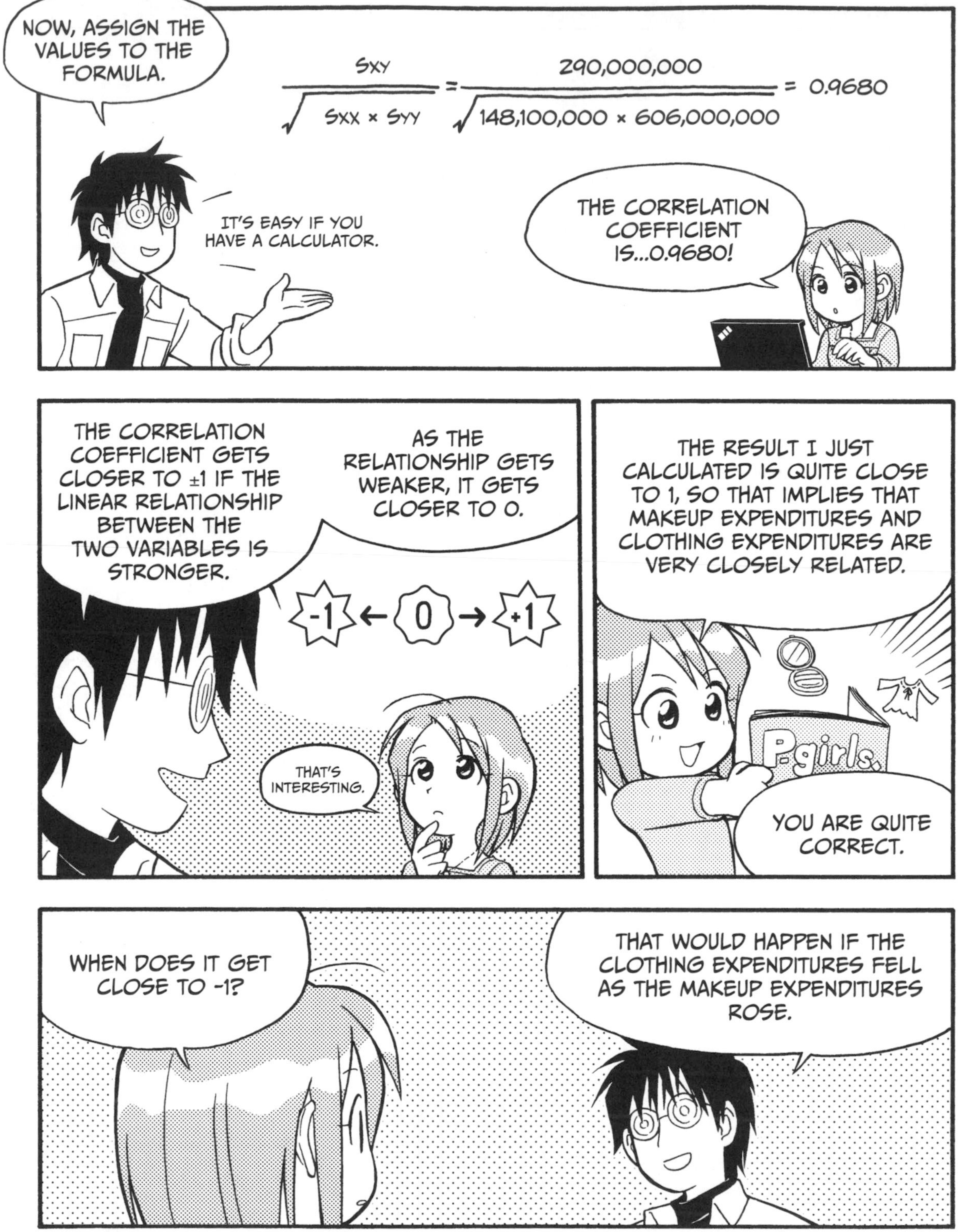
NOW, ASSIGN THE VALUES TO THE FORMULA.
$$\frac{S_{xy}}{\sqrt{S_{xx} \times S_{yy}}} = \frac{290,000,000}{\sqrt{148,100,000 \times 606,000,000}} = 0.9680$$
IT'S EASY IF YOU HAVE A CALCULATOR.
THE CORRELATION COEFFICIENT IS...0.9680!
THE CORRELATION COEFFICIENT GETS CLOSER TO ±1 IF THE LINEAR RELATIONSHIP BETWEEN THE TWO VARIABLES IS STRONGER.
AS THE RELATIONSHIP GETS WEAKER, IT GETS CLOSER TO 0.
-1 ← 0 → +1
THAT'S INTERESTING.
THE RESULT I JUST CALCULATED IS QUITE CLOSE TO 1, SO THAT IMPLIES THAT MAKEUP EXPENDITURES AND CLOTHING EXPENDITURES ARE VERY CLOSELY RELATED.
P-girls.
YOU ARE QUITE CORRECT.
WHEN DOES IT GET CLOSE TO -1?
THAT WOULD HAPPEN IF THE CLOTHING EXPENDITURES FELL AS THE MAKEUP EXPENDITURES ROSE.

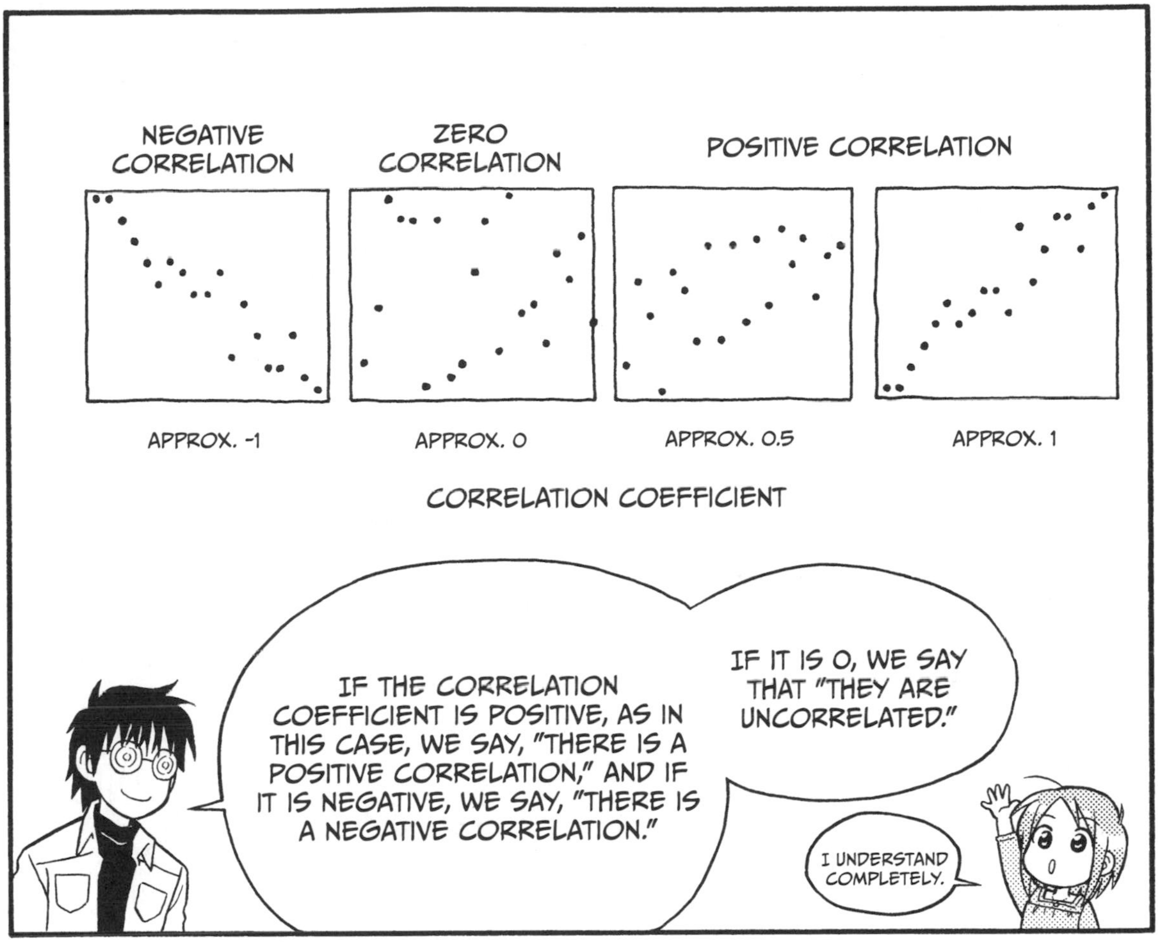
NEGATIVE CORRELATION
ZERO CORRELATION
POSITIVE CORRELATION
APPROX. -1
APPROX. 0
APPROX. 0.5
APPROX. 1
CORRELATION COEFFICIENT
IF THE CORRELATION COEFFICIENT IS POSITIVE, AS IN THIS CASE, WE SAY, "THERE IS A POSITIVE CORRELATION," AND IF IT IS NEGATIVE, WE SAY, "THERE IS A NEGATIVE CORRELATION."
IF IT IS 0, WE SAY THAT "THEY ARE UNCORRELATED."
I UNDERSTAND COMPLETELY.

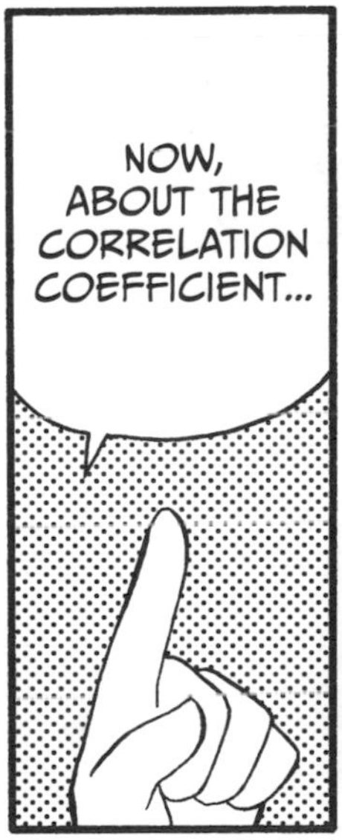
NOW, ABOUT THE CORRELATION COEFFICIENT...

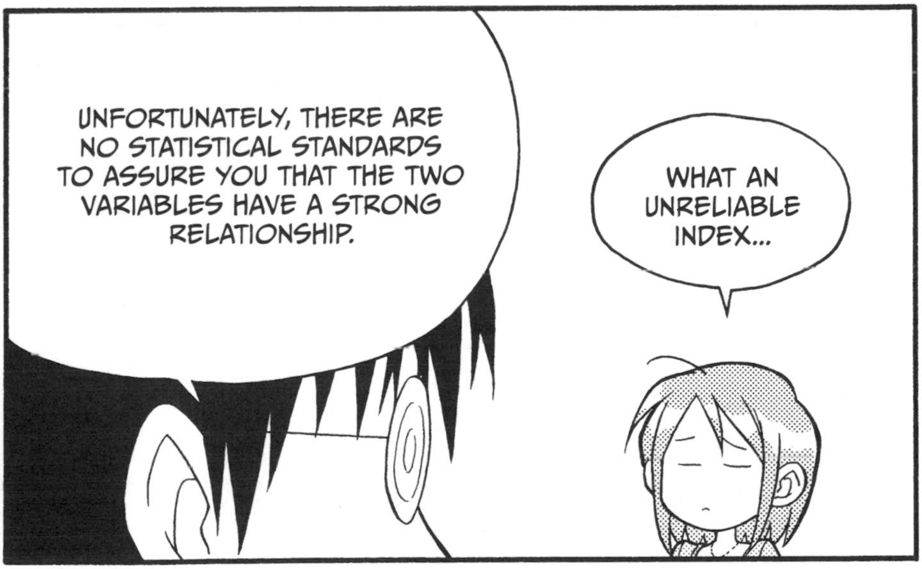
UNFORTUNATELY, THERE ARE NO STATISTICAL STANDARDS TO ASSURE YOU THAT THE TWO VARIABLES HAVE A STRONG RELATIONSHIP.
WHAT AN UNRELIABLE INDEX...

INFORMAL STANDARDS OF THE CORRELATION COEFFICIENT

Absolute value of the correlation coefficient		Detailed description	Rough description
1.0–0.9	⇨	Very strongly related	
0.9–0.7	⇨	Fairly strongly related	Related
0.7–0.5	⇨	Fairly weakly related	
Below 0.5	⇨	Very weakly related	Not related

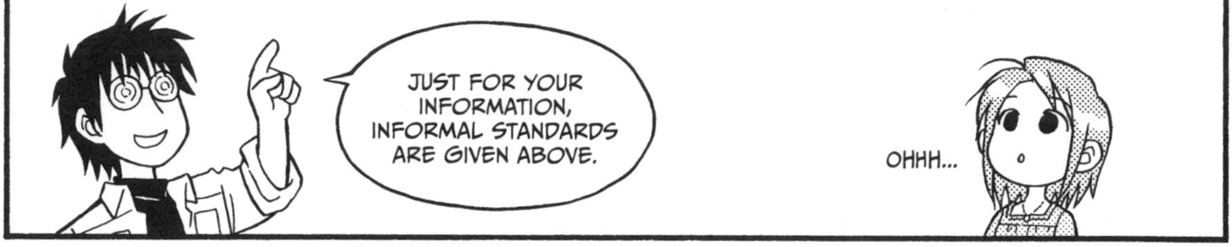
JUST FOR YOUR INFORMATION, INFORMAL STANDARDS ARE GIVEN ABOVE.
OHHH...

WARNING

I mentioned earlier that the correlation coefficient is an index that shows the degree of *linear* relation between two numerical variables.

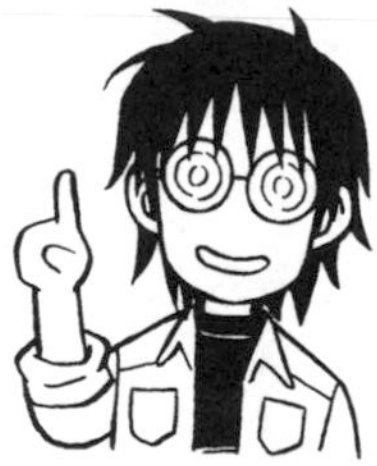

SAMPLE OF DATA UNSUITABLE FOR CORRELATION COEFFICIENT

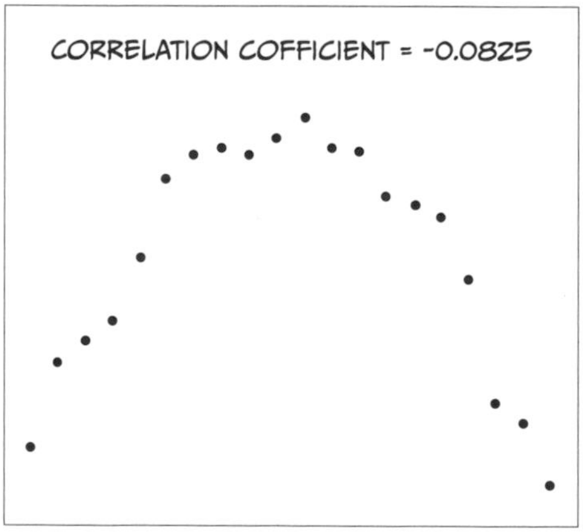
CORRELATION COFFICIENT = -0.0825

For example, the two variables are obviously related in this chart. However, the correlation coefficient is almost 0 because the relationship is *non-linear*.

2. CORRELATION RATIO

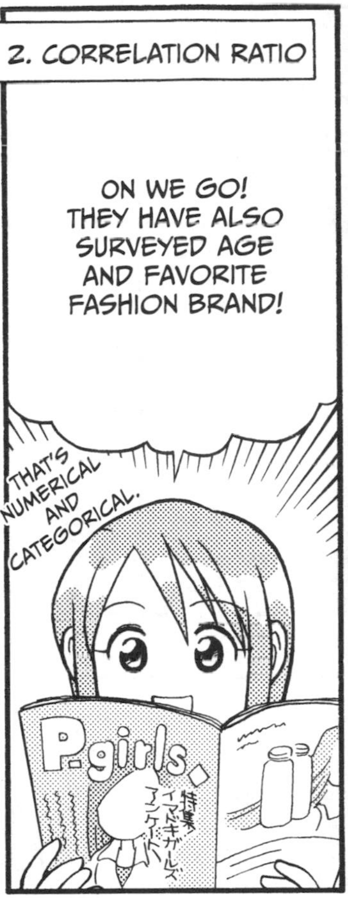
ON WE GO! THEY HAVE ALSO SURVEYED AGE AND FAVORITE FASHION BRAND!
THAT'S NUMERICAL AND CATEGORICAL.
P.girls

Street Survey in Everyhills

Age and Favorite Fashion Brand

Respondent	Age	Brand
Ms. A	27	Theremes
Ms. B	33	Channelior
Ms. C	16	Bureperry
Ms. D	29	Bureperry
Ms. E	32	Channelior
Ms. F	23	Theremes
Ms. G	25	Channelior
Ms. H	28	Theremes
Ms. I	22	Bureperry
Ms. J	18	Bureperry
Ms. K	26	Channelior
Ms. L	26	Theremes
Ms. M	15	Bureperry
Ms. N	29	Channelior
Ms. O	26	Bureperry

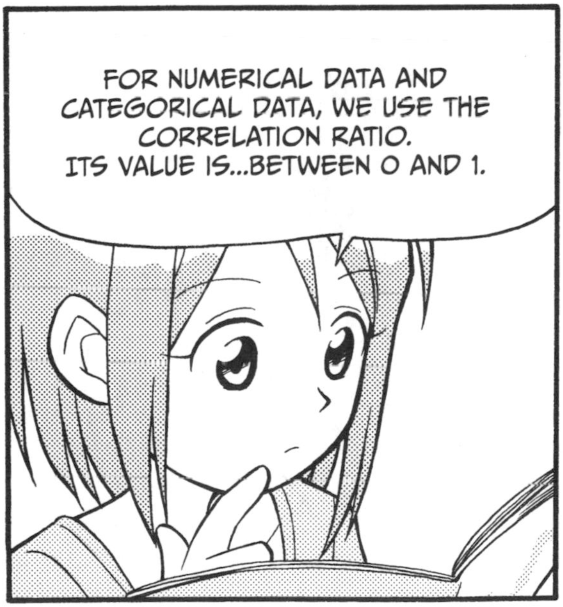
FOR NUMERICAL DATA AND CATEGORICAL DATA, WE USE THE CORRELATION RATIO. ITS VALUE IS...BETWEEN 0 AND 1.

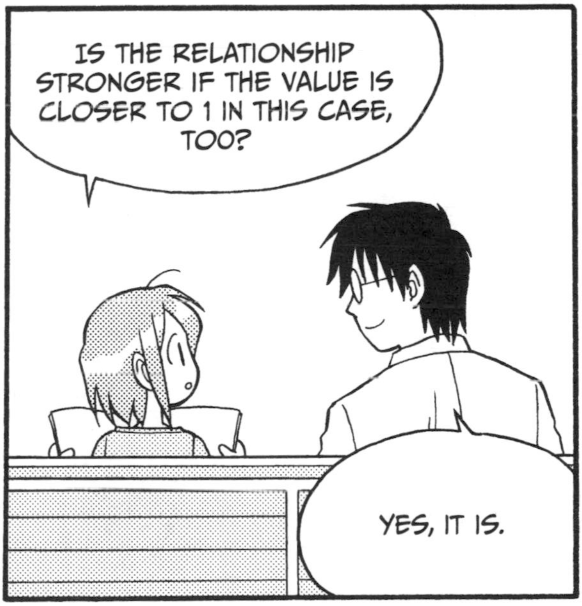
IS THE RELATIONSHIP STRONGER IF THE VALUE IS CLOSER TO 1 IN THIS CASE, TOO?
YES, IT IS.

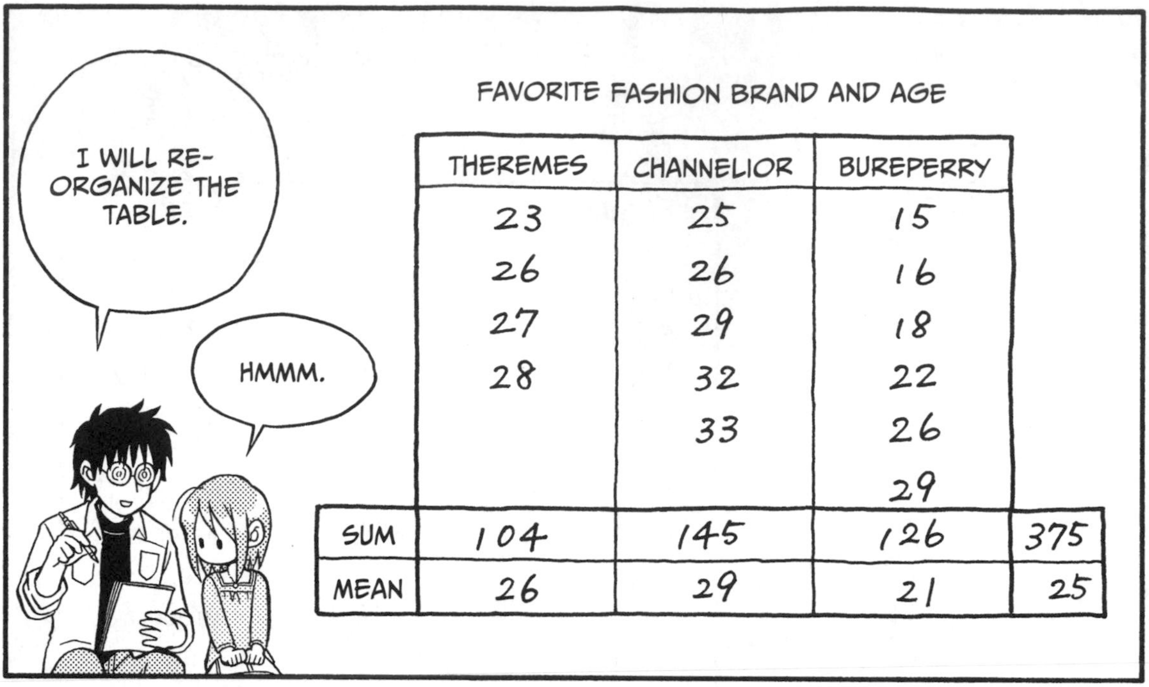
I WILL RE-ORGANIZE THE TABLE.
HMMM.
FAVORITE FASHION BRAND AND AGE

	THEREMES	CHANNELIOR	BUREPERRY	
	23	25	15	
	26	26	16	
	27	29	18	
	28	32	22	
		33	26	
			29	
SUM	104	145	126	375
MEAN	26	29	21	25

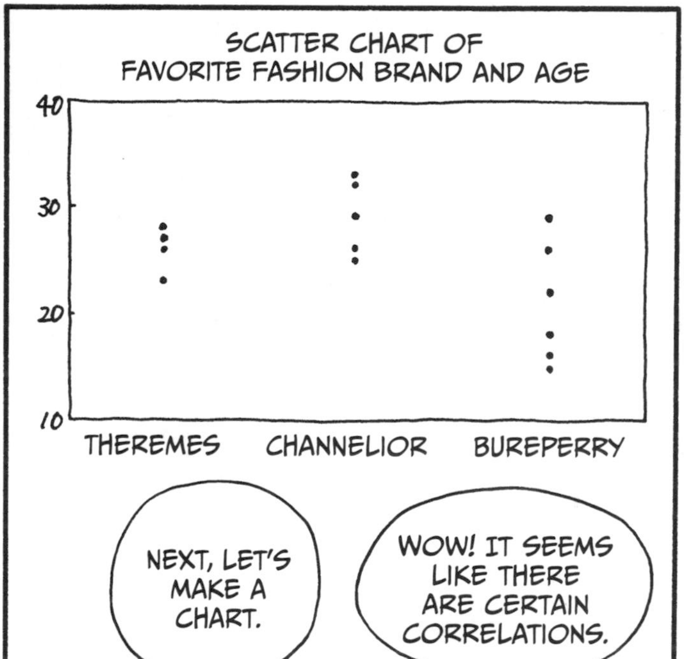
SCATTER CHART OF
FAVORITE FASHION BRAND AND AGE
40
30
20
10
THEREMES
CHANNELIOR
BUREPERRY
NEXT, LET'S MAKE A CHART.
WOW! IT SEEMS LIKE THERE ARE CERTAIN CORRELATIONS.

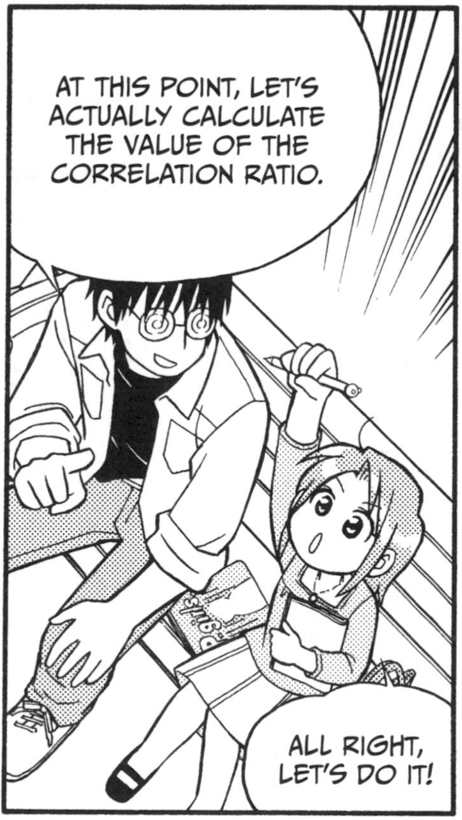
AT THIS POINT, LET'S ACTUALLY CALCULATE THE VALUE OF THE CORRELATION RATIO.
ALL RIGHT, LET'S DO IT!

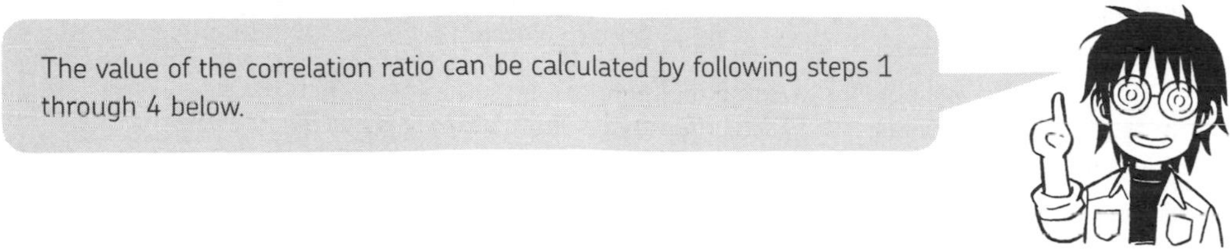

The value of the correlation ratio can be calculated by following steps 1 through 4 below.

Step 1

Do the calculations in the table below.

		Sum	
(Theremes – average for Theremes)²	$(23 - 26)^2 = (-3)^2 = 9$ $(26 - 26)^2 = 0^2 = 0$ $(27 - 26)^2 = 1^2 = 1$ $(28 - 26)^2 = 2^2 = 4$	14	S_{TT}
(Channelior – average for Channelior)²	$(25 - 29)^2 = (-4)^2 = 16$ $(26 - 29)^2 = (-3)^2 = 9$ $(29 - 29)^2 = 0^2 = 0$ $(32 - 29)^2 = 3^2 = 9$ $(33 - 29)^2 = 4^2 = 16$	50	S_{CC}
(Bureperry – average for Bureperry)²	$(15 - 21)^2 = (-6)^2 = 36$ $(16 - 21)^2 = (-5)^2 = 25$ $(18 - 21)^2 = (-3)^2 = 9$ $(22 - 21)^2 = 1^2 = 1$ $(26 - 21)^2 = 5^2 = 25$ $(29 - 21)^2 = 8^2 = 64$	160	S_{BB}

Step 2

Calculate the intraclass variance ($S_{TT} + S_{CC} + S_{BB}$ = how much the data within each category varies).

$$S_{TT} + S_{CC} + S_{BB} = 14 + 50 + 160 = 224$$

Step 3

Calculate the interclass variance, or how different the categories are from each other.

(number of votes for Theremes) × (average for Theremes − average for all data)^2
+ (number of votes for Channelior) × (average for Channelior − average for all data)^2
+ (number of votes for Bureperry) × (average for Bureperry − average for all data)^2

$$\begin{aligned} & 4 \times (26 - 25)^2 + 5 \times (29 - 25)^2 + 6 \times (21 - 25)^2 \\ &= 4 \times 1 + 5 \times 16 + 6 \times 16 \\ &= 4 + 80 + 96 \\ &= 180 \end{aligned}$$

Step 4

Calculate the value of the correlation ratio.

$$\frac{\text{interclass variance}}{\text{intraclass variance} + \text{interclass variance}}$$

$$\frac{180}{224 + 180} = \frac{180}{404} = 0.4455$$

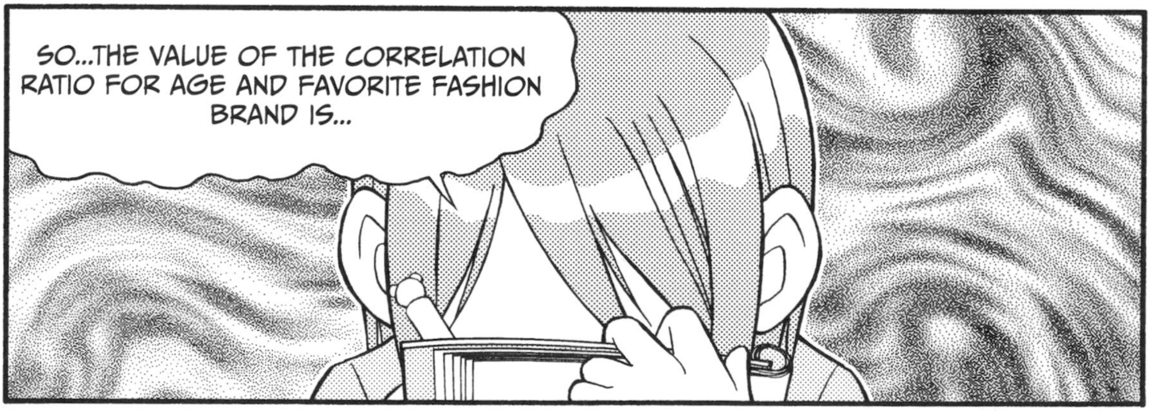
SO...THE VALUE OF THE CORRELATION RATIO FOR AGE AND FAVORITE FASHION BRAND IS...

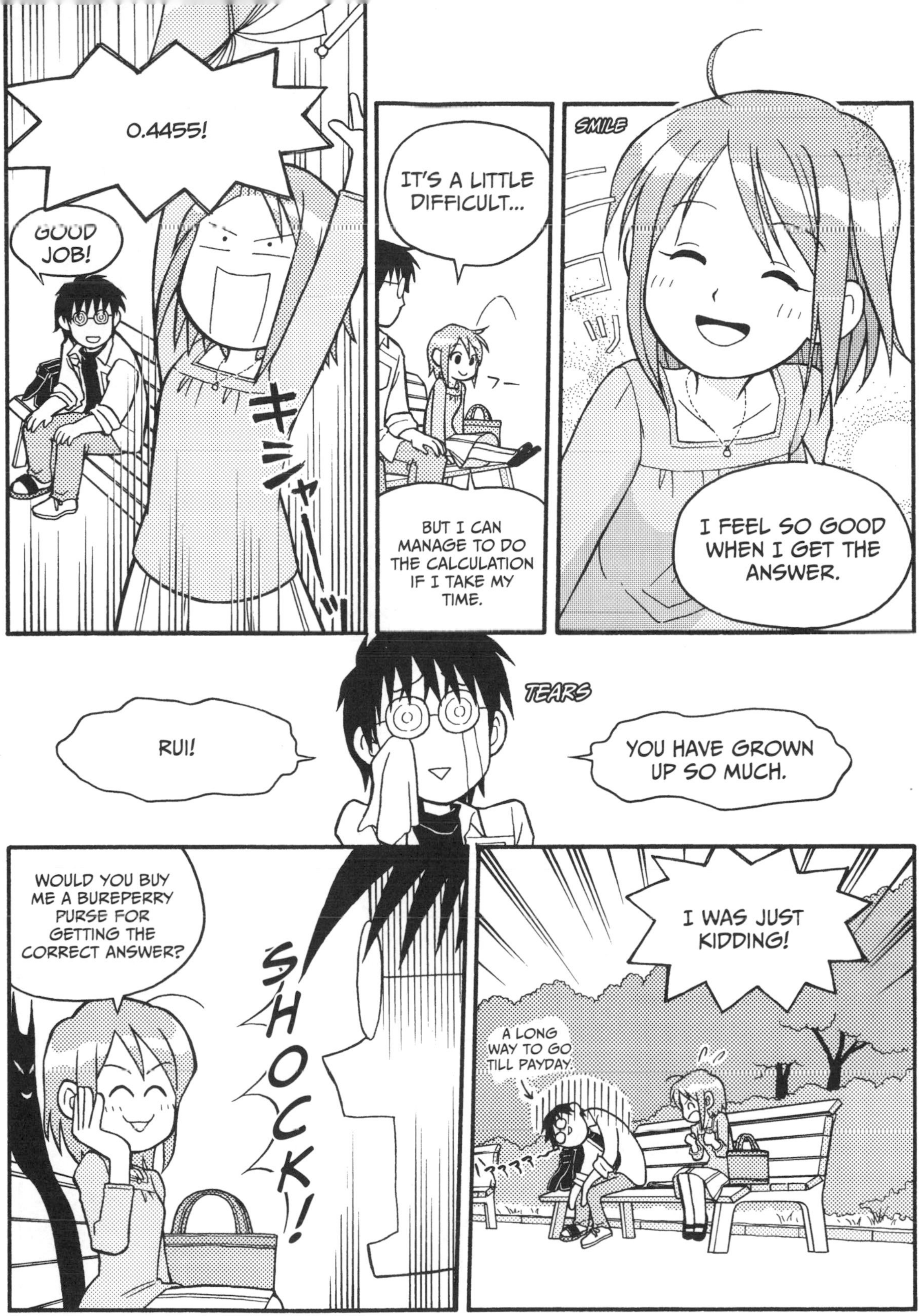
0.4455!
GOOD JOB!
IT'S A LITTLE DIFFICULT...
BUT I CAN MANAGE TO DO THE CALCULATION IF I TAKE MY TIME.
SMILE
I FEEL SO GOOD WHEN I GET THE ANSWER.
RUI!
TEARS
YOU HAVE GROWN UP SO MUCH.
WOULD YOU BUY ME A BUREPERRY PURSE FOR GETTING THE CORRECT ANSWER?
SHOCK!
I WAS JUST KIDDING!
A LONG WAY TO GO TILL PAYDAY.

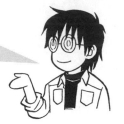

As explained earlier, the value of the correlation ratio is between 0 and 1. The stronger the correlation is between the two variables, the closer the value is to 1, and the weaker the correlation is between two variables, the closer the value is to 0. Refer to the charts below for more details.

Here is a scatter chart of favorite fashion brand and age (when the correlation ratio is 1).

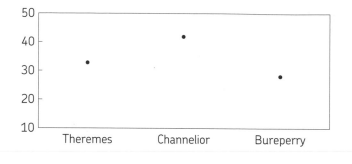

50
40
30
20
10
Theremes
Channelior
Bureperry

correlation ratio is 1 ⇔ data included in each group is the same ⇔ intraclass variance is 0

Here is a scatter chart of favorite fashion brand and age (when the correlation ratio is 0).

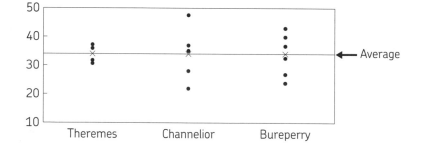

50
40
30
20
10
Average
Theremes
Channelior
Bureperry

correlation ratio is 0 ⇔ average for each group is the same ⇔ intraclass variance is 0

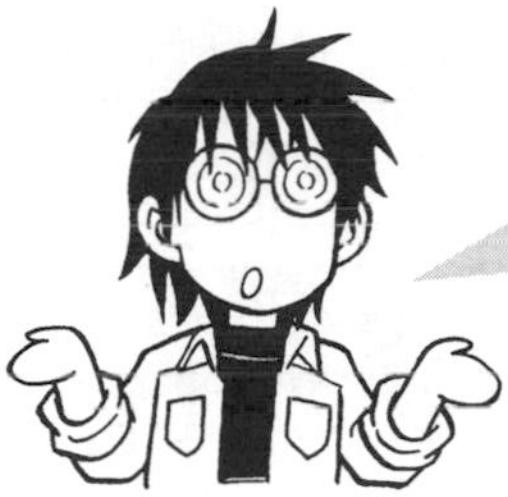

Unfortunately, there are no statistical standards such as "the two variables have a strong correlation if the correlation ratio is above a certain benchmark." However, informal standards are given below.

INFORMAL STANDARDS OF THE CORRELATION RATIO

Correlation ratio		Detailed description	Rough description
1.0–0.8	⇨	Very strongly related	
0.8–0.5	⇨	Fairly strongly related	Related
0.5–0.25	⇨	Fairly weakly related	
Below 0.25	⇨	Very weakly related	Not related

The result of the calculation for the case in question was 0.4455, so the variables are fairly weakly related!

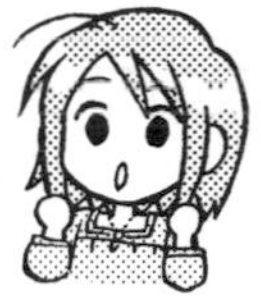

3. CRAMER'S COEFFICIENT

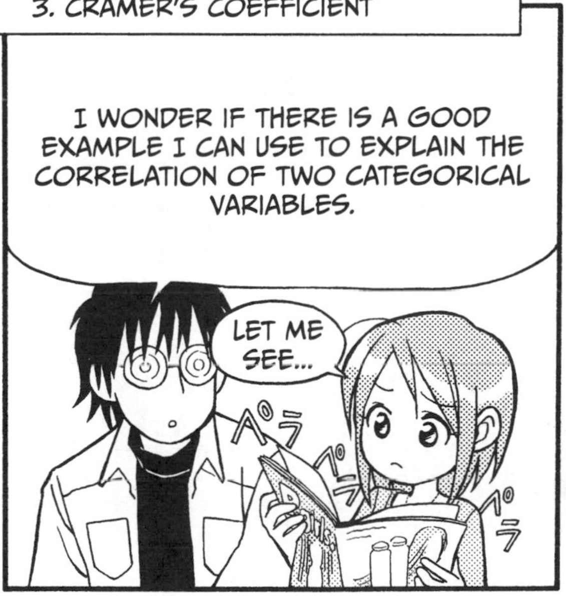
I WONDER IF THERE IS A GOOD EXAMPLE I CAN USE TO EXPLAIN THE CORRELATION OF TWO CATEGORICAL VARIABLES.
LET ME SEE...
ペラ
ペラ
ペラ

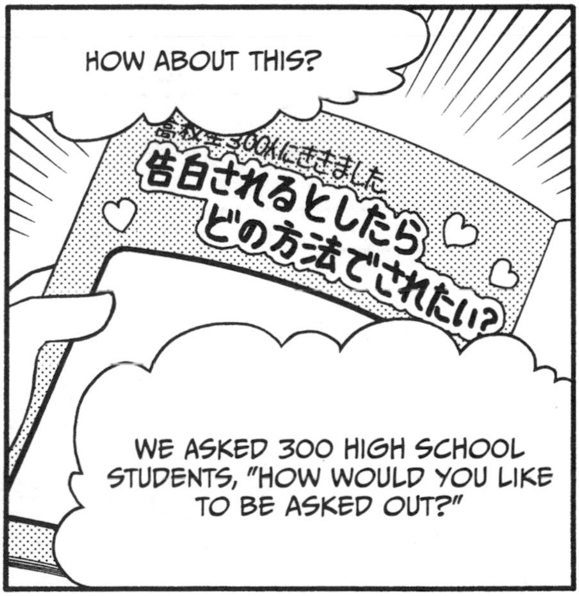
HOW ABOUT THIS?
告白されるとしたら
どの方法でされたい?
WE ASKED 300 HIGH SCHOOL STUDENTS, "HOW WOULD YOU LIKE TO BE ASKED OUT?"

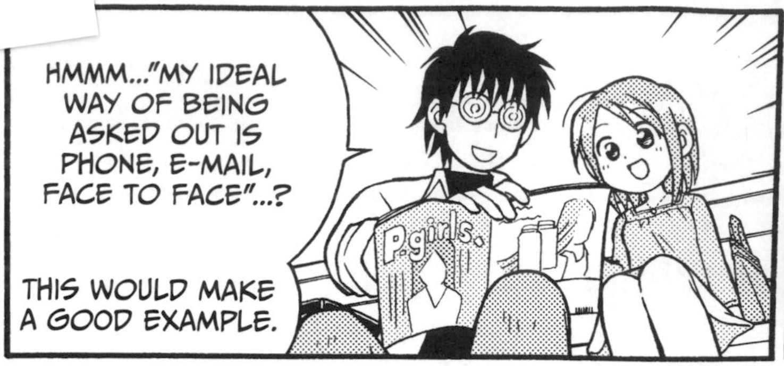
HMMM..."MY IDEAL WAY OF BEING ASKED OUT IS PHONE, E-MAIL, FACE TO FACE"...?
THIS WOULD MAKE A GOOD EXAMPLE.

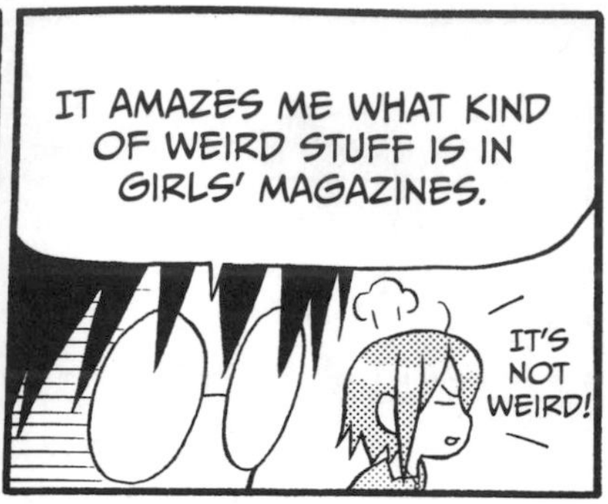
IT AMAZES ME WHAT KIND OF WEIRD STUFF IS IN GIRLS' MAGAZINES.
IT'S NOT WEIRD!

CROSS TABULATION OF SEX AND DESIRED WAY OF BEING ASKED OUT

		DESIRED WAY OF BEING ASKED OUT			SUM
		PHONE	E-MAIL	FACE TO FACE	
SEX	FEMALE	34	61	53	148
	MALE	38	40	74	152
SUM		72	101	127	300

This indicates that 74 out of 152 males answered that they'd like to be asked out directly.

CROSS TABULATION OF SEX AND DESIRED WAY OF BEING ASKED OUT (HORIZONTAL PERCENTAGE TABLE)

		DESIRED WAY OF BEING ASKED OUT			SUM
		PHONE	E-MAIL	FACE TO FACE	
SEX	FEMALE	23%	41%	36%	100%
	MALE	25%	26%	49%	100%
SUM		24%	34%	42%	100%

This shows that 49% ($\frac{74}{152} \times 100$) of the 152 males would like to be asked out directly.

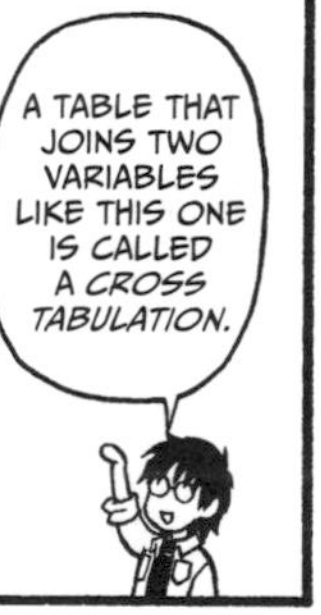
A TABLE THAT JOINS TWO VARIABLES LIKE THIS ONE IS CALLED A *CROSS TABULATION*.

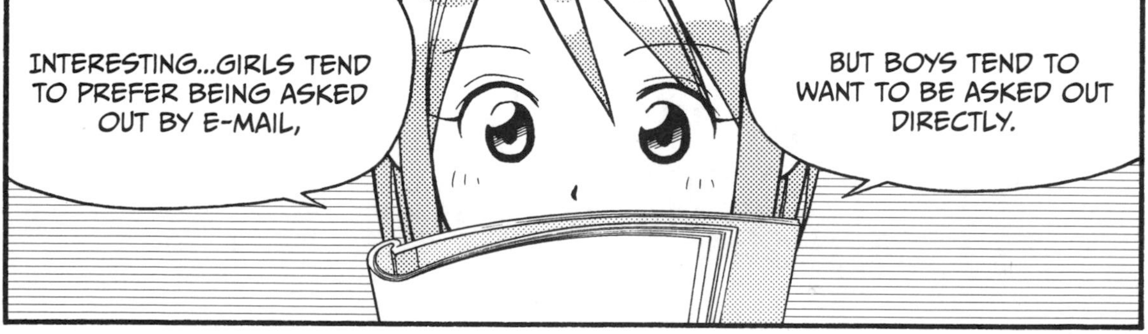
INTERESTING...GIRLS TEND TO PREFER BEING ASKED OUT BY E-MAIL,
BUT BOYS TEND TO WANT TO BE ASKED OUT DIRECTLY.

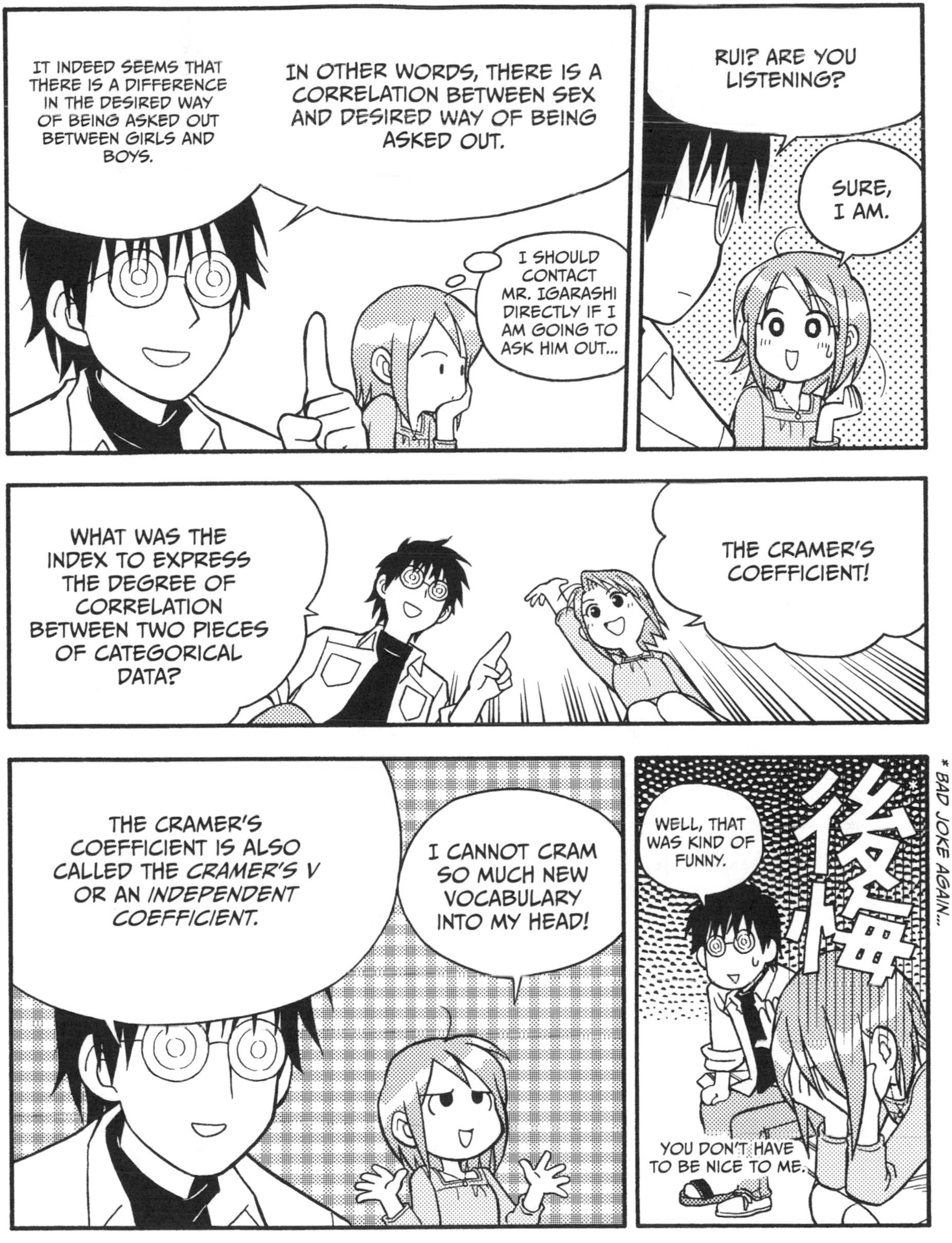
IT INDEED SEEMS THAT THERE IS A DIFFERENCE IN THE DESIRED WAY OF BEING ASKED OUT BETWEEN GIRLS AND BOYS.
IN OTHER WORDS, THERE IS A CORRELATION BETWEEN SEX AND DESIRED WAY OF BEING ASKED OUT.
I SHOULD CONTACT MR. IGARASHI DIRECTLY IF I AM GOING TO ASK HIM OUT...
RUI? ARE YOU LISTENING?
SURE, I AM.
WHAT WAS THE INDEX TO EXPRESS THE DEGREE OF CORRELATION BETWEEN TWO PIECES OF CATEGORICAL DATA?
THE CRAMER'S COEFFICIENT!
THE CRAMER'S COEFFICIENT IS ALSO CALLED THE *CRAMER'S V* OR AN *INDEPENDENT COEFFICIENT*.
I CANNOT CRAM SO MUCH NEW VOCABULARY INTO MY HEAD!
後悔*
WELL, THAT WAS KIND OF FUNNY.
YOU DON'T HAVE TO BE NICE TO ME.
* BAD JOKE AGAIN...

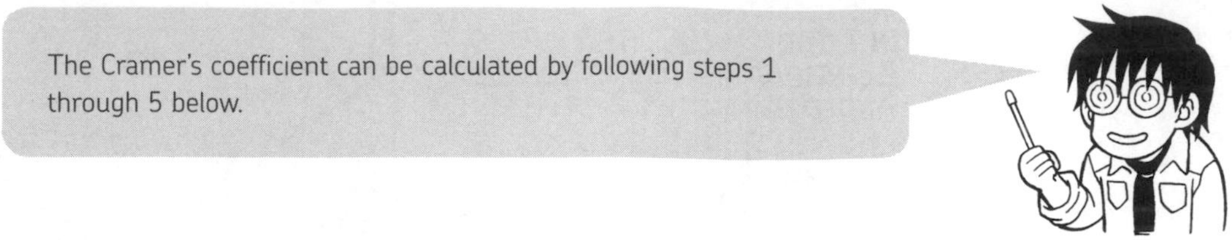
The Cramer's coefficient can be calculated by following steps 1 through 5 below.

Step 1

Prepare a cross tabulation. The values surrounded by the bold frame are called *actual measurement frequencies*.

		Desired way of being asked out			Sum
		Phone	E-mail	Face to face	
Sex	Female	34	61	53	148
	Male	38	40	74	152
Sum		72	101	127	300

Step 2

Do the calculations in the table below. The values surrounded by the bold frame are called *expected frequencies*.

		Desired way of being asked out			Sum
		Phone	E-mail	Face to face	
Sex	Female	$\frac{148 \times 72}{300}$	$\frac{148 \times 101}{300}$	$\frac{148 \times 127}{300}$	148
	Male	$\frac{152 \times 72}{300}$	$\frac{152 \times 101}{300}$	$\frac{152 \times 127}{300}$	152
Sum		72	101	127	300

$$\frac{\text{sum of male} \times \text{sum of face to face}}{\text{total number of values}} \qquad \text{Formula A}$$

If sex and desired way of being asked out have no relationship, the ratio between phone, e-mail, and face to face should be

$$72 : 101 : 127 = \frac{72}{72 + 101 + 127} : \frac{101}{72 + 101 + 127} : \frac{127}{72 + 101 + 127}$$

$$= \frac{72}{300} : \frac{101}{300} : \frac{127}{300}$$

for both males and females, according to the sum column in the table in step 2. Thus, our expected frequency (Formula A) shows the predicted number of males who wish to be asked out directly when there is no relationship between sex and desired way of being asked out is 152 × (127 ÷ 300) = (152 × 127) ÷ 300, or

$$152 \times \frac{127}{300} = \frac{152 \times 127}{300} = 64.3$$

Step 3

Calculate $\dfrac{(\text{actual frequency} - \text{expected frequency})^2}{\text{expected frequency}}$ for each square.

		Desired way of being asked out			Sum
		Phone	E-mail	Face to face	
Sex	Female	$\dfrac{\left(34 - \dfrac{148 \times 72}{300}\right)^2}{\dfrac{148 \times 72}{300}}$	$\dfrac{\left(61 - \dfrac{148 \times 101}{300}\right)^2}{\dfrac{148 \times 101}{300}}$	$\dfrac{\left(53 - \dfrac{148 \times 127}{300}\right)^2}{\dfrac{148 \times 127}{300}}$	148
	Male	$\dfrac{\left(38 - \dfrac{152 \times 72}{300}\right)^2}{\dfrac{152 \times 72}{300}}$	$\dfrac{\left(40 - \dfrac{152 \times 101}{300}\right)^2}{\dfrac{152 \times 101}{300}}$	$\dfrac{\left(74 - \dfrac{152 \times 127}{300}\right)^2}{\dfrac{152 \times 127}{300}}$	152
Sum		72	101	127	300

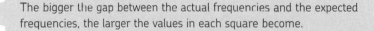

The bigger the gap between the actual frequencies and the expected frequencies, the larger the values in each square become.

Step 4

Calculate the sum of the value inside the bold frame in the table of step 3. This value is called *Pearson's chi-square test statistic*. It will be written as χ_0^2 from now on.

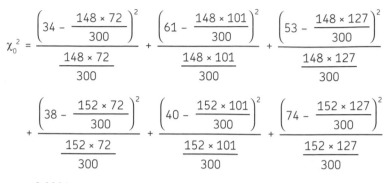

$$\chi_0^2 = \frac{\left(34 - \dfrac{148 \times 72}{300}\right)^2}{\dfrac{148 \times 72}{300}} + \frac{\left(61 - \dfrac{148 \times 101}{300}\right)^2}{\dfrac{148 \times 101}{300}} + \frac{\left(53 - \dfrac{148 \times 127}{300}\right)^2}{\dfrac{148 \times 127}{300}}$$

$$+ \frac{\left(38 - \dfrac{152 \times 72}{300}\right)^2}{\dfrac{152 \times 72}{300}} + \frac{\left(40 - \dfrac{152 \times 101}{300}\right)^2}{\dfrac{152 \times 101}{300}} + \frac{\left(74 - \dfrac{152 \times 127}{300}\right)^2}{\dfrac{152 \times 127}{300}}$$

$$= 8.0091$$

As can be understood from step 3, the more the actual measurements diverge from their expected frequencies, or the greater the correlation between sex and desired way of being asked out, the larger Pearson's chi-square test statistic (χ_0^2) becomes.

Step 5

Calculate the Cramer's coefficient.

$$\sqrt{\dfrac{\chi_0^{\,2}}{\text{the total number of values} \times (\min\{\text{the number of lines in the cross tabulation, the number of rows in the cross tabulation}\} - 1)}}$$

$\min\{a,b\}$ means "whichever is smaller, a or b."

$$\sqrt{\dfrac{8.0091}{300 \times \min\{2,3\} - 1}} = \sqrt{\dfrac{8.0091}{300 \times (2 - 1)}} = \sqrt{\dfrac{8.0091}{300}} = 0.1634$$

THUS, THE CRAMER'S COEFFICIENT IS 0.1634.

HELP. I'M FEELING DIZZY.

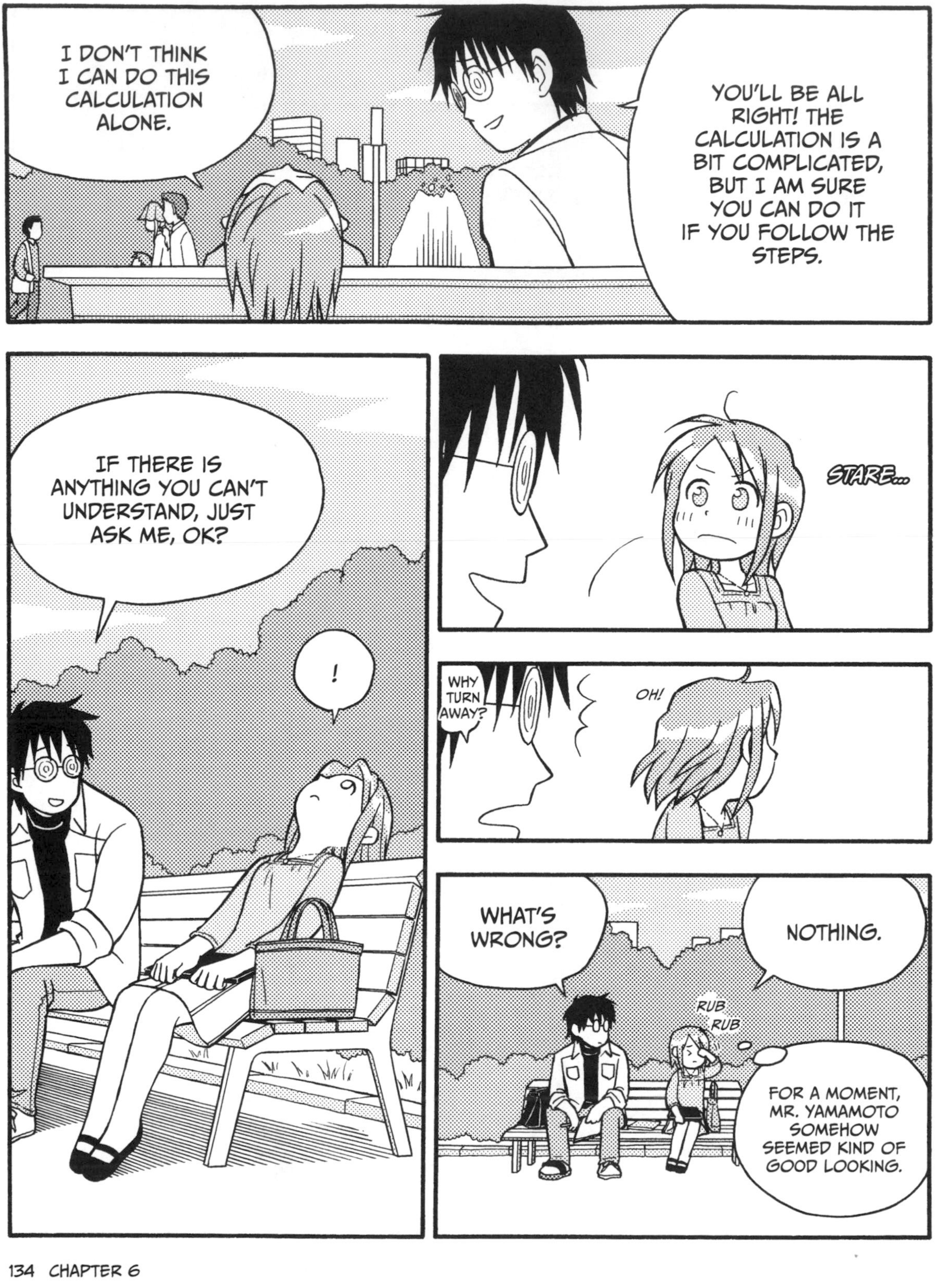
I DON'T THINK I CAN DO THIS CALCULATION ALONE.
YOU'LL BE ALL RIGHT! THE CALCULATION IS A BIT COMPLICATED, BUT I AM SURE YOU CAN DO IT IF YOU FOLLOW THE STEPS.
IF THERE IS ANYTHING YOU CAN'T UNDERSTAND, JUST ASK ME, OK?
!
STARE...
WHY TURN AWAY?
OH!
WHAT'S WRONG?
NOTHING.
RUB RUB
FOR A MOMENT, MR. YAMAMOTO SOMEHOW SEEMED KIND OF GOOD LOOKING.

As explained earlier, the Cramer's coefficient is between 0 and 1. The stronger the correlation between two variables, the closer the coefficient gets to 1, and the weaker the correlation, the closer the coefficient gets to 0. See the cross tabulation (horizontal percentage table) below for more details.

Here is the cross tabulation of sex and desired way of being asked out (horizontal percentage table) when the value of the Cramer's coefficient is 1.

		Desired way of being asked out			Sum
		Phone	E-mail	Face to face	
Sex	Female	17%	83%	0%	100%
	Male	0%	0%	100%	100%

Cramer's coefficient is 1 ⇔ the preferences of female and male are completely different

Here is the cross tabulation of sex and desired way of being asked out (horizontal percentage table) when the value of the Cramer's coefficient is 0.

		Desired way of being asked out			Sum
		Phone	E-mail	Face to face	
Sex	Female	17%	48%	35%	100%
	Male	17%	48%	35%	100%

Cramer's coefficient is 0 ⇔ the preferences of female and male are the same

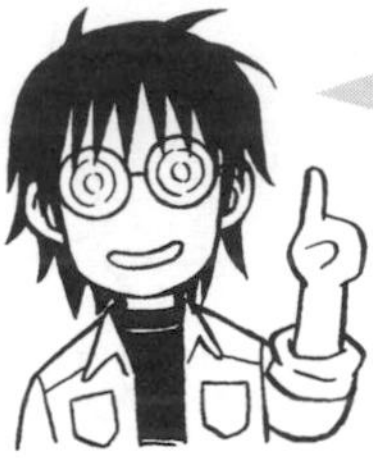

Unfortunately, there are no statistical standards such as "the two variables have a strong correlation if the Cramer's coefficient is above a certain benchmark." However, informal standards are given below.

INFORMAL STANDARDS OF THE CRAMER'S COEFFICIENT

Cramer's coefficient		Detailed description	Rough description
1.0–0.8	⇨	Very strongly related	
0.8–0.5	⇨	Fairly strongly related	Related
0.5–0.25	⇨	Fairly weakly related	
Below 0.25	⇨	Very weakly related	Not related

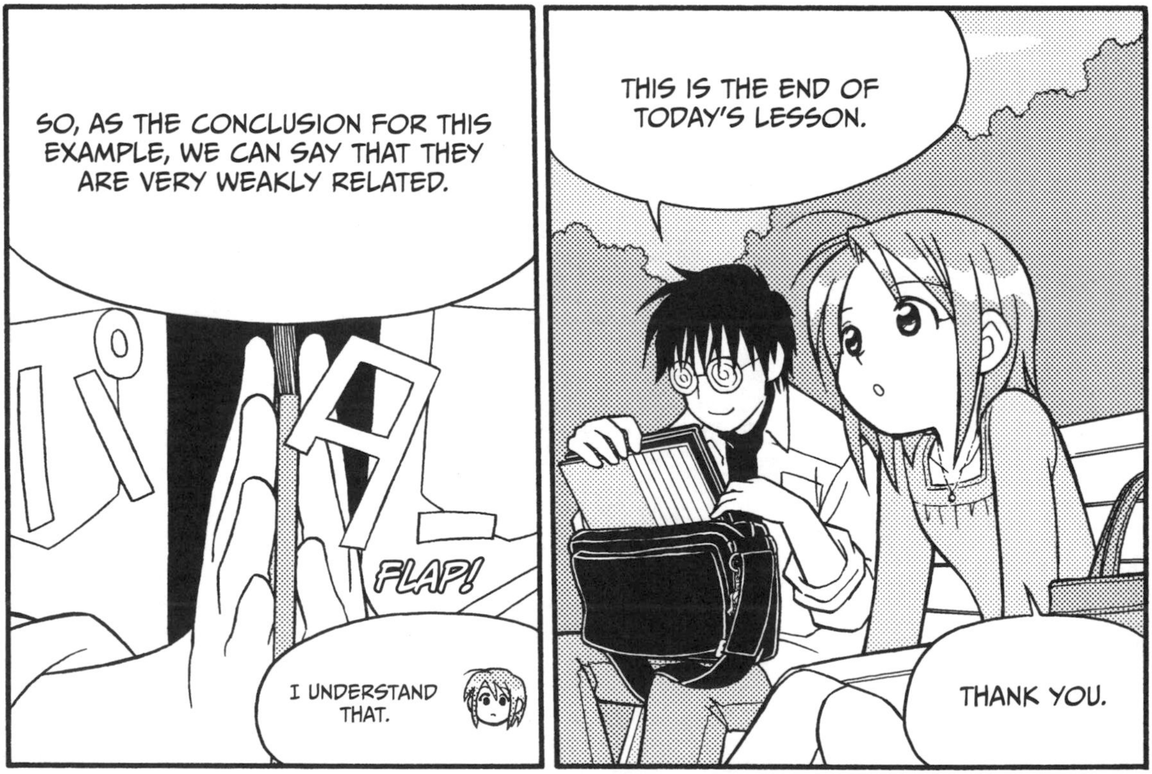
SO, AS THE CONCLUSION FOR THIS EXAMPLE, WE CAN SAY THAT THEY ARE VERY WEAKLY RELATED.
FLAP!
I UNDERSTAND THAT.
THIS IS THE END OF TODAY'S LESSON.
THANK YOU.

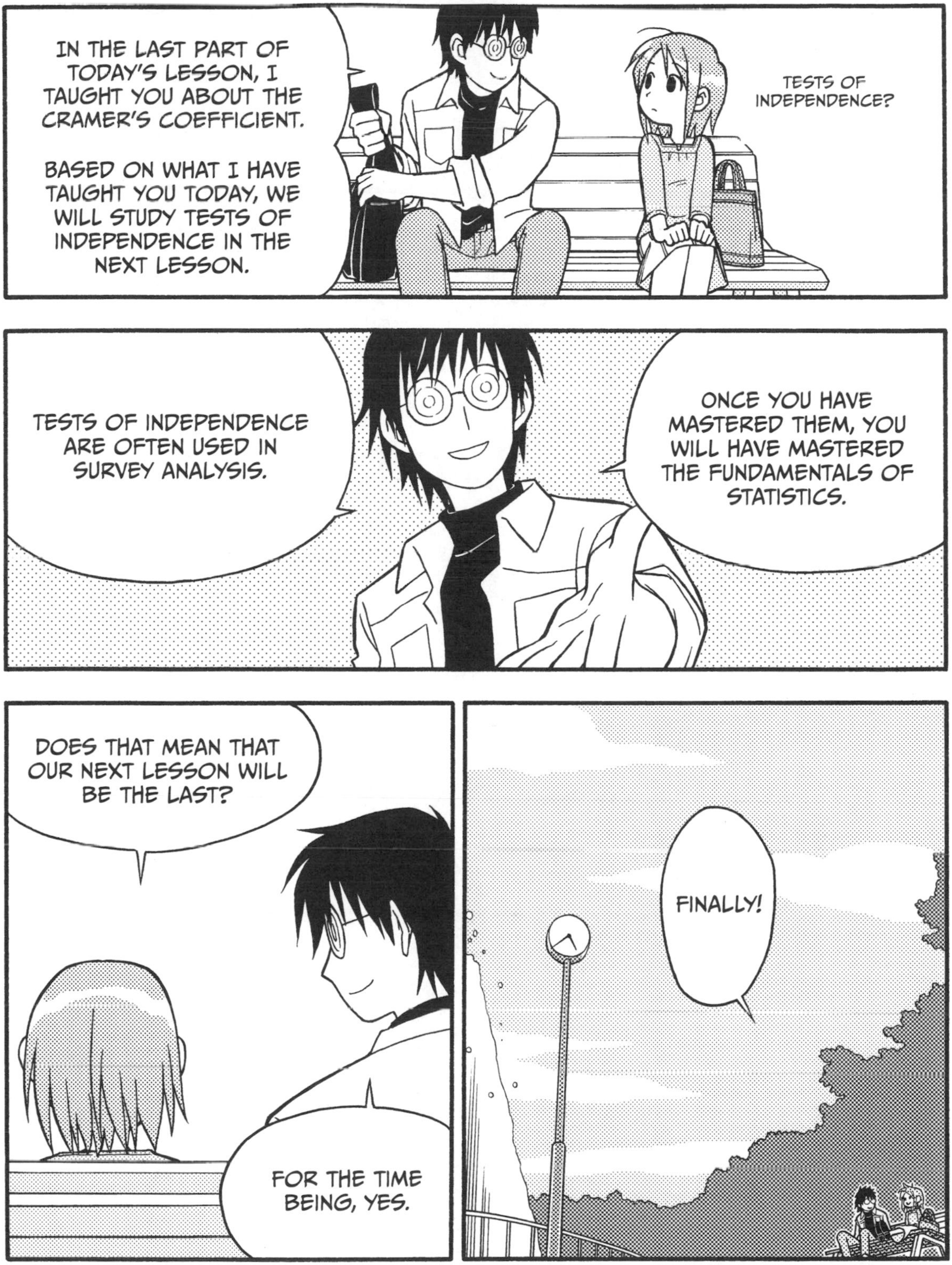
IN THE LAST PART OF TODAY'S LESSON, I TAUGHT YOU ABOUT THE CRAMER'S COEFFICIENT.
BASED ON WHAT I HAVE TAUGHT YOU TODAY, WE WILL STUDY TESTS OF INDEPENDENCE IN THE NEXT LESSON.
TESTS OF INDEPENDENCE?
TESTS OF INDEPENDENCE ARE OFTEN USED IN SURVEY ANALYSIS.
ONCE YOU HAVE MASTERED THEM, YOU WILL HAVE MASTERED THE FUNDAMENTALS OF STATISTICS.
DOES THAT MEAN THAT OUR NEXT LESSON WILL BE THE LAST?
FOR THE TIME BEING, YES.
FINALLY!

EXERCISE AND ANSWER

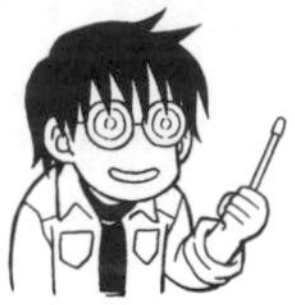

EXERCISE

Company *X* runs a casual dining restaurant. Its financial status was declining recently. Thus, Company *X* decided to study its customers' needs and conducted a survey of randomly chosen people, age 20 or older, residing in Japan. The table below shows the results of this survey.

Respondent	**What food do you often have in a casual dining restaurant?**	**If a free drink is to be served after a meal, which would you prefer? Coffee or tea?**
1	Chinese	Coffee
2	European	Coffee
...	...	...
250	Japanese	Tea

Below is a cross tabulation made using the table above.

		Preference for coffee or tea		Sum
		Coffee	Tea	
Type of food often ordered	Japanese	43	33	76
	European	51	53	104
	Chinese	29	41	70
Sum		123	127	250

Calculate the Cramer's coefficient for the food often ordered in casual dining restaurants and the preferred free drink of either coffee or tea.

ANSWER

Step 1

Prepare a cross tabulation.

		Preference for coffee or tea		Sum
		Coffee	Tea	
Type of food often ordered	Japanese	43	33	76
	European	51	53	104
	Chinese	29	41	70
Sum		123	127	250

Step 2

Calculate the expected frequency.

		Preference for coffee or tea		Sum
		Coffee	Tea	
Type of food often ordered	Japanese	$\dfrac{76 \times 123}{250}$	$\dfrac{76 \times 127}{250}$	76
	European	$\dfrac{104 \times 123}{250}$	$\dfrac{104 \times 127}{250}$	104
	Chinese	$\dfrac{70 \times 123}{250}$	$\dfrac{70 \times 127}{250}$	70
Sum		123	127	250

Step 3

Calculate

$$\frac{(\text{actual measurement frequency} - \text{expected frequency})^2}{\text{expected frequency}}$$

for each square.

		Preference for coffee or tea		Sum
		Coffee	Tea	
Type of food often ordered	Japanese	$\dfrac{\left(43 - \dfrac{76 \times 123}{250}\right)^2}{\dfrac{76 \times 123}{250}}$	$\dfrac{\left(33 - \dfrac{76 \times 127}{250}\right)^2}{\dfrac{76 \times 127}{250}}$	76
	European	$\dfrac{\left(51 - \dfrac{104 \times 123}{250}\right)^2}{\dfrac{104 \times 123}{250}}$	$\dfrac{\left(53 - \dfrac{104 \times 127}{250}\right)^2}{\dfrac{104 \times 127}{250}}$	104
	Chinese	$\dfrac{\left(29 - \dfrac{70 \times 123}{250}\right)^2}{\dfrac{70 \times 123}{250}}$	$\dfrac{\left(41 - \dfrac{70 \times 127}{250}\right)^2}{\dfrac{70 \times 127}{250}}$	70
Sum		123	127	250

Step 4

Calculate the sum of the value inside the bold frame in the table in step 3, which is the value of Pearson's chi-square test statistic (χ_0^2).

$$\chi_0^2 = \frac{\left(43 - \dfrac{76 \times 123}{250}\right)^2}{\dfrac{76 \times 123}{250}} + \frac{\left(33 - \dfrac{76 \times 127}{250}\right)^2}{\dfrac{76 \times 127}{250}}$$

$$+ \frac{\left(51 - \dfrac{104 \times 123}{250}\right)^2}{\dfrac{104 \times 123}{250}} + \frac{\left(53 - \dfrac{104 \times 127}{250}\right)^2}{\dfrac{104 \times 127}{250}}$$

$$+ \frac{\left(29 - \dfrac{70 \times 123}{250}\right)^2}{\dfrac{70 \times 123}{250}} + \frac{\left(41 - \dfrac{70 \times 127}{250}\right)^2}{\dfrac{70 \times 127}{250}}$$

$$= 3.3483$$

Step 5

Calculate the Cramer's coefficient.

$$\sqrt{\frac{\chi_0^2}{\text{the total number of values} \times \left(\min\left\{\begin{array}{c}\text{the number of lines} \\ \text{in the cross tabulation}\end{array}, \begin{array}{c}\text{the number of rows} \\ \text{in the cross tabulation}\end{array}\right\} - 1\right)}}$$

$$\sqrt{\frac{3.3483}{250 \times (\min\{3,2\} - 1)}} = \sqrt{\frac{3.3483}{250 \times (2 - 1)}} = \sqrt{\frac{3.3483}{250}} = 0.1157$$

SUMMARY

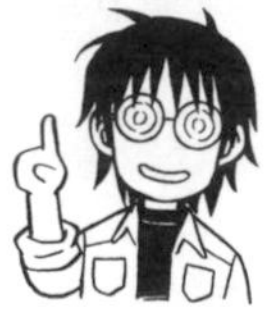

- The index used to describe the degree of correlation between numerical data and numerical data is the *correlation coefficient*.
- The index used to describe the degree of correlation between numerical data and categorical data is the *correlation ratio*.
- The index used to describe the degree of correlation between categorical data and categorical data is the *Cramer's coefficient* (sometimes called the *Cramer's V* or an *independent coefficient*).
- The characteristics of the correlation coefficient, correlation ratio, and Cramer's coefficient are shown in the table below.

	Minimum	**Maximum**	**The value when the two variables are not correlated at all**	**The value when the two variables are most strongly correlated**
Correlation coefficient	–1	1	0	–1 or 1
Correlation ratio	0	1	0	1
Cramer's coefficient	0	1	0	1

- There are no statistical standards for the correlation coefficient, correlation ratio, and Cramer's coefficient, such as "the two variables have a strong correlation if the value is above a certain benchmark."

7

LET'S EXPLORE
THE HYPOTHESIS TESTS

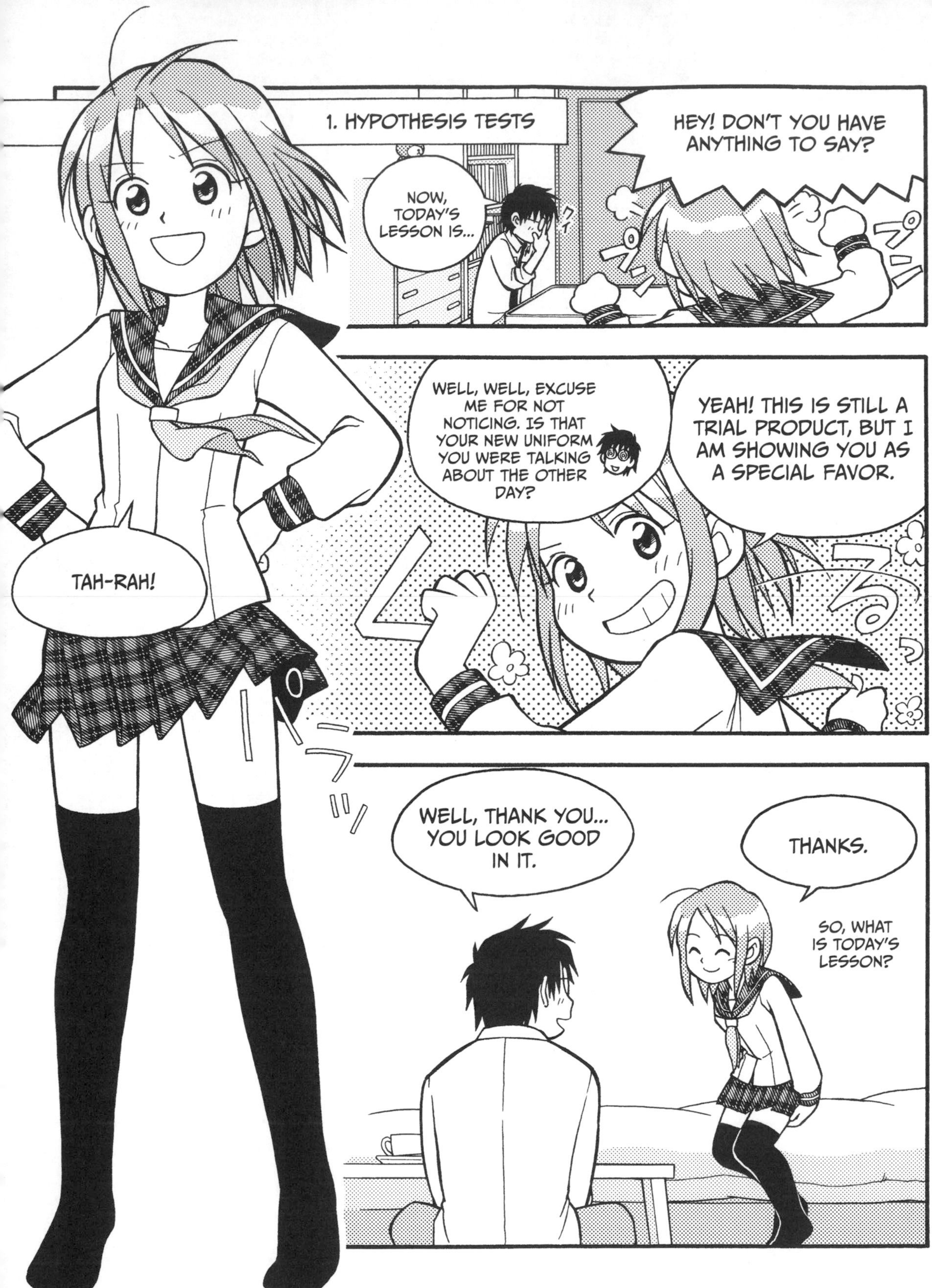
TAH-RAH!
1. HYPOTHESIS TESTS
NOW, TODAY'S LESSON IS...
HEY! DON'T YOU HAVE ANYTHING TO SAY?
WELL, WELL, EXCUSE ME FOR NOT NOTICING. IS THAT YOUR NEW UNIFORM YOU WERE TALKING ABOUT THE OTHER DAY?
YEAH! THIS IS STILL A TRIAL PRODUCT, BUT I AM SHOWING YOU AS A SPECIAL FAVOR.
WELL, THANK YOU... YOU LOOK GOOD IN IT.
THANKS.
SO, WHAT IS TODAY'S LESSON?

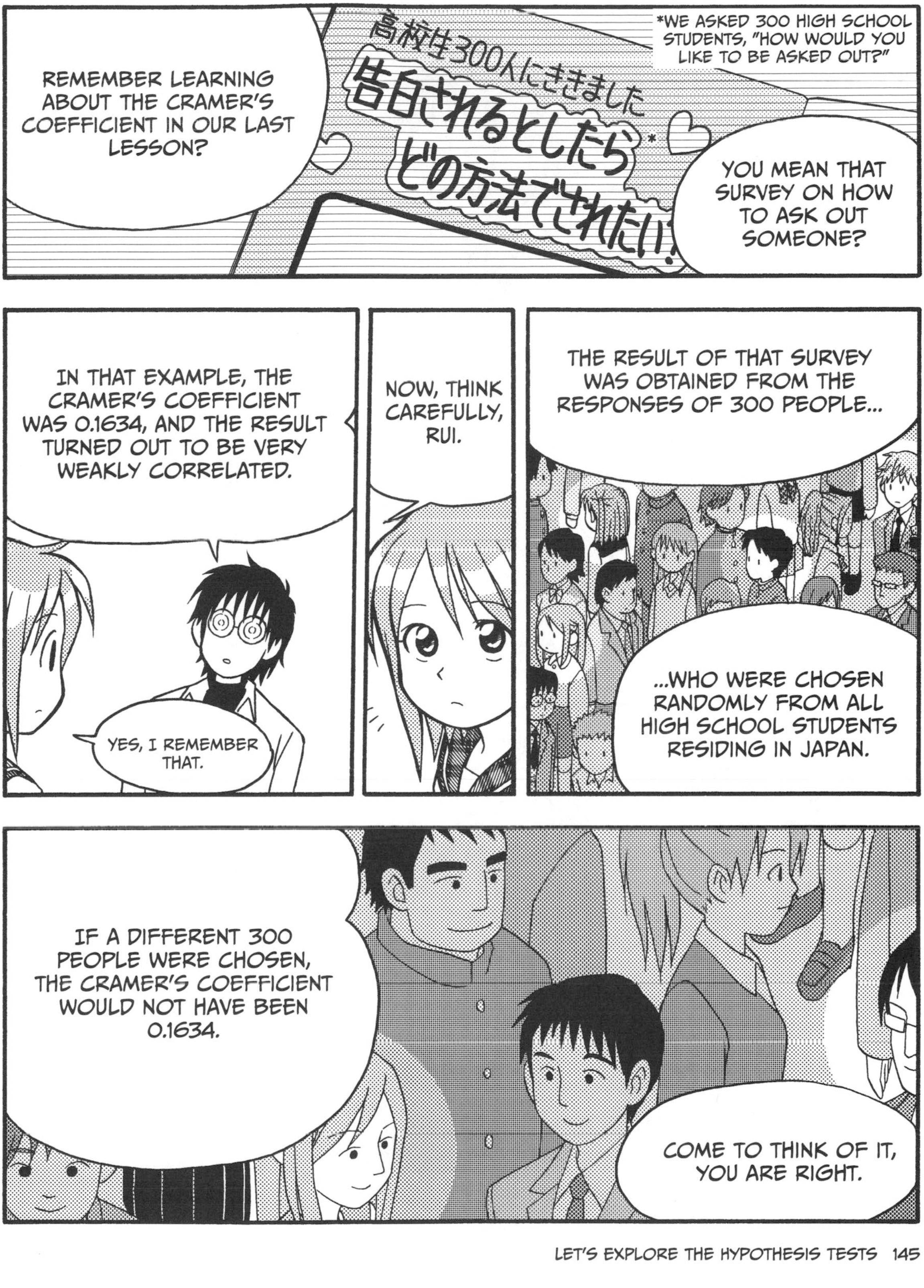
REMEMBER LEARNING ABOUT THE CRAMER'S COEFFICIENT IN OUR LAST LESSON?
高校生300人にききました
告白されるとしたら*
どの方法でされたい!?
*WE ASKED 300 HIGH SCHOOL STUDENTS, "HOW WOULD YOU LIKE TO BE ASKED OUT?"
YOU MEAN THAT SURVEY ON HOW TO ASK OUT SOMEONE?
IN THAT EXAMPLE, THE CRAMER'S COEFFICIENT WAS 0.1634, AND THE RESULT TURNED OUT TO BE VERY WEAKLY CORRELATED.
YES, I REMEMBER THAT.
NOW, THINK CAREFULLY, RUI.
THE RESULT OF THAT SURVEY WAS OBTAINED FROM THE RESPONSES OF 300 PEOPLE...
...WHO WERE CHOSEN RANDOMLY FROM ALL HIGH SCHOOL STUDENTS RESIDING IN JAPAN.
IF A DIFFERENT 300 PEOPLE WERE CHOSEN, THE CRAMER'S COEFFICIENT WOULD NOT HAVE BEEN 0.1634.
COME TO THINK OF IT, YOU ARE RIGHT.

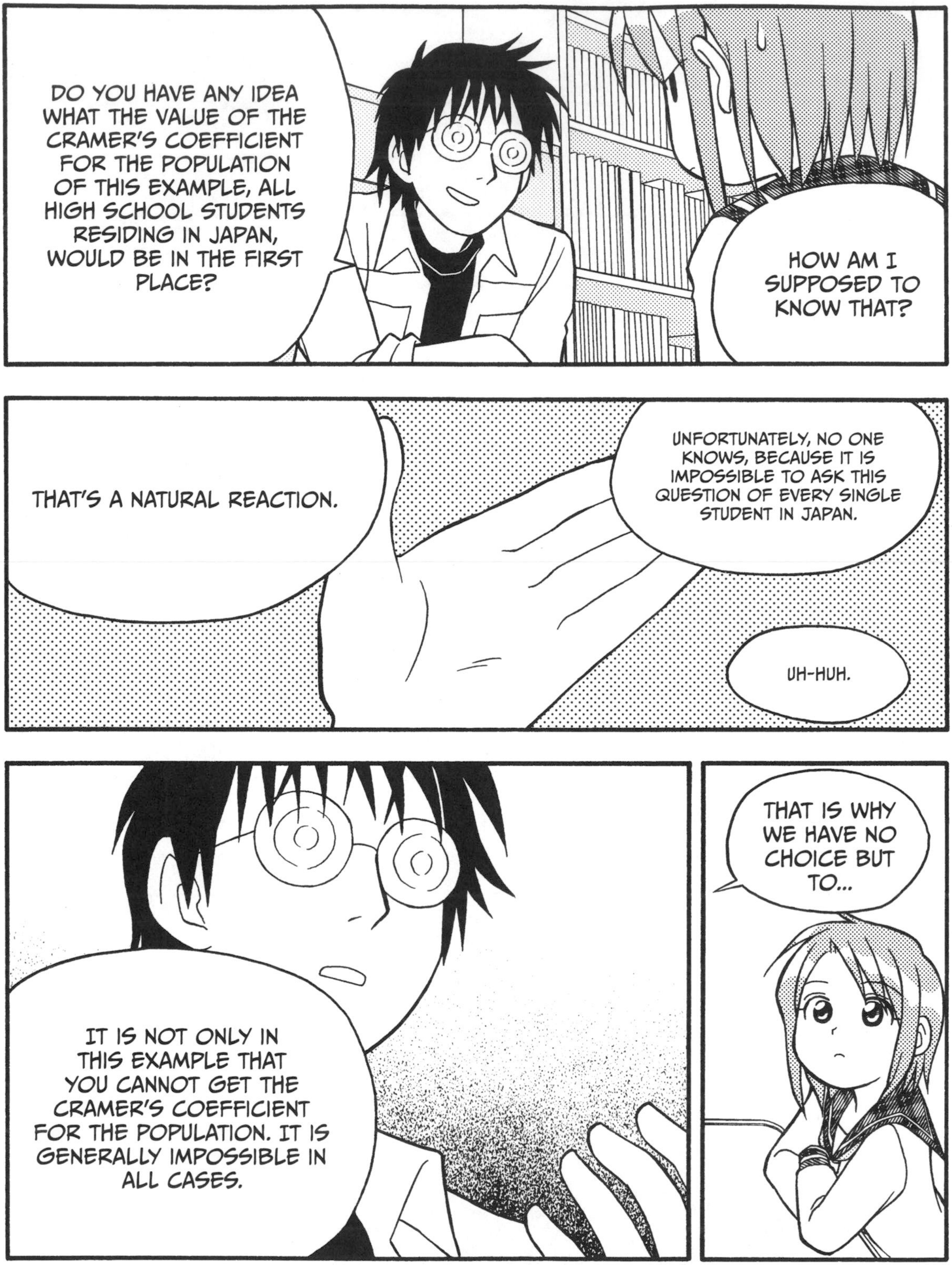
DO YOU HAVE ANY IDEA WHAT THE VALUE OF THE CRAMER'S COEFFICIENT FOR THE POPULATION OF THIS EXAMPLE, ALL HIGH SCHOOL STUDENTS RESIDING IN JAPAN, WOULD BE IN THE FIRST PLACE?
HOW AM I SUPPOSED TO KNOW THAT?
THAT'S A NATURAL REACTION.
UNFORTUNATELY, NO ONE KNOWS, BECAUSE IT IS IMPOSSIBLE TO ASK THIS QUESTION OF EVERY SINGLE STUDENT IN JAPAN.
UH-HUH.
IT IS NOT ONLY IN THIS EXAMPLE THAT YOU CANNOT GET THE CRAMER'S COEFFICIENT FOR THE POPULATION. IT IS GENERALLY IMPOSSIBLE IN ALL CASES.
THAT IS WHY WE HAVE NO CHOICE BUT TO...

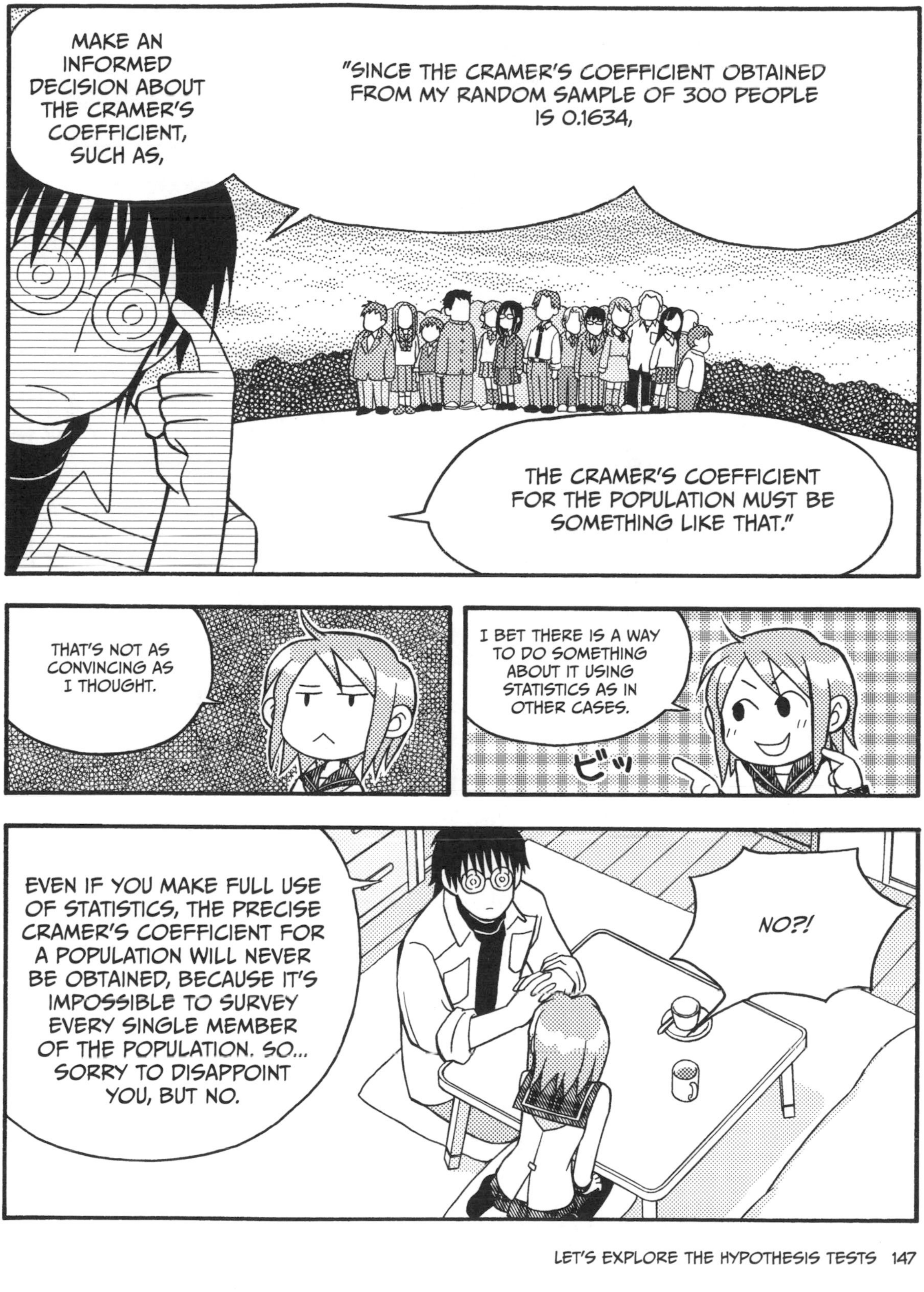
MAKE AN INFORMED DECISION ABOUT THE CRAMER'S COEFFICIENT, SUCH AS,
"SINCE THE CRAMER'S COEFFICIENT OBTAINED FROM MY RANDOM SAMPLE OF 300 PEOPLE IS 0.1634,
THE CRAMER'S COEFFICIENT FOR THE POPULATION MUST BE SOMETHING LIKE THAT."
THAT'S NOT AS CONVINCING AS I THOUGHT.
I BET THERE IS A WAY TO DO SOMETHING ABOUT IT USING STATISTICS AS IN OTHER CASES.
ビッ
EVEN IF YOU MAKE FULL USE OF STATISTICS, THE PRECISE CRAMER'S COEFFICIENT FOR A POPULATION WILL NEVER BE OBTAINED, BECAUSE IT'S IMPOSSIBLE TO SURVEY EVERY SINGLE MEMBER OF THE POPULATION. SO... SORRY TO DISAPPOINT YOU, BUT NO.
NO?!

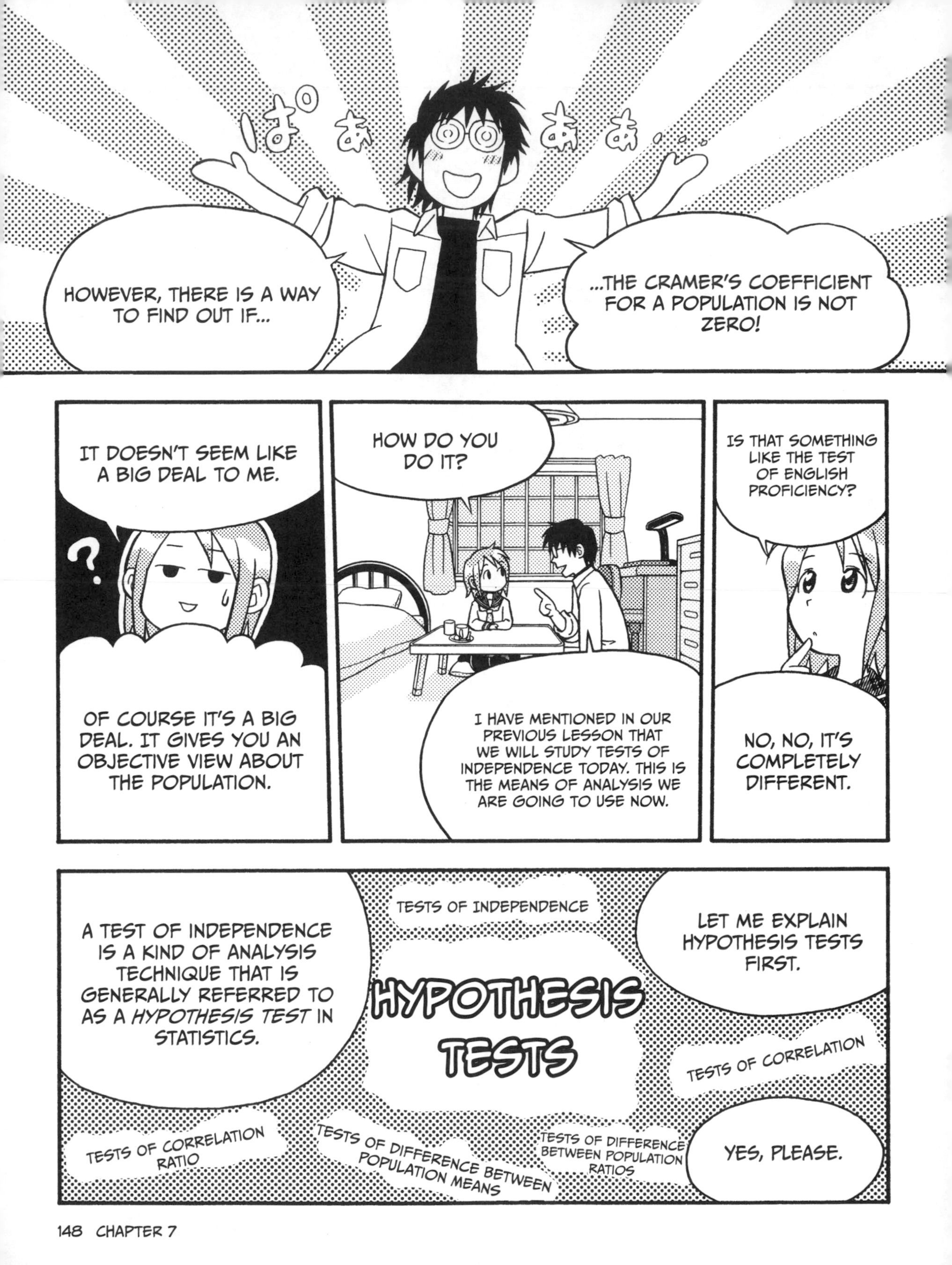
ぱあああ
HOWEVER, THERE IS A WAY TO FIND OUT IF...
...THE CRAMER'S COEFFICIENT FOR A POPULATION IS NOT ZERO!
IT DOESN'T SEEM LIKE A BIG DEAL TO ME.
?
OF COURSE IT'S A BIG DEAL. IT GIVES YOU AN OBJECTIVE VIEW ABOUT THE POPULATION.
HOW DO YOU DO IT?
I HAVE MENTIONED IN OUR PREVIOUS LESSON THAT WE WILL STUDY TESTS OF INDEPENDENCE TODAY. THIS IS THE MEANS OF ANALYSIS WE ARE GOING TO USE NOW.
IS THAT SOMETHING LIKE THE TEST OF ENGLISH PROFICIENCY?
NO, NO, IT'S COMPLETELY DIFFERENT.
A TEST OF INDEPENDENCE IS A KIND OF ANALYSIS TECHNIQUE THAT IS GENERALLY REFERRED TO AS A *HYPOTHESIS TEST* IN STATISTICS.
TESTS OF INDEPENDENCE
HYPOTHESIS TESTS
LET ME EXPLAIN HYPOTHESIS TESTS FIRST.
TESTS OF CORRELATION
TESTS OF CORRELATION RATIO
TESTS OF DIFFERENCE BETWEEN POPULATION MEANS
TESTS OF DIFFERENCE BETWEEN POPULATION RATIOS
YES, PLEASE.

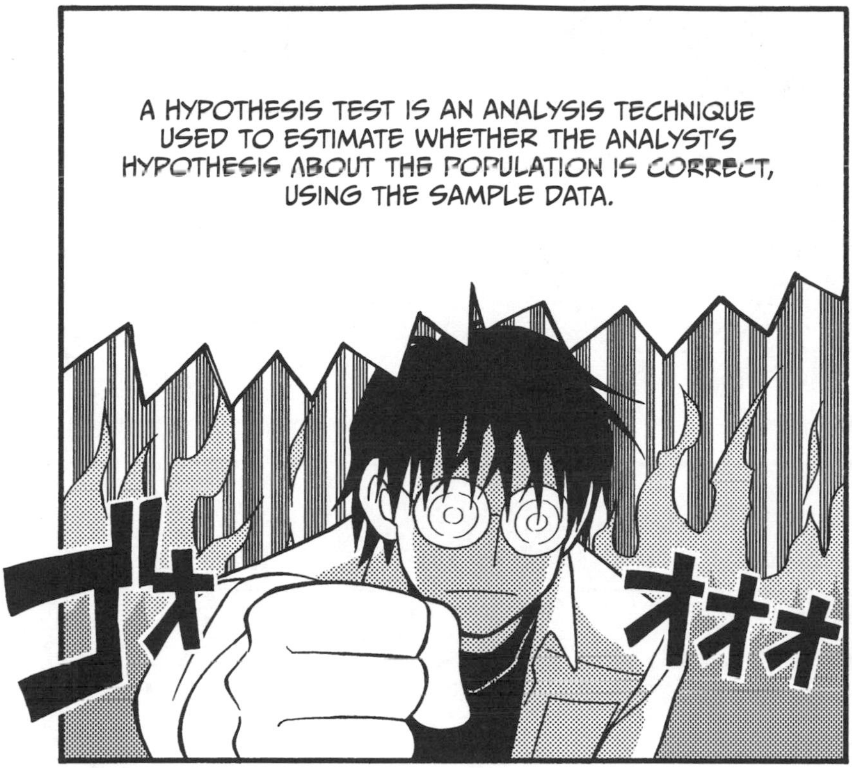
A HYPOTHESIS TEST IS AN ANALYSIS TECHNIQUE USED TO ESTIMATE WHETHER THE ANALYST'S HYPOTHESIS ABOUT THE POPULATION IS CORRECT, USING THE SAMPLE DATA.
ゴオ
オオオ

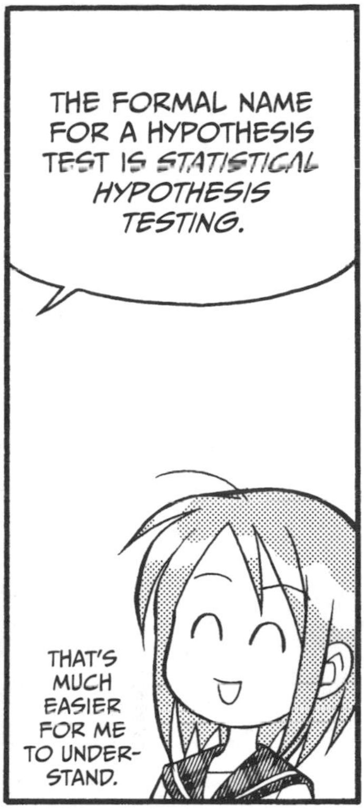
THE FORMAL NAME FOR A HYPOTHESIS TEST IS *STATISTICAL HYPOTHESIS TESTING*.
THAT'S MUCH EASIER FOR ME TO UNDER-STAND.

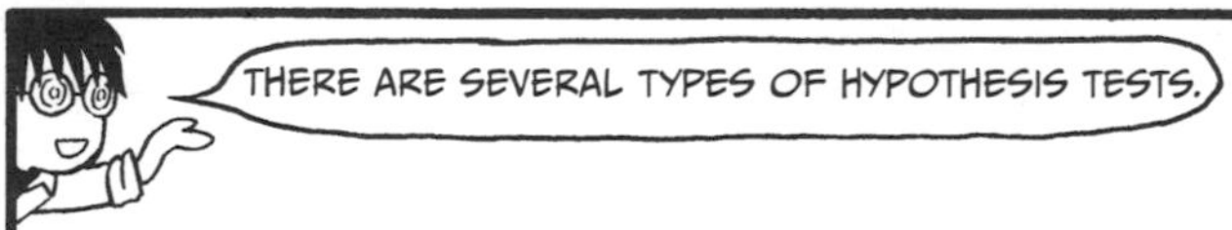
THERE ARE SEVERAL TYPES OF HYPOTHESIS TESTS.

EXAMPLES OF HYPOTHESIS TESTS

Name	Example of use
Tests of independence	Estimates whether the value of the Cramer's coefficient for sex and desired way of being asked out is zero for a population
Tests of correlation ratio	Estimates whether the value of the correlation ratio for favorite fashion brand and age is zero for a population
Tests of correlation	Estimates whether the correlation coefficient for amount spent on makeup and amount spent on clothes is zero for a population
Tests of difference between population means	Estimates whether allowances are different between high school girls in Tokyo and Osaka*
Tests of difference between population ratios	Estimates whether the approval rating of cabinet *X* is different between voters residing in urban areas and rural areas*

* Note that two populations are being considered.

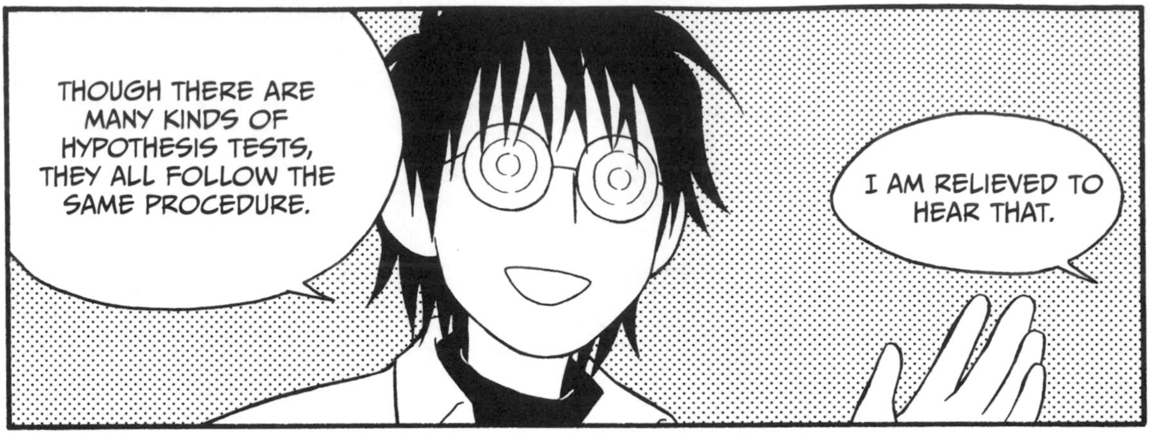
THOUGH THERE ARE MANY KINDS OF HYPOTHESIS TESTS, THEY ALL FOLLOW THE SAME PROCEDURE.
I AM RELIEVED TO HEAR THAT.

PROCEDURE FOR A HYPOTHESIS TEST

Step 1	Define the population.
Step 2	Set up a null hypothesis and an alternative hypothesis.
Step 3	Select which hypothesis test to conduct.
Step 4	Determine the significance level.
Step 5	Obtain the test statistic from the sample data.
Step 6	Determine whether the test statistic obtained in step 5 is in the critical region.
Step 7	If the test statistic is in the critical region, you must reject the null hypothesis. If not, you fail to reject the null hypothesis.

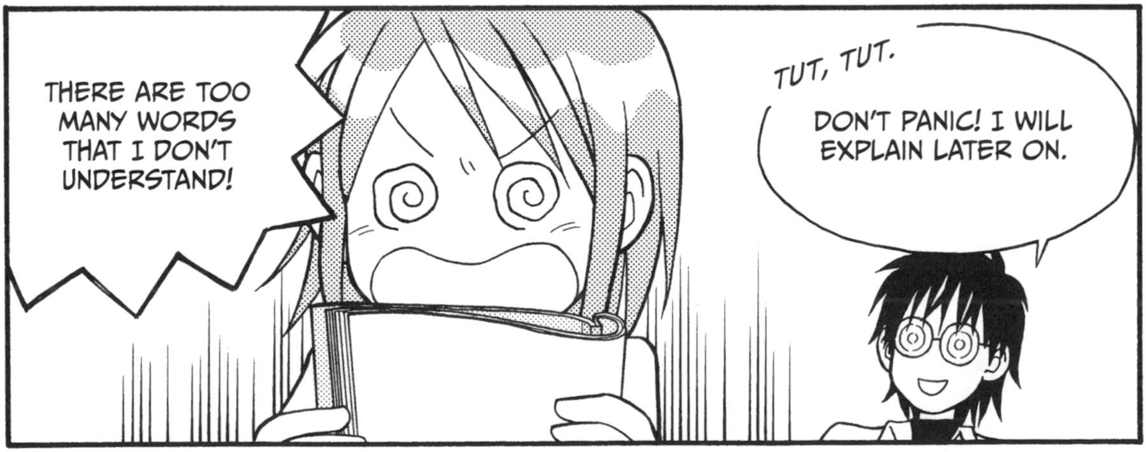
THERE ARE TOO MANY WORDS THAT I DON'T UNDERSTAND!
TUT, TUT.
DON'T PANIC! I WILL EXPLAIN LATER ON.

2. THE CHI-SQUARE TEST OF INDEPENDENCE

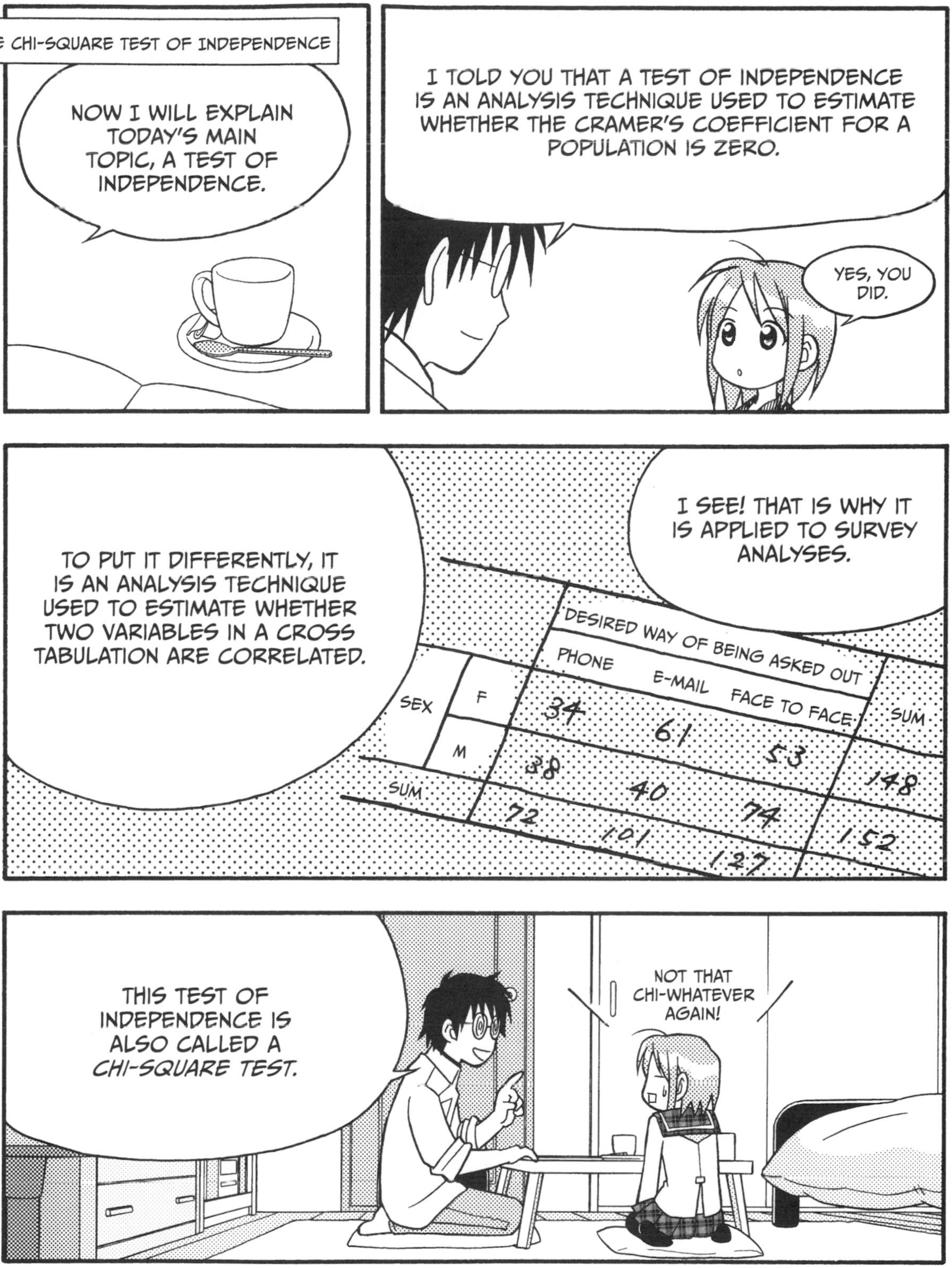

NOW I WILL EXPLAIN TODAY'S MAIN TOPIC, A TEST OF INDEPENDENCE.
I TOLD YOU THAT A TEST OF INDEPENDENCE IS AN ANALYSIS TECHNIQUE USED TO ESTIMATE WHETHER THE CRAMER'S COEFFICIENT FOR A POPULATION IS ZERO.
YES, YOU DID.
TO PUT IT DIFFERENTLY, IT IS AN ANALYSIS TECHNIQUE USED TO ESTIMATE WHETHER TWO VARIABLES IN A CROSS TABULATION ARE CORRELATED.
I SEE! THAT IS WHY IT IS APPLIED TO SURVEY ANALYSES.
DESIRED WAY OF BEING ASKED OUT
PHONE
E-MAIL
FACE TO FACE
SUM
SEX
F
M
SUM
34
61
53
148
38
40
74
152
72
101
127
THIS TEST OF INDEPENDENCE IS ALSO CALLED A *CHI-SQUARE TEST*.
NOT THAT CHI-WHATEVER AGAIN!

EXPLANATION

Pearson's chi-square test statistic (χ_0^2) and chi-square distribution

> Before giving an actual example of a test of independence, I would like to explain an important fact that is fundamental to tests of independence. Though it is impossible to do this in reality, suppose the below experiment is conducted.

Step 1

Take a random sample of 300 students from the population "all high school students residing in Japan."

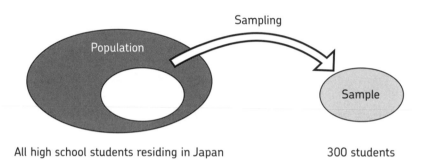

Sampling

Population

Sample

All high school students residing in Japan   300 students

Step 2

Conduct the survey on page 127 with the 300 people chosen in step 1 to obtain the chi-square statistic (χ_0^2).

Step 3

Put the 300 people back into the population.

Step 4

Repeat steps 1 through 3 over and over.

In this experiment, if the value of the Cramer's coefficient for the population "all high school students residing in Japan" is 0, the graph of Pearson's chi-square test statistic (χ_0^2) turns out to be a chi-square distribution with 2 degrees of freedom. In other words, if the value of the Cramer's coefficient for the population "all high school students residing in Japan" is 0, then Pearson's chi-square test statistic (χ_0^2) follows a chi-square distribution with 2 degrees of freedom.

*   See pages 130–133 for information on how to obtain Pearson's chi-square test statistic (χ_0^2).
*   See page 100 for information on a chi-square distribution with 2 degrees of freedom.

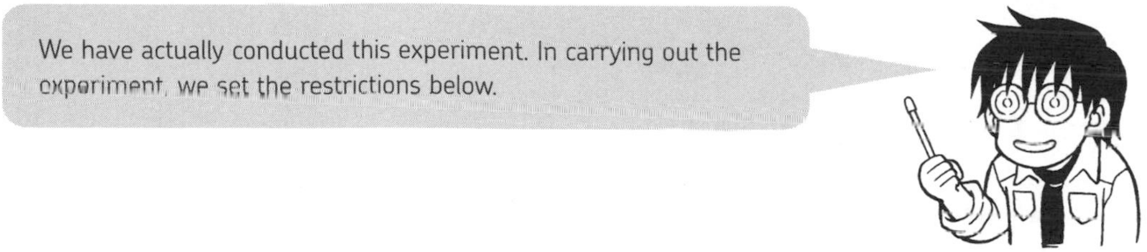

We have actually conducted this experiment. In carrying out the experiment, we set the restrictions below.

- As it is impossible to experiment with the actual population of "all high school students residing in Japan," the group of 10,000 people in Table 7-1 will be regarded as "all high school students residing in Japan" instead.
- We assume that the Cramer's coefficient for "all high school students residing in Japan" is 0. This means that the ratio of those who prefer being asked out by phone to those who prefer being asked out by e-mail to those who prefer being asked out directly is equal for girls and boys (see page 135). The cross tabulation for Table 7-1 is Table 7-2.
- Since it is otherwise endless, we will stop repeating steps 1 through 3 after 10,000 times.

TABLE 7-1: DESIRED WAY OF BEING ASKED OUT
(ALL HIGH SCHOOL STUDENTS RESIDING IN JAPAN)

Respondent	Sex	Desired way of being asked out
1	Female	Face to face
2	Female	Phone
...	...	...
10,000	Male	E-mail

TABLE 7-2: CROSS TABULATION OF SEX AND DESIRED WAY OF BEING ASKED OUT

		Desired way of being asked out			Sum
		Phone	E-mail	Face to face	
Sex	Female	400	1,600	2,000	4,000
	Male	600	2,400	3,000	6,000
Sum		1,000	4,000	5,000	10,000

Table 7-3 shows the result of the experiment. Figure 7-1 is a histogram made according to Table 7-3.

TABLE 7-3: RESULT OF EXPERIMENT

Experiment	Pearson's chi-square test statistic (χ_0^2)
1	0.8598
2	0.7557
...	...
10,000	2.7953

FIGURE 7-1: A HISTOGRAM BASED ON TABLE 7-3 (RANGE OF CLASS = 1)

0.5
0.4
0.3
0.2
0.1
0
Ratio of experiments
0.5
5.5
10.5
15.5
Chi-square value

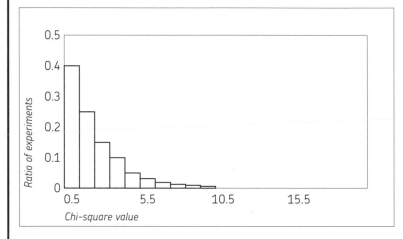

Figure 7-1 indeed looks similar to the graph on page 100, "2 Degrees of Freedom." It seems to be correct that Pearson's chi-square test statistic (χ_0^2) follows a chi-square distribution with 2 degrees of freedom. Though this has nothing to do with the experiment itself, here is one point to note. Two degrees of freedom comes from:

$$(2 - 1) \times (3 - 1) = 1 \times 2 = 2$$

↑ 2 patterns: female and male

↑ 3 patterns: phone, e-mail, and face to face

I will not go into why such a strange calculation is applied, as it is a topic much too advanced for the level of this book. But don't worry—even if you don't fully understand the calculation, you won't be at any disadvantage.

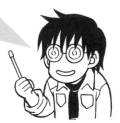

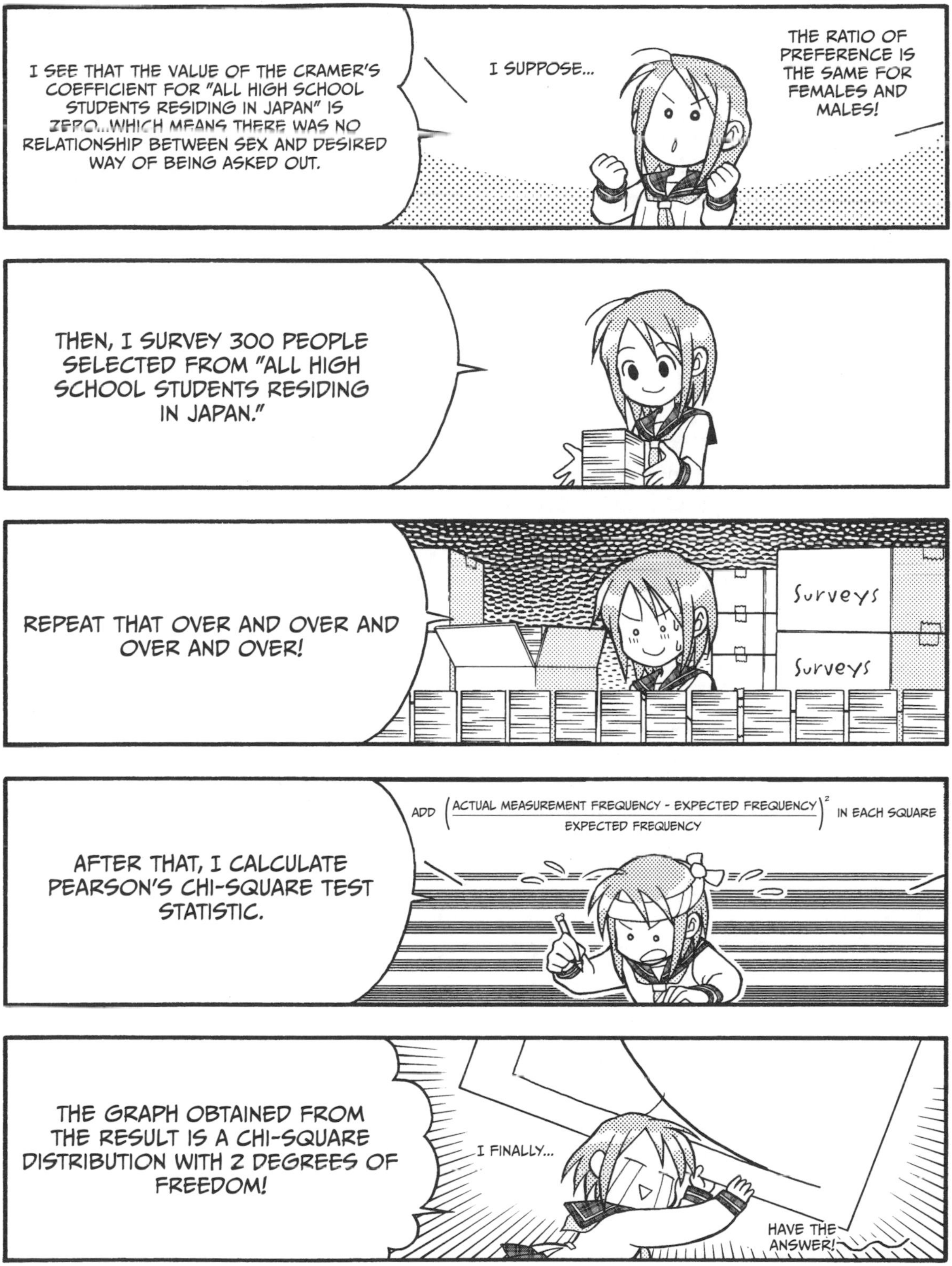

I SEE THAT THE VALUE OF THE CRAMER'S COEFFICIENT FOR "ALL HIGH SCHOOL STUDENTS RESIDING IN JAPAN" IS ZERO...WHICH MEANS THERE WAS NO RELATIONSHIP BETWEEN SEX AND DESIRED WAY OF BEING ASKED OUT.
I SUPPOSE...
THE RATIO OF PREFERENCE IS THE SAME FOR FEMALES AND MALES!
THEN, I SURVEY 300 PEOPLE SELECTED FROM "ALL HIGH SCHOOL STUDENTS RESIDING IN JAPAN."
REPEAT THAT OVER AND OVER AND OVER AND OVER!
Surveys
Surveys
AFTER THAT, I CALCULATE PEARSON'S CHI-SQUARE TEST STATISTIC.
ADD $\left(\frac{\text{ACTUAL MEASUREMENT FREQUENCY} - \text{EXPECTED FREQUENCY}}{\text{EXPECTED FREQUENCY}}\right)^2$ IN EACH SQUARE
THE GRAPH OBTAINED FROM THE RESULT IS A CHI-SQUARE DISTRIBUTION WITH 2 DEGREES OF FREEDOM!
I FINALLY...
HAVE THE ANSWER!

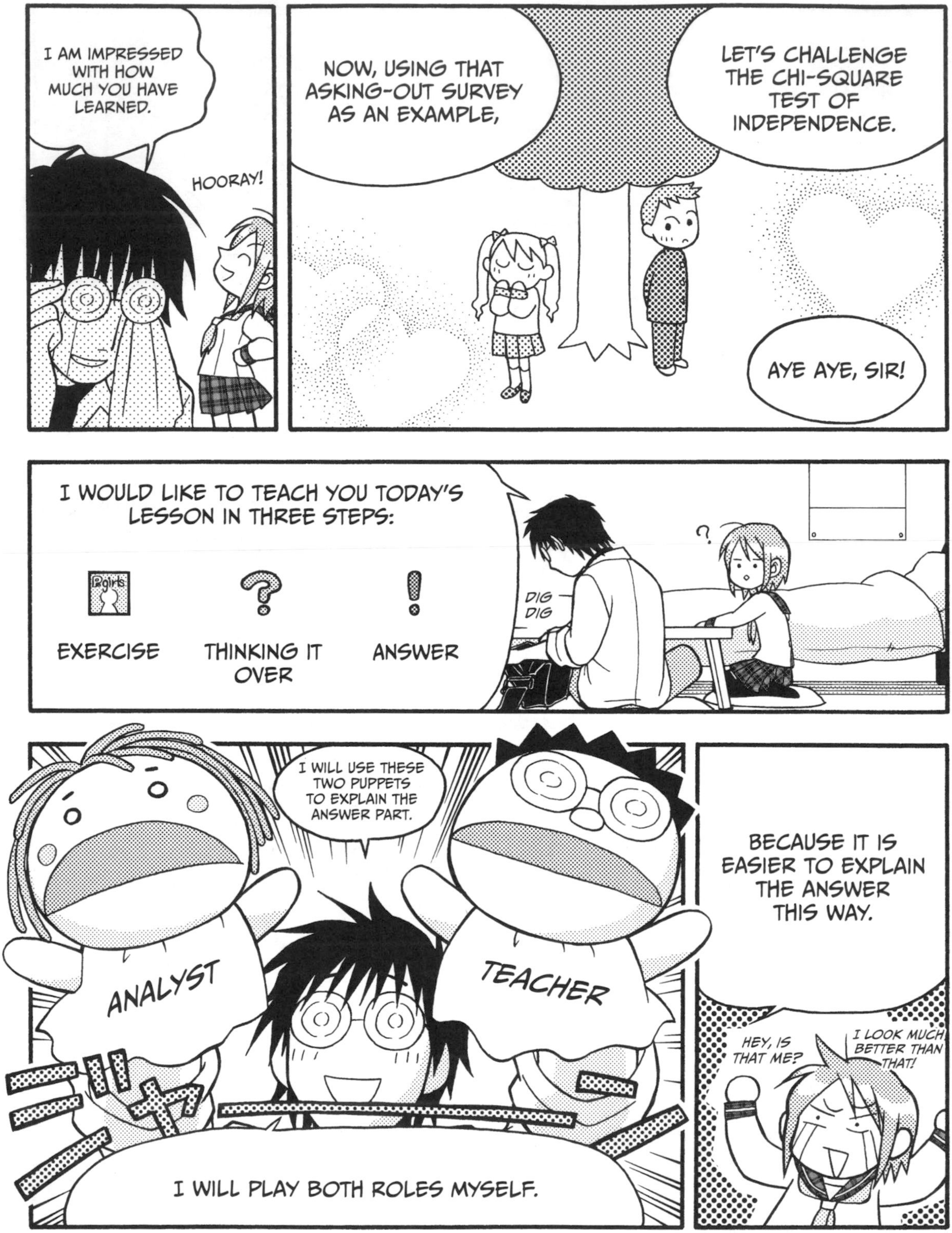
I AM IMPRESSED WITH HOW MUCH YOU HAVE LEARNED.
HOORAY!
NOW, USING THAT ASKING-OUT SURVEY AS AN EXAMPLE,
LET'S CHALLENGE THE CHI-SQUARE TEST OF INDEPENDENCE.
AYE AYE, SIR!
I WOULD LIKE TO TEACH YOU TODAY'S LESSON IN THREE STEPS:
EXERCISE
THINKING IT OVER
ANSWER
DIG DIG
?
I WILL USE THESE TWO PUPPETS TO EXPLAIN THE ANSWER PART.
ANALYST
TEACHER
I WILL PLAY BOTH ROLES MYSELF.
BECAUSE IT IS EASIER TO EXPLAIN THE ANSWER THIS WAY.
HEY, IS THAT ME?
I LOOK MUCH BETTER THAN THAT!

EXERCISE

P-Girls Magazine decided to publish an article titled "We Asked 300 High School Students, 'How Would You Like to Be Asked Out?'" In order to prepare the article, a journalist randomly chose 300 people from all the high school students residing in Japan and took a survey. The table below is the result of this survey.

Respondent	Desired way of being asked out	Age	Sex
1	Face to face	17	Female
2	Phone	15	Female
...	...	...	...
300	E-mail	18	Male

The table below is the cross tabulation of sex and desired way of being asked out.

		Desired way of being asked out			Sum
		Phone	E-mail	Face to face	
Sex	Female	34	61	53	148
	Male	38	40	74	152
Sum		72	101	127	300

Using the chi-square test of independence, estimate if the Cramer's coefficient for sex and desired way of being asked out in the population "all high school students residing in Japan" is greater than 0. This is the same as estimating with a test of independence whether sex and desired way of being asked out are correlated. Remember that the significance level (explained later) is 0.05.

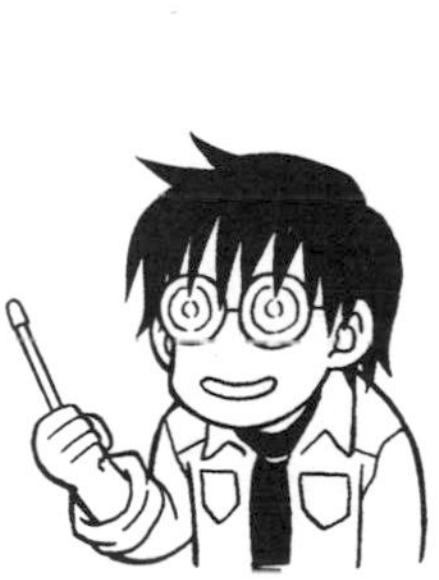

As explained on pages 152–154, Pearson's chi-square test statistic (χ_0^2) follows a chi-square distribution with 2 degrees of freedom if the null hypothesis states that the value of the Cramer's coefficient for the population "all high school students residing in Japan" is 0. If that's true, then the probability that χ_0^2 obtained from the 300 people who have been chosen randomly is 5.9915 or more is 0.05.

FIGURE 7-2: PROBABILITY THAT χ_0^2 IS 5.9915 OR MORE

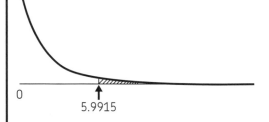

This is clear from the table of chi-square distribution on page 103.

χ_0^2 for this exercise has already been calculated on page 132. It is 8.0091. True, this figure has been calculated based on data from 300 randomly chosen people, but doesn't this seem too large? Taking into consideration the comment on page 132, isn't it natural to assume that the Cramer's coefficient for the population "all high school students residing in Japan" is greater than 0?

Remember that the process for a chi-square test of independence (not limited to this exercise) goes like this:

1. Assume a null hypothesis that "the Cramer's coefficient for the population is 0" for the time being.

2. Calculate χ_0^2 from the sample data.

3. If χ_0^2 is too large, reject the null hypothesis and conclude that "the Cramer's coefficient for the population is greater than 0."

As χ_0^2 becomes larger, the probability shown as the shaded area in Figure 7-3 naturally becomes smaller.

FIGURE 7-3: PROBABILITY IN CORRESPONDENCE TO $\chi_0^{\;2}$

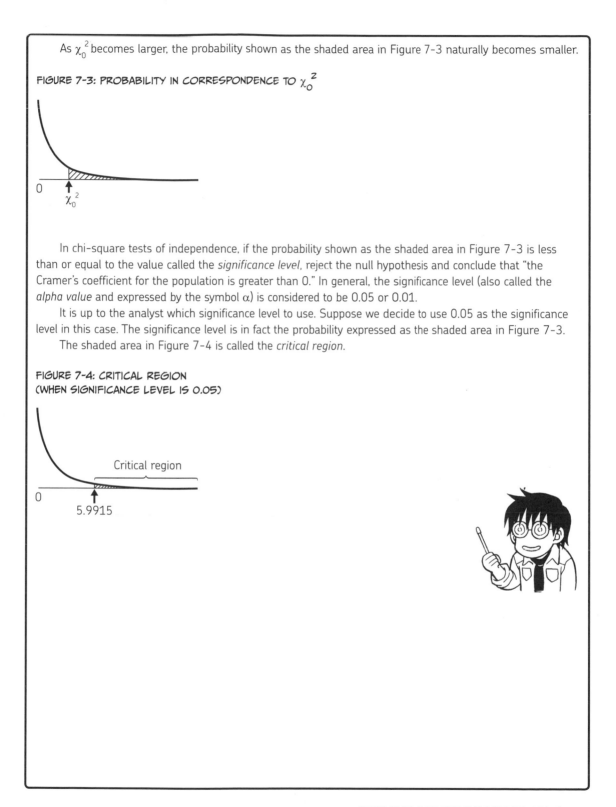

In chi-square tests of independence, if the probability shown as the shaded area in Figure 7-3 is less than or equal to the value called the *significance level*, reject the null hypothesis and conclude that "the Cramer's coefficient for the population is greater than 0." In general, the significance level (also called the *alpha value* and expressed by the symbol α) is considered to be 0.05 or 0.01.

It is up to the analyst which significance level to use. Suppose we decide to use 0.05 as the significance level in this case. The significance level is in fact the probability expressed as the shaded area in Figure 7-3.

The shaded area in Figure 7-4 is called the *critical region*.

FIGURE 7-4: CRITICAL REGION
(WHEN SIGNIFICANCE LEVEL IS 0.05)

! ANSWER

Step 1

Define the population.

The population is:

ALL HIGH SCHOOL STUDENTS RESIDING IN JAPAN

ANALYST

TEACHER

In this exercise, the population was defined as "all high school students residing in Japan." Thus, in this particular exercise, step 1 is unnecessary.

However, for "Tests of difference between population ratios" in the table on page 149, the populations in question are "voters residing in urban areas" and "voters residing in rural areas." Where are the urban areas exactly? Are they Tokyo and Osaka? Or are they the capitals of the prefectures? This must be specified by the analyst.

I repeat: When you are actually doing a hypothesis test, you must determine the population. No matter which hypothesis test you are trying to carry out, you must not fail to properly define the population.

Otherwise, you might fall into a situation in which you are lost, wondering, "What was I trying to estimate?" Lots of statisticians fall into traps like this. Take great care about this point.

Step 2

Set up a null hypothesis and an alternative hypothesis.

The null hypothesis is: "The Cramer's coefficient for the population is 0. In other words, sex and desired way of being asked out are not correlated."

The alternative hypothesis is: "The Cramer's coefficient for the population is greater than 0. In other words, sex and desired way of being asked out are correlated."

ANALYST

TEACHER

An explanation of null hypotheses and alternative hypotheses is given on page 170.

Step 3

Choose which hypothesis test to do.

I am going to do a chi-square test of independence.

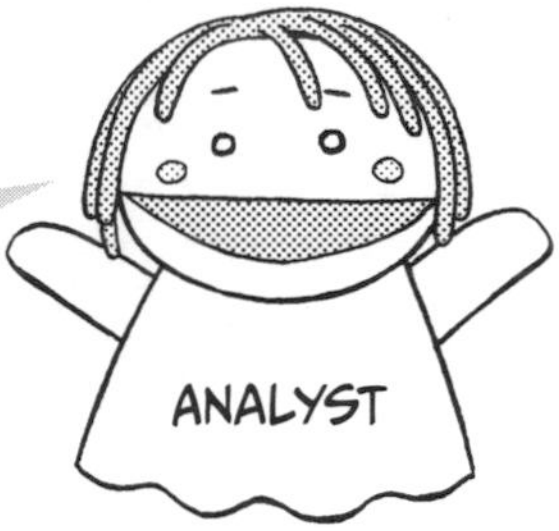
ANALYST

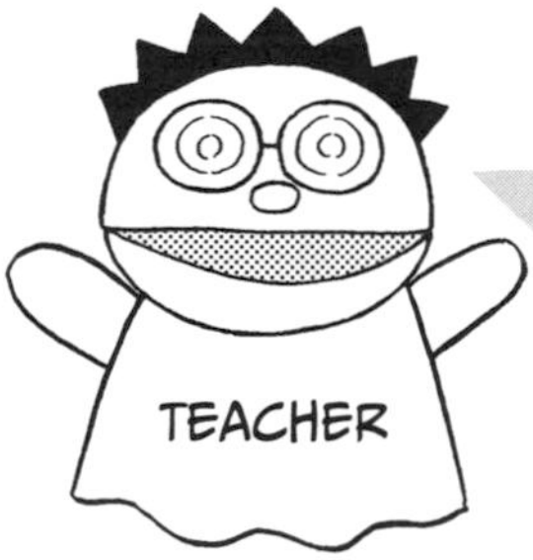
TEACHER

This exercise asks you to do a chi-square test of independence. So in this particular exercise, step 3 is unnecessary.

(When you are actually doing a hypothesis test and not an exercise, you must select the hypothesis test suitable for the objective of analysis on your own.)

Step 4

Determine the significance level.

I will use 0.05 as the significance level.

ANALYST

TEACHER

The exercise assigns 0.05 as the significance level, so in this particular exercise, step 4 is unnecessary. When you are actually doing a hypothesis test and not an exercise, you must determine the significance level. As mentioned earlier, normally either 0.05 or 0.01 is used. The smaller the P-value computed from the sample data, the stronger the evidence is against the null hypothesis. In general, the symbol α is used to express the significance level (alpha value).

Step 5

Calculate the test statistic from the sample data.

I am trying to do a chi-square test of independence. Thus the test statistic is Pearson's chi-square test statistic (χ_0^2). The value of χ_0^2 for this exercise has already been calculated on page 132: $\chi_0^2 = 8.0091$.

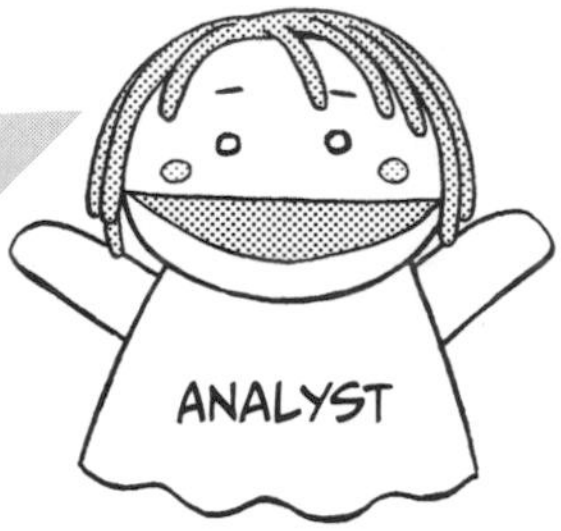
ANALYST

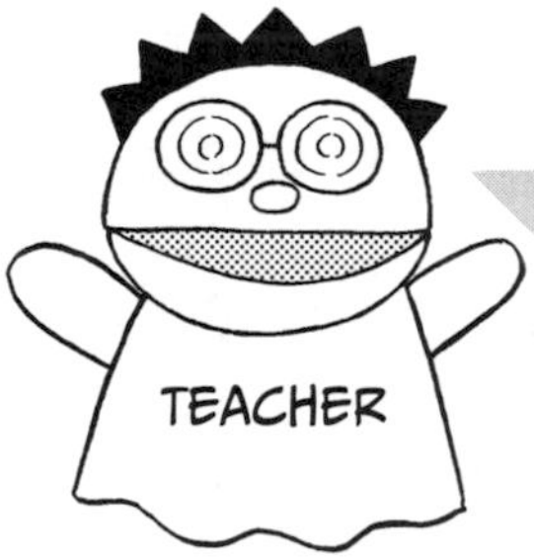
TEACHER

The test statistic is obtained from a function that calculates a single value from the sample data. Different kinds of hypothesis tests have different test statistics. As mentioned above, the value for a test of independence is χ_0^2, and in the case of tests of correlation (see page 149), the test statistic is as below.

$$\frac{\text{correlation coefficient}^2 \times \sqrt{\text{number of values} - 2}}{1 - \sqrt{\text{correlation coefficient}^2}}$$

Step 6

Determine whether or not the test statistic from step 5 is in the critical region.

Pearson's chi-square test statistic (χ_0^2) is 8.0091. As the significance level (α) is 0.05, the critical region is 5.9915 or above, according to the table of chi-square distribution on page 103. As shown in the chart below, the test statistic is within the critical region.

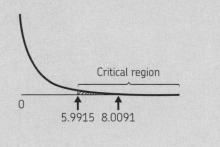

ANALYST

TEACHER

The critical region changes depending on the significance level (α). If α in this exercise was 0.01 instead of 0.05, the critical region would be 9.2104 or above, according to the table of chi-square distribution on page 103.

Step 7

If the test statistic is in the critical region in step 6, you reject the null hypothesis. If not, you fail to reject the null hypothesis. In this case, the test statistic was in the critical region.

Thus the alternative hypothesis, "the Cramer's coefficient for the population is greater than 0," is correct!

ANALYST

TEACHER

You cannot conclude that the alternative hypothesis is absolutely correct in a hypothesis test, even if the test statistic is within the critical region. The only conclusion you can make is, "I would like to say that the alternative hypothesis is 'absolutely' correct . . . but there is, at most, a $(\alpha \times 100)\%$ possibility that the null hypothesis is correct."

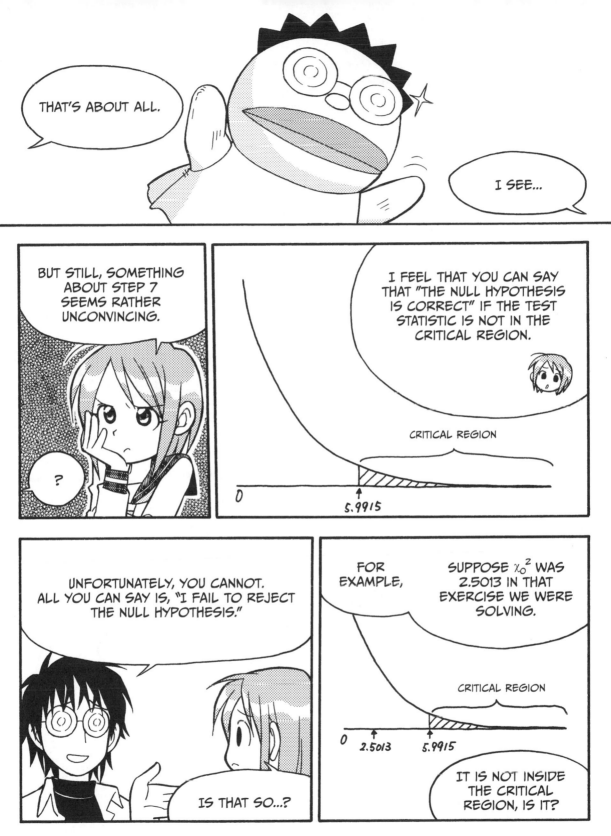

THAT'S ABOUT ALL.

I SEE...

BUT STILL, SOMETHING ABOUT STEP 7 SEEMS RATHER UNCONVINCING.

?

I FEEL THAT YOU CAN SAY THAT "THE NULL HYPOTHESIS IS CORRECT" IF THE TEST STATISTIC IS NOT IN THE CRITICAL REGION.

CRITICAL REGION

0   5.9915

UNFORTUNATELY, YOU CANNOT. ALL YOU CAN SAY IS, "I FAIL TO REJECT THE NULL HYPOTHESIS."

IS THAT SO...?

FOR EXAMPLE,

SUPPOSE χ_0^2 WAS 2.5013 IN THAT EXERCISE WE WERE SOLVING.

CRITICAL REGION

0   2.5013   5.9915

IT IS NOT INSIDE THE CRITICAL REGION, IS IT?

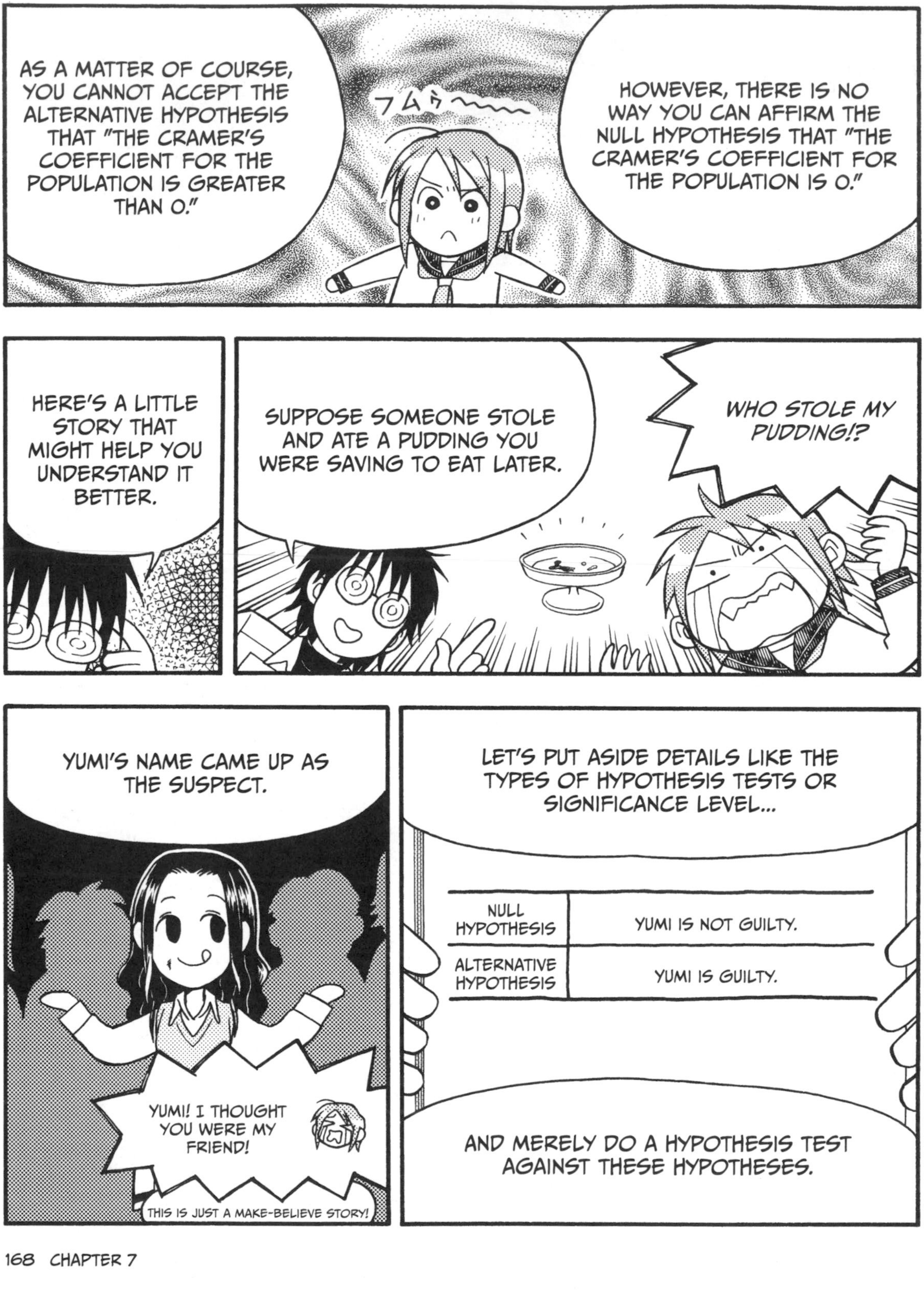
AS A MATTER OF COURSE, YOU CANNOT ACCEPT THE ALTERNATIVE HYPOTHESIS THAT "THE CRAMER'S COEFFICIENT FOR THE POPULATION IS GREATER THAN 0."
フムぅ～～～
HOWEVER, THERE IS NO WAY YOU CAN AFFIRM THE NULL HYPOTHESIS THAT "THE CRAMER'S COEFFICIENT FOR THE POPULATION IS 0."
HERE'S A LITTLE STORY THAT MIGHT HELP YOU UNDERSTAND IT BETTER.
SUPPOSE SOMEONE STOLE AND ATE A PUDDING YOU WERE SAVING TO EAT LATER.
WHO STOLE MY PUDDING!?
YUMI'S NAME CAME UP AS THE SUSPECT.
YUMI! I THOUGHT YOU WERE MY FRIEND!
THIS IS JUST A MAKE-BELIEVE STORY!
LET'S PUT ASIDE DETAILS LIKE THE TYPES OF HYPOTHESIS TESTS OR SIGNIFICANCE LEVEL...
NULL HYPOTHESIS
YUMI IS NOT GUILTY.
ALTERNATIVE HYPOTHESIS
YUMI IS GUILTY.
AND MERELY DO A HYPOTHESIS TEST AGAINST THESE HYPOTHESES.

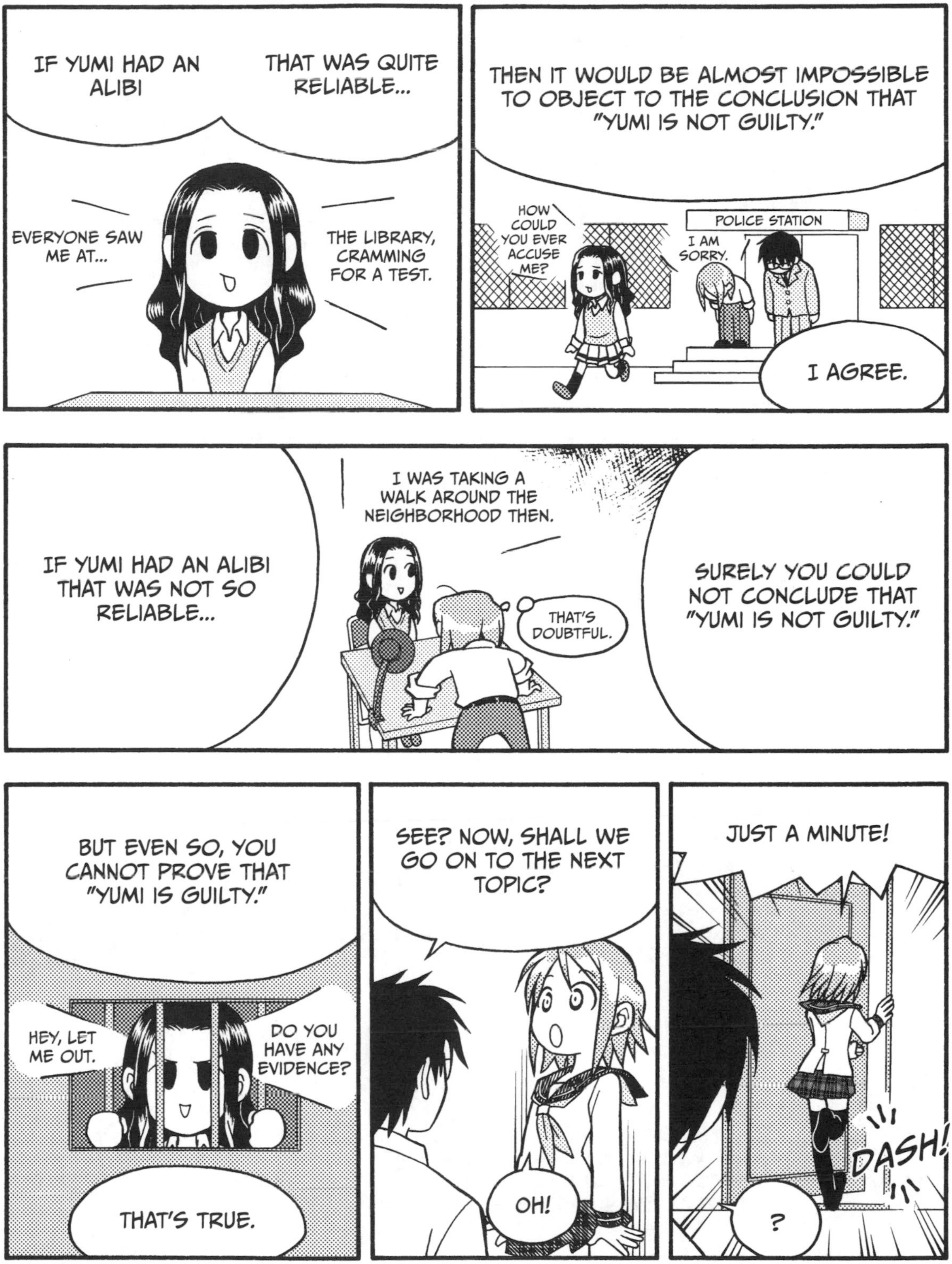
IF YUMI HAD AN ALIBI
THAT WAS QUITE RELIABLE...
EVERYONE SAW ME AT...
THE LIBRARY, CRAMMING FOR A TEST.
THEN IT WOULD BE ALMOST IMPOSSIBLE TO OBJECT TO THE CONCLUSION THAT "YUMI IS NOT GUILTY."
HOW COULD YOU EVER ACCUSE ME?
POLICE STATION
I AM SORRY.
I AGREE.
IF YUMI HAD AN ALIBI THAT WAS NOT SO RELIABLE...
I WAS TAKING A WALK AROUND THE NEIGHBORHOOD THEN.
THAT'S DOUBTFUL.
SURELY YOU COULD NOT CONCLUDE THAT "YUMI IS NOT GUILTY."
BUT EVEN SO, YOU CANNOT PROVE THAT "YUMI IS GUILTY."
HEY, LET ME OUT.
DO YOU HAVE ANY EVIDENCE?
THAT'S TRUE.
SEE? NOW, SHALL WE GO ON TO THE NEXT TOPIC?
OH!
JUST A MINUTE!
DASH!
?

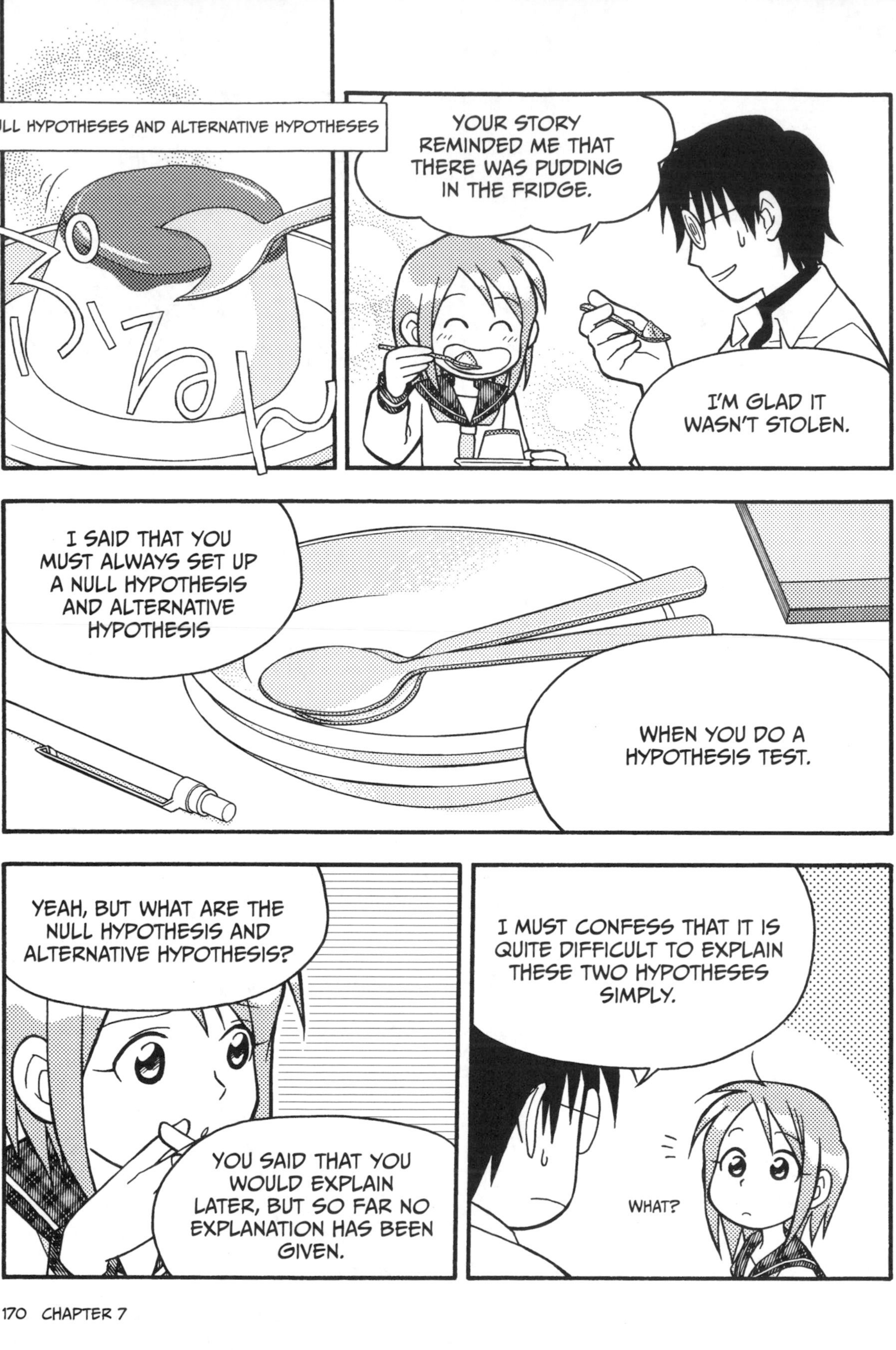
3. NULL HYPOTHESES AND ALTERNATIVE HYPOTHESES
ぷるん
YOUR STORY REMINDED ME THAT THERE WAS PUDDING IN THE FRIDGE.
I'M GLAD IT WASN'T STOLEN.
I SAID THAT YOU MUST ALWAYS SET UP A NULL HYPOTHESIS AND ALTERNATIVE HYPOTHESIS
WHEN YOU DO A HYPOTHESIS TEST.
YEAH, BUT WHAT ARE THE NULL HYPOTHESIS AND ALTERNATIVE HYPOTHESIS?
YOU SAID THAT YOU WOULD EXPLAIN LATER, BUT SO FAR NO EXPLANATION HAS BEEN GIVEN.
I MUST CONFESS THAT IT IS QUITE DIFFICULT TO EXPLAIN THESE TWO HYPOTHESES SIMPLY.
WHAT?

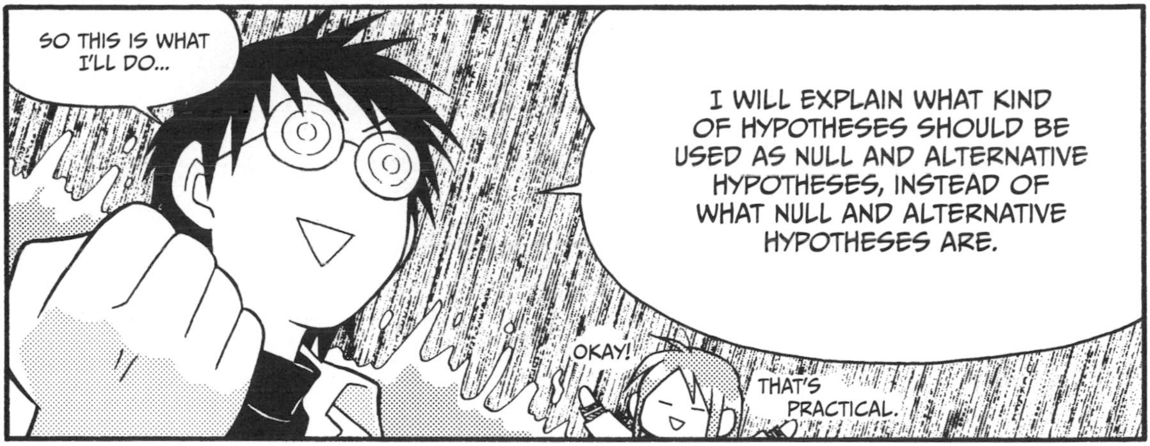
SO THIS IS WHAT I'LL DO...
I WILL EXPLAIN WHAT KIND OF HYPOTHESES SHOULD BE USED AS NULL AND ALTERNATIVE HYPOTHESES, INSTEAD OF WHAT NULL AND ALTERNATIVE HYPOTHESES ARE.
OKAY!
THAT'S PRACTICAL.

EXAMPLES OF HYPOTHESIS TESTS

Name	Example of use
Tests of independence	Estimates whether the value of the Cramer's coefficient for sex and desired way of being asked out is zero for a population
Tests of correlation ratio	Estimates whether the value of the correlation ratio for favorite fashion brand and age is zero for a population
Tests of correlation	Estimates whether the correlation coefficient for amount spent on makeup and amount spent on clothes is zero for a population
Tests of difference between population means	Estimates whether allowances are different between high school girls in Tokyo and Osaka*
Tests of difference between population ratios	Estimates whether the approval rating of cabinet *X* is different between voters residing in urban areas and rural areas*

* Note that two populations are being considered.

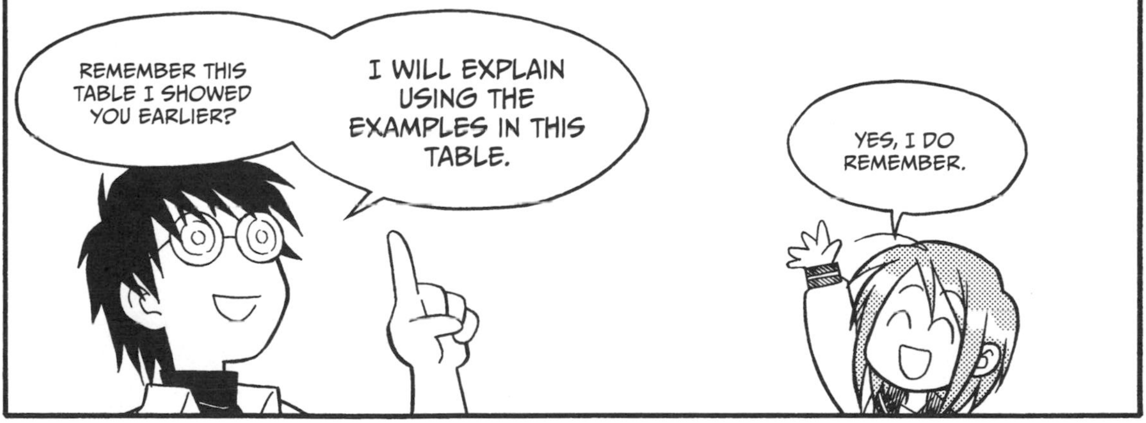
REMEMBER THIS TABLE I SHOWED YOU EARLIER?
I WILL EXPLAIN USING THE EXAMPLES IN THIS TABLE.
YES, I DO REMEMBER.

TESTS OF INDEPENDENCE

Null hypothesis	The Cramer's coefficient for sex and desired way of being asked out is 0 for a population.
Alternative hypothesis	The Cramer's coefficient for sex and desired way of being asked out is greater than 0 for a population.

TESTS OF CORRELATION RATIO

Null hypothesis	The correlation ratio for favorite fashion brand and age is 0 for a population.
Alternative hypothesis	The correlation ratio for favorite fashion brand and age is greater than 0 for a population.

TESTS OF CORRELATION

Null hypothesis	The correlation coefficient for amount spent on makeup and amount spent on clothes is 0 for a population.
Alternative hypothesis	The correlation coefficient for amount spent on makeup and amount spent on clothes is not 0 for a population. or The correlation coefficient for amount spent on makeup and amount spent on clothes is greater than 0 for a population. or The correlation coefficient for amount spent on makeup and amount spent on clothes is less than 0 for a population.

TESTS OF DIFFERENCE BETWEEN POPULATION MEANS

Null hypothesis	The allowances of high school girls in Tokyo and Osaka are the same.
Alternative hypothesis	The allowances of high school girls in Tokyo and Osaka are not the same.
	or
	The allowances of high school girls in Tokyo are larger than those of high school girls in Osaka.
	or
	The allowances of high school girls in Tokyo are smaller than those of high school girls in Osaka.

TESTS OF DIFFERENCE BETWEEN POPULATION RATIOS

Null hypothesis	The approval ratings of cabinet X for voters residing in urban areas and rural areas are the same.
Alternative hypothesis	The approval ratings of cabinet X for voters residing in urban areas and rural areas are not the same.
	or
	The approval rating of cabinet X for voters residing in urban areas is higher than that of voters residing in rural areas.
	or
	The approval rating of cabinet X for voters residing in urban areas is lower than that of voters residing in rural areas.

GOOD
EXPLANATION!

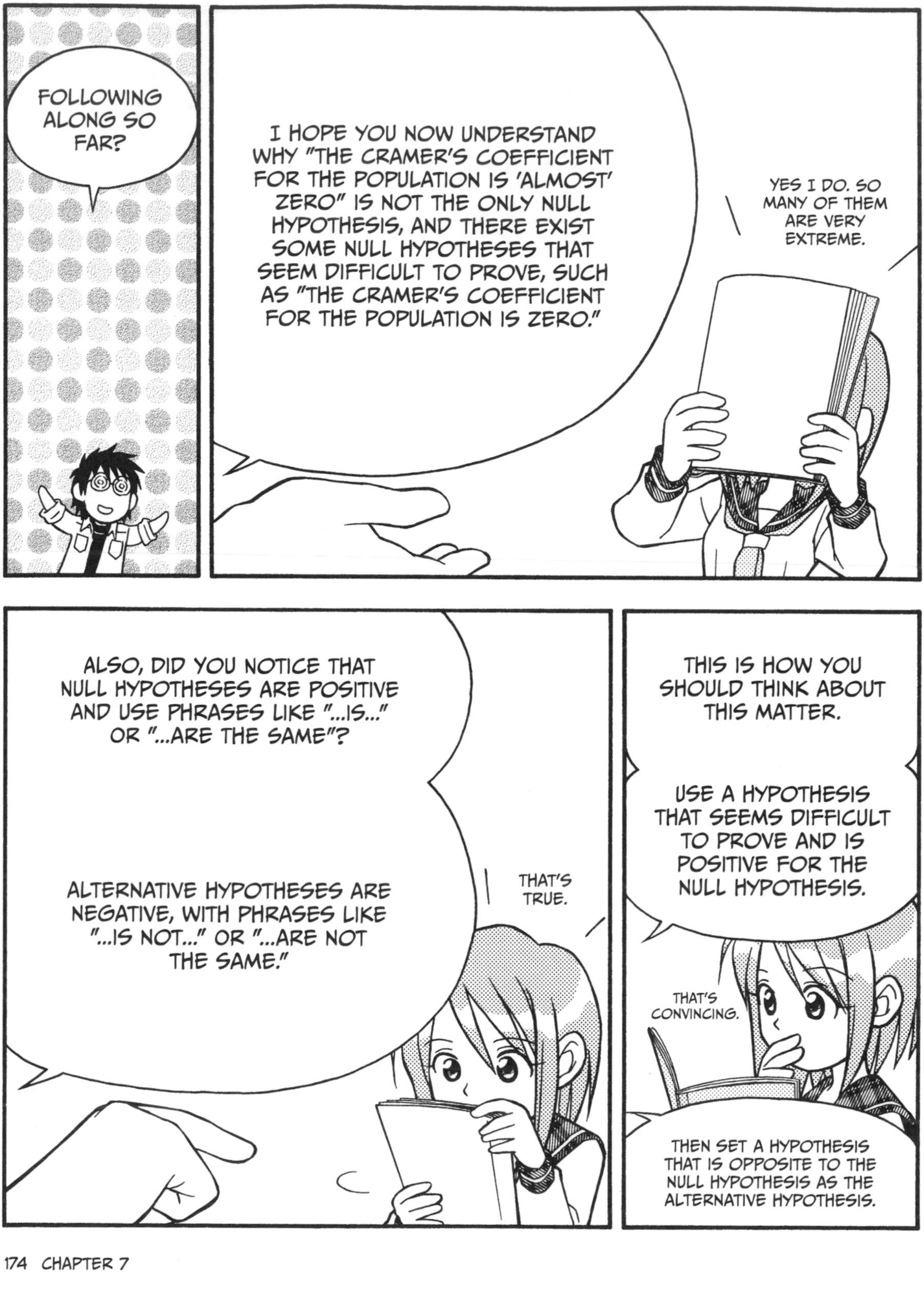
FOLLOWING ALONG SO FAR?
I HOPE YOU NOW UNDERSTAND WHY "THE CRAMER'S COEFFICIENT FOR THE POPULATION IS 'ALMOST' ZERO" IS NOT THE ONLY NULL HYPOTHESIS, AND THERE EXIST SOME NULL HYPOTHESES THAT SEEM DIFFICULT TO PROVE, SUCH AS "THE CRAMER'S COEFFICIENT FOR THE POPULATION IS ZERO."
YES I DO. SO MANY OF THEM ARE VERY EXTREME.
ALSO, DID YOU NOTICE THAT NULL HYPOTHESES ARE POSITIVE AND USE PHRASES LIKE "...IS..." OR "...ARE THE SAME"?
ALTERNATIVE HYPOTHESES ARE NEGATIVE, WITH PHRASES LIKE "...IS NOT..." OR "...ARE NOT THE SAME."
THAT'S TRUE.
THIS IS HOW YOU SHOULD THINK ABOUT THIS MATTER.
USE A HYPOTHESIS THAT SEEMS DIFFICULT TO PROVE AND IS POSITIVE FOR THE NULL HYPOTHESIS.
THAT'S CONVINCING.
THEN SET A HYPOTHESIS THAT IS OPPOSITE TO THE NULL HYPOTHESIS AS THE ALTERNATIVE HYPOTHESIS.

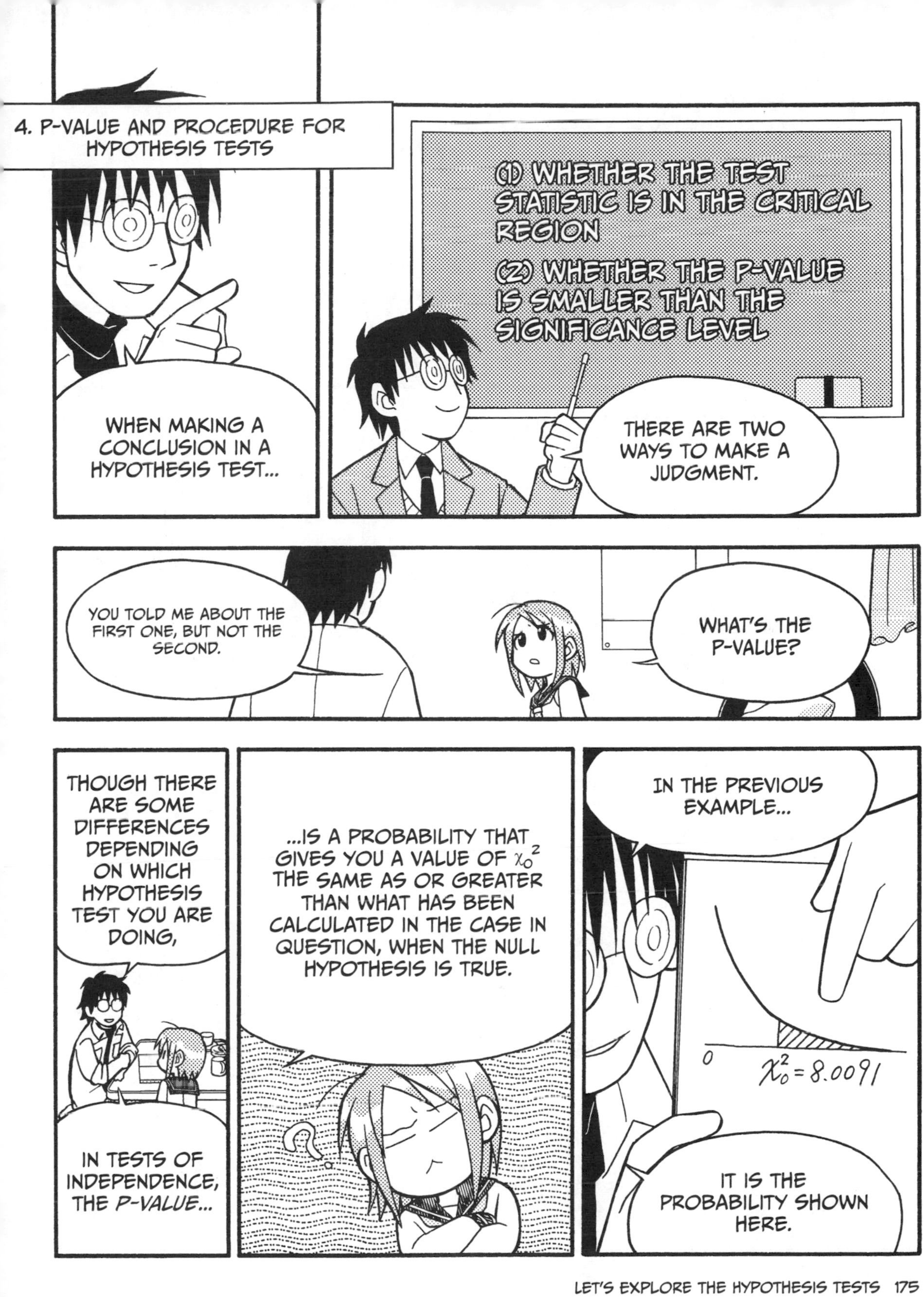
4. P-VALUE AND PROCEDURE FOR HYPOTHESIS TESTS
WHEN MAKING A CONCLUSION IN A HYPOTHESIS TEST...
(1) WHETHER THE TEST STATISTIC IS IN THE CRITICAL REGION
(2) WHETHER THE P-VALUE IS SMALLER THAN THE SIGNIFICANCE LEVEL
THERE ARE TWO WAYS TO MAKE A JUDGMENT.
YOU TOLD ME ABOUT THE FIRST ONE, BUT NOT THE SECOND.
WHAT'S THE P-VALUE?
THOUGH THERE ARE SOME DIFFERENCES DEPENDING ON WHICH HYPOTHESIS TEST YOU ARE DOING,
IN TESTS OF INDEPENDENCE, THE *P-VALUE*...
...IS A PROBABILITY THAT GIVES YOU A VALUE OF χ_0^2 THE SAME AS OR GREATER THAN WHAT HAS BEEN CALCULATED IN THE CASE IN QUESTION, WHEN THE NULL HYPOTHESIS IS TRUE.
IN THE PREVIOUS EXAMPLE...
0
$\chi_0^2 = 8.0091$
IT IS THE PROBABILITY SHOWN HERE.

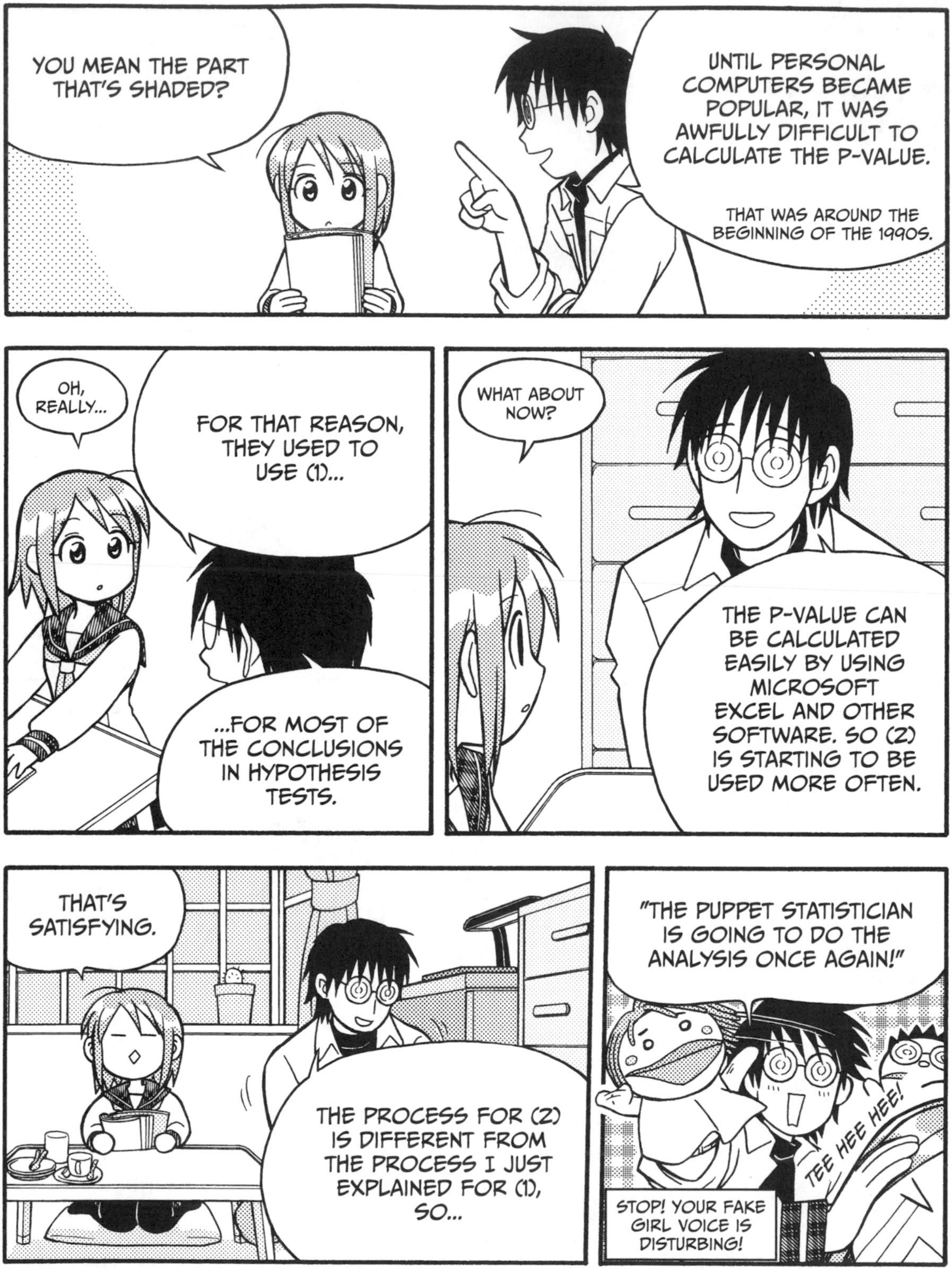

YOU MEAN THE PART THAT'S SHADED?
UNTIL PERSONAL COMPUTERS BECAME POPULAR, IT WAS AWFULLY DIFFICULT TO CALCULATE THE P-VALUE.
THAT WAS AROUND THE BEGINNING OF THE 1990S.
OH, REALLY...
FOR THAT REASON, THEY USED TO USE (1)...
...FOR MOST OF THE CONCLUSIONS IN HYPOTHESIS TESTS.
WHAT ABOUT NOW?
THE P-VALUE CAN BE CALCULATED EASILY BY USING MICROSOFT EXCEL AND OTHER SOFTWARE. SO (2) IS STARTING TO BE USED MORE OFTEN.
THAT'S SATISFYING.
THE PROCESS FOR (2) IS DIFFERENT FROM THE PROCESS I JUST EXPLAINED FOR (1), SO...
"THE PUPPET STATISTICIAN IS GOING TO DO THE ANALYSIS ONCE AGAIN!"
TEE HEE HEE!
STOP! YOUR FAKE GIRL VOICE IS DISTURBING!

Step 6p

Determine whether or not the P-value corresponding to the test statistic obtained in step 5 is smaller than the significance level.

The significance level is 0.05. Since Pearson's chi-square test statistic (χ_0^2, which is the test statistic in this case) is 8.0091, the P-value is 0.0182.

$$0.0182 < 0.05$$

Thus the P-value is smaller.

ANALYST

TEACHER

As mentioned before, you can calculate the P-value using Excel (though this depends on what type of hypothesis test you are doing). See page 208 for details.

Step 7p

If the P-value is smaller than the significance level in step 6p, you reject the null hypothesis. If not, you fail to reject the null hypothesis.

The P-value was smaller than the significance level. Therefore, you conclude in favor of the alternative hypothesis, "the Cramer's coefficient for the population is greater than 0."

ANALYST

TEACHER

Even if the P-value was smaller than the significance level, you cannot really conclude that the alternative hypothesis is "absolutely" correct in a hypothesis test. The only conclusion you can make is: "I would like to say that the alternative hypothesis is 'absolutely' correct . . . but there is a ($\alpha \times 100$)% possibility that the null hypothesis is correct."

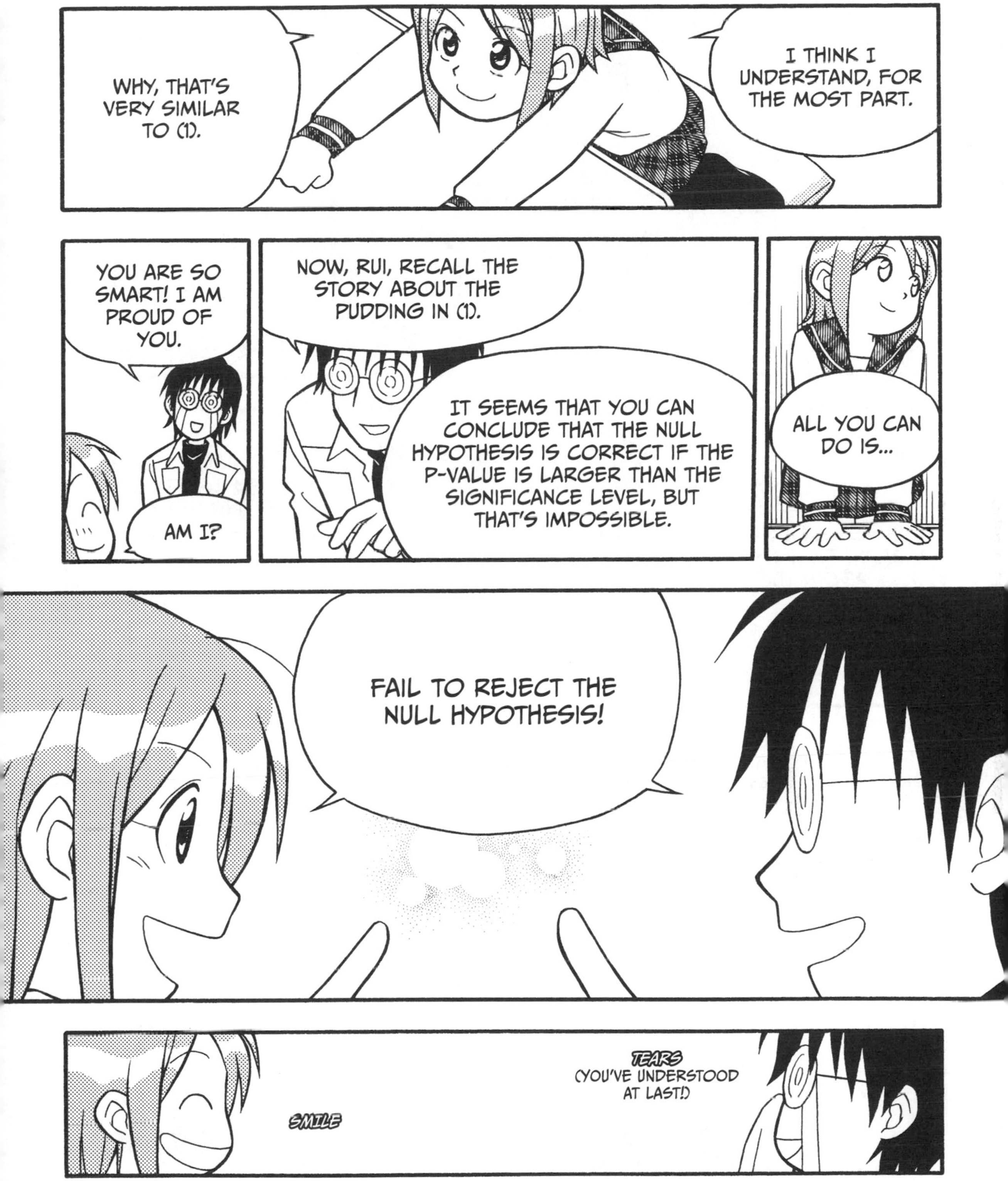
WHY, THAT'S VERY SIMILAR TO (1).
I THINK I UNDERSTAND, FOR THE MOST PART.
YOU ARE SO SMART! I AM PROUD OF YOU.
AM I?
NOW, RUI, RECALL THE STORY ABOUT THE PUDDING IN (1).
IT SEEMS THAT YOU CAN CONCLUDE THAT THE NULL HYPOTHESIS IS CORRECT IF THE P-VALUE IS LARGER THAN THE SIGNIFICANCE LEVEL, BUT THAT'S IMPOSSIBLE.
ALL YOU CAN DO IS...
FAIL TO REJECT THE NULL HYPOTHESIS!
SMILE
TEARS
(YOU'VE UNDERSTOOD AT LAST!)

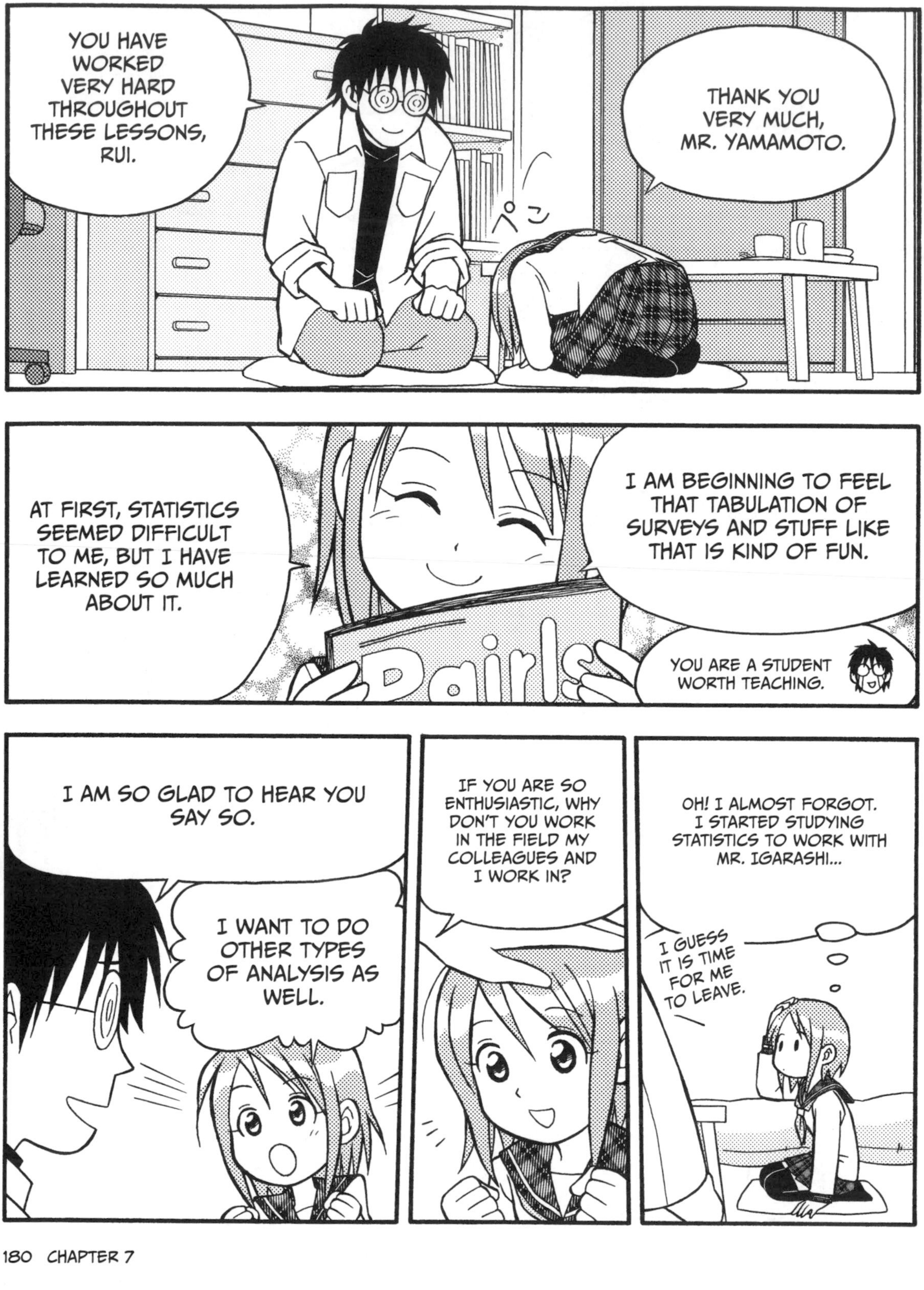
YOU HAVE WORKED VERY HARD THROUGHOUT THESE LESSONS, RUI.
ペコ
THANK YOU VERY MUCH, MR. YAMAMOTO.
AT FIRST, STATISTICS SEEMED DIFFICULT TO ME, BUT I HAVE LEARNED SO MUCH ABOUT IT.
I AM BEGINNING TO FEEL THAT TABULATION OF SURVEYS AND STUFF LIKE THAT IS KIND OF FUN.
Pairls
YOU ARE A STUDENT WORTH TEACHING.
I AM SO GLAD TO HEAR YOU SAY SO.
I WANT TO DO OTHER TYPES OF ANALYSIS AS WELL.
IF YOU ARE SO ENTHUSIASTIC, WHY DON'T YOU WORK IN THE FIELD MY COLLEAGUES AND I WORK IN?
OH! I ALMOST FORGOT. I STARTED STUDYING STATISTICS TO WORK WITH MR. IGARASHI...
I GUESS IT IS TIME FOR ME TO LEAVE.

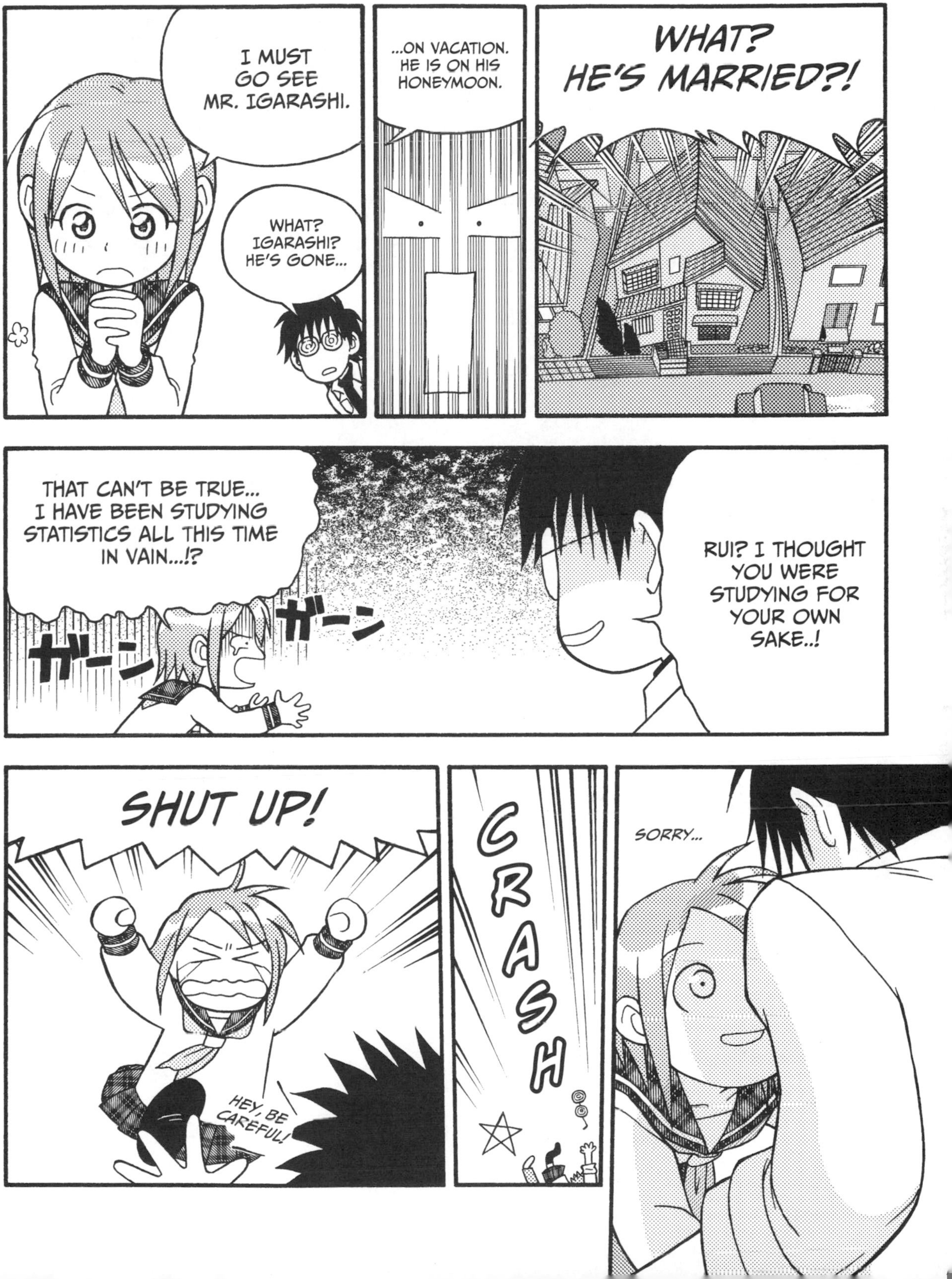
I MUST GO SEE MR. IGARASHI.
WHAT? IGARASHI? HE'S GONE...
...ON VACATION. HE IS ON HIS HONEYMOON.
WHAT? HE'S MARRIED?!
THAT CAN'T BE TRUE... I HAVE BEEN STUDYING STATISTICS ALL THIS TIME IN VAIN...!?
ガーン
ガーン
RUI? I THOUGHT YOU WERE STUDYING FOR YOUR OWN SAKE..!
SHUT UP!
HEY, BE CAREFUL!
CRASH
SORRY...

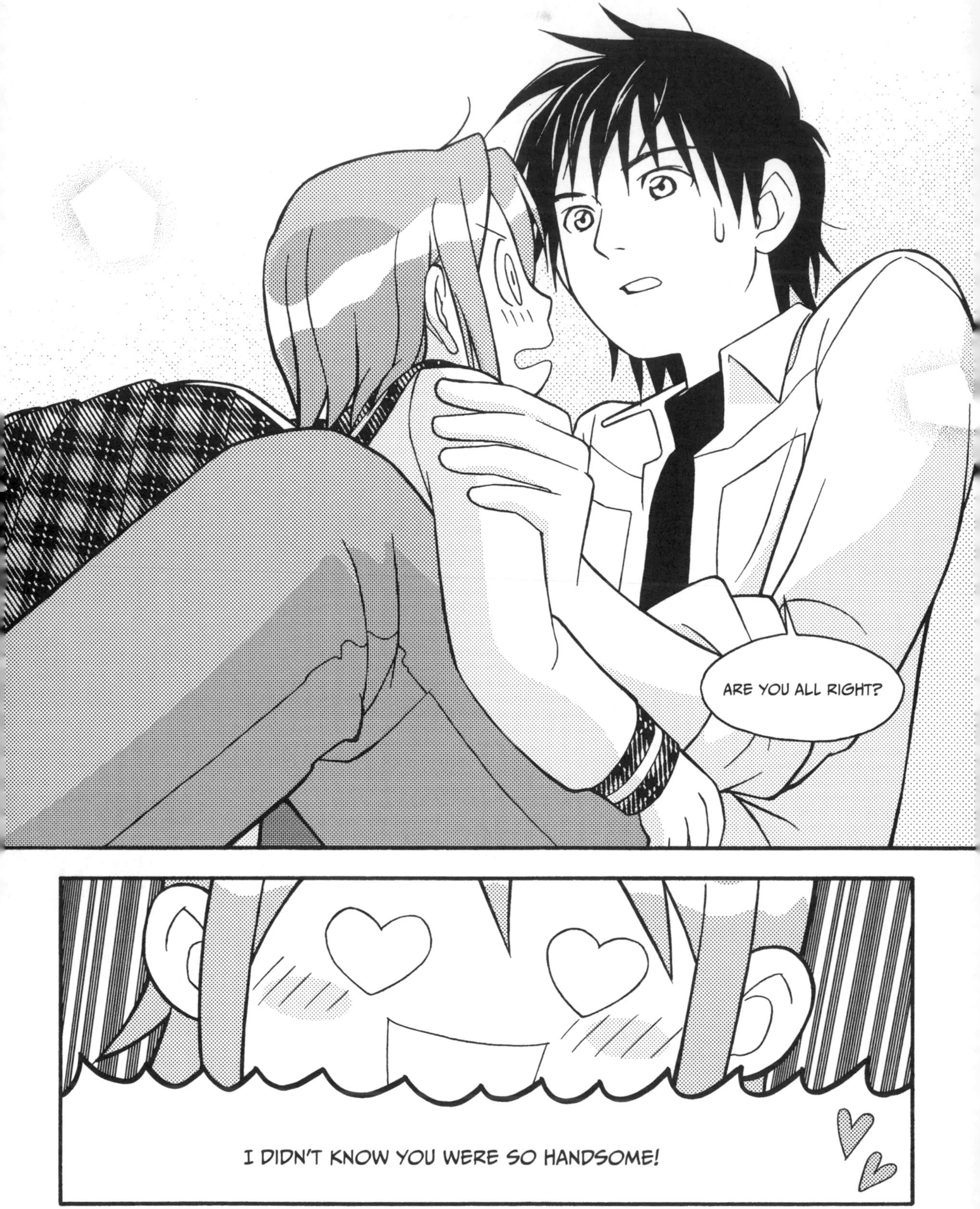
ARE YOU ALL RIGHT?
I DIDN'T KNOW YOU WERE SO HANDSOME!

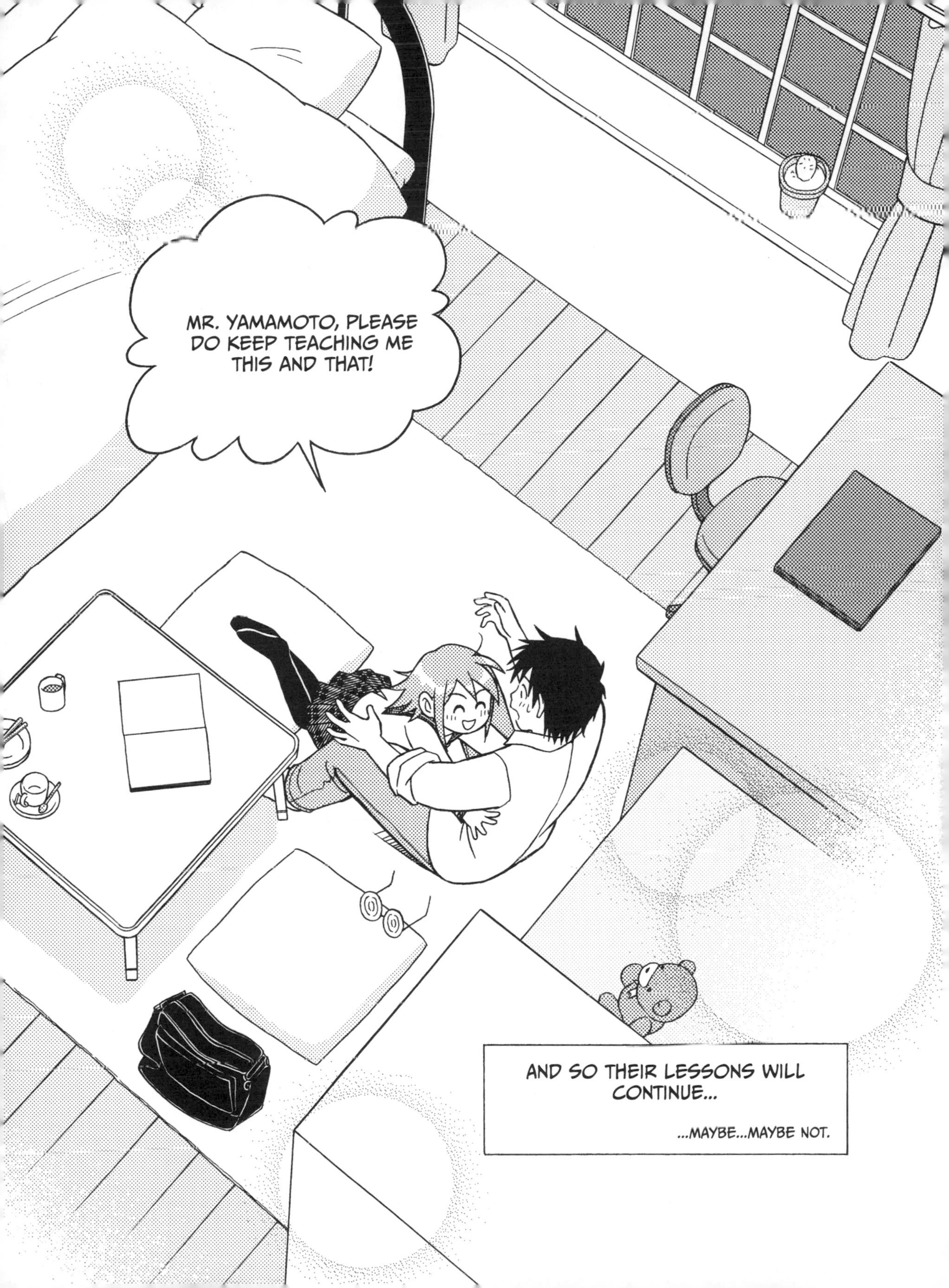
MR. YAMAMOTO, PLEASE DO KEEP TEACHING ME THIS AND THAT!
AND SO THEIR LESSONS WILL CONTINUE...
...MAYBE...MAYBE NOT.

5. TESTS OF INDEPENDENCE AND TESTS OF HOMOGENEITY

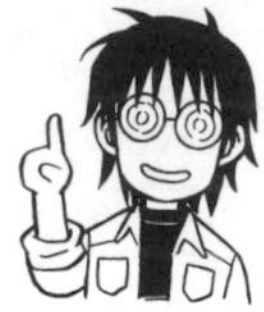

There is a hypothesis test very similar to a test of independence called a *test of homogeneity*. Below is an example of a test of homogeneity. As you read it, think about how it is different from a test of independence.

EXAMPLE

P-Girls Magazine published an article titled, "We Asked 300 High School Students, 'How Would You Like to Be Asked Out?'" The choices were phone, e-mail, or face to face.

> HYPOTHESIS: THE RATIO OF PHONE TO E-MAIL TO FACE-TO-FACE IS DIFFERENT BETWEEN HIGH SCHOOL GIRLS AND BOYS.

To find out if this hypothesis is true or not, a journalist actually conducted a survey by randomly choosing respondents from each of the two groups, "all high school girls residing in Japan" and "all high school boys residing in Japan." The table below is the result.

Respondent	Desired way of being asked out	Age	Sex
1	Face to face	17	Female
...	...	...	...
148	E-mail	16	Female
149	Phone	15	Male
...	...	...	...
300	E-mail	18	Male

The cross tabulation of sex and desired way of being asked out is the table below.

		Desired way of being asked out			
		Phone	E-mail	Face to face	Sum
Sex	Female	34	61	53	148
	Male	38	40	74	152
Sum		72	101	127	300

Estimate whether or not the hypothesis stated above is correct using a test of homogeneity. Use 0.05 as the significance level.

PROCEDURE

Step 1	Define the population.	The population in this case is "all high school girls residing in Japan" and "all high school boys residing in Japan."
Step 2	Set up a null hypothesis and an alternative hypothesis.	The null hypothesis is "the ratio of phone to e-mail to face to face is the same for high school girls and boys." The alternative hypothesis is "the ratio of phone to e-mail to face to face is different between high school girls and boys."
Step 3	Choose which hypothesis test to do.	A test of homogeneity will be applied.
Step 4	Determine the significance level.	The significance level is 0.05.
Step 5	Calculate the test statistic from the sample data.	A test of homogeneity is being used in this exercise. Therefore, the test statistic is Pearson's chi-square test statistic. The value of χ_0^2 in this exercise has already been calculated on page 132. $$\chi_0^2 = 8.0091$$ Pearson's chi-square test statistic (χ_0^2) in this exercise follows a chi-square distribution of degrees of freedom $(2 - 1) \times (3 - 1) = 1 \times 2 = 2$, if the null hypothesis is true.
Step 6	Determine whether the test statistic in step 5 is in the critical region.	The test statistic χ_0^2 is 8.0091. Since the significance level is 0.05, the critical region is 5.9915 or more, according to the table of chi-square distribution on page 103. The test statistic is within the critical region.
Step 7	If the test statistic is in the critical region in step 6, reject the null hypothesis and conclude in favor of the alternative. If not, fail to reject the null hypothesis.	The test statistic was within the critical region. Thus, you conclude in favor of the alternative hypothesis, "the ratio of phone to e-mail to face to face is different between high school girls and boys."

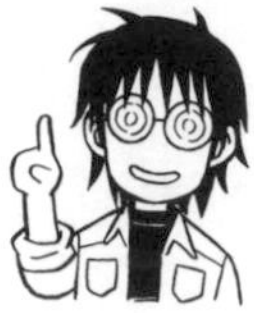

Don't you think that both the exercise and procedure are quite similar to those for a test of independence? Let's now look at the differences between tests of independence and tests of homogeneity. There are three things to note.

First, the population defined is different. There is only one population ("all high school students residing in Japan") in the former. In the latter, there are two populations: "all high school girls residing in Japan" and "all high school boys residing in Japan."

Also, the hypotheses are different. In the former,

Null hypothesis	The Cramer's coefficient for the population is 0. In other words, sex and desired way of being asked out are not correlated.
Alternative hypothesis	The Cramer's coefficient for the population is greater than 0. In other words, sex and desired way of being asked out are correlated.

In the latter,

Null hypothesis	The ratio of phone to e-mail to face to face is the same for high school girls and boys.
Alternative hypothesis	The ratio of phone to e-mail to face to face is different between high school girls and boys.

Finally, the order of procedure is different. In the former, the hypothesis is set after the data is collected, whereas the hypothesis is set before collecting the data in the latter.

As confirmed in the previous paragraph, tests of independence and tests of homogeneity have obvious differences. However, in practice, people tend to do tests of homogeneity when they are actually intending to do tests of independence, or vice versa. Be careful.

6. HYPOTHESIS TEST CONCLUSIONS

Up to this point, we have expressed the conclusion of a hypothesis test as follows:

> IF THE TEST STATISTIC IS IN THE CRITICAL REGION, YOU CAN
> CONCLUDE, "I REJECT THE NULL HYPOTHESIS." IF NOT, YOU CONCLUDE,
> "I FAIL TO REJECT THE NULL HYPOTHESIS."

But there are other ways to express the conclusions of hypothesis tests. They are summarized below.

TABLE 7-4: EXPRESSIONS OF HYPOTHESIS TEST CONCLUSIONS

When the test statistic is in the critical region	When the test statistic is not in the critical region
• Conclude in favor of the alternative hypothesis • Conclude that the result is statistically significant • Reject the null hypothesis	• Fail to reject the null hypothesis • Conclude that the result is not statistically significant • Accept the null hypothesis

The expressions "it is statistically significant" and "it is not statistically significant" seem to be popular in introductions to statistics. So why did we use an unpopular expression on purpose? I recognize that many beginners to hypothesis tests use the expression "it is significant" without actually understanding the meaning of the phrase. They seem to be merely confirming the test statistic or P-value. If you do not set a proper null and alternative hypothesis, the meaning of *significant* will be ambiguous. Beginners' definitions of their populations are frequently unclear as well.

I used to think I shouldn't be so strict with beginners. But it's impossible to make an accurate conclusion with uncertain null and alternative hypotheses. So in this book, I use the expressions "reject the null hypothesis" and "fail to reject the null hypothesis" so that you will get into the habit of thinking hard about your hypotheses.

EXERCISE

The table below is the same as the cross tabulation found on page 138.

		Preference for coffee or tea		Sum
		Coffee	Tea	
Type of food often ordered	Japanese	43	33	76
	European	51	53	104
	Chinese	29	41	70
Sum		123	127	250

Using a chi-square test of independence, estimate if the Cramer's coefficient for type of food often ordered and preference for coffee or tea in the population "people of age 20 or older residing in Japan" is greater than 0. This is the same as estimating whether there is a correlation between type of food often ordered and preference for coffee or tea. Use 0.01 as the significance level.

ANSWER

Step 1	Define the population.	The population in this case is "people of age 20 or older residing in Japan."
Step 2	Set up a null hypothesis and an alternative hypothesis.	The null hypothesis is "type of food often ordered and preference for coffee or tea are not correlated." The alternative hypothesis is "type of food often ordered and preference for coffee or tea are correlated."
Step 3	Choose which hypothesis test to do.	A chi-square test of independence will be applied.
Step 4	Determine the significance level.	The significance level is 0.01.
Step 5	Calculate the test statistic from the sample data.	A chi-square test of independence is being used in this exercise. Therefore, the test statistic is Pearson's chi-square test statistic (χ_0^2). The value of χ_0^2 in this exercise has already been calculated on page 141. $\chi_0^2 = 3.3483$
Step 6	Determine whether the test statistic obtained in step 5 is in the critical region.	The test statistic χ_0^2 is 3.3483. Because the significance level (α) is 0.01, the critical region is 9.2104 or above, according to the table of chi-square distribution on page 103. The test statistic is not within the critical region.
Step 7	If the test statistic is in the critical region in step 6, reject the null hypothesis. If not, fail to reject the null hypothesis.	The test statistic was not within the critical region. Thus, the null hypothesis "type of food often ordered and preference for coffee or tea are not correlated" cannot be rejected.

SUMMARY

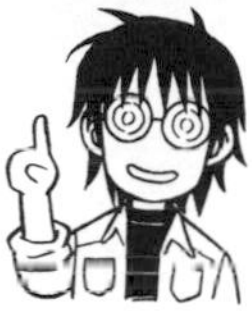

- A *hypothesis test* is an analysis technique used to estimate whether the analyst's hypothesis about the population is correct using the sample data.
- The formal name for a hypothesis test is *statistical hypothesis testing*.
- Test statistics are obtained from a function that calculates a single value from the sample data.
- In general, 0.05 or 0.01 is used as the significance level.
- The *critical region* is an area that corresponds to the significance level (also called the *alpha value* and expressed by the symbol α).
- A *chi-square test of independence* is an analysis technique used to estimate whether the Cramer's coefficient for a population is 0. It can also be said that it is an analysis technique used to estimate whether the two variables in a cross tabulation are correlated.
- If the Cramer's coefficient for a population is 0, Pearson's chi-square test statistic follows a chi-square distribution.
- The *P-value* in a test of independence is a probability that gives a Pearson's chi-square test statistic equal to or greater than the value earned in the case when the null hypothesis is true.
- When making a conclusion in a hypothesis test, there are two bases of judgment:

  1. Whether the test statistic is in the critical region
  2. Whether the P-value is smaller than the significance level

- The process of analysis in any hypothesis test is the same as the process for the test of independence or any other kind of test. The actual procedure is:

Step 1	Define the population.
Step 2	Set up a null hypothesis and an alternative hypothesis.
Step 3	Choose which hypothesis test to do.
Step 4	Determine the significance level.
Step 5	Calculate the value of the test statistic from the sample data.
Step 6	Determine whether the test statistic obtained in step 5 is in the critical region.
Step 7	If the test statistic is in the critical region in step 6, reject the null hypothesis. If not, fail to reject the null hypothesis.
Step 6p	Determine whether the P-value corresponding to the test statistic obtained in step 5 is smaller than the significance level.
Step 7p	If the P-value is smaller than the significance level in step 6p, reject the null hypothesis. If not, fail to reject the null hypothesis.

LET'S CALCULATE USING EXCEL

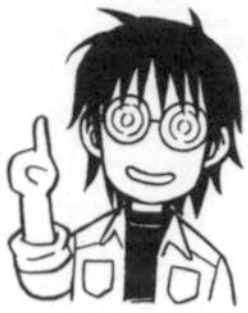

This appendix contains instructions for calculating various statistics using Microsoft Excel. You'll learn how to do the following things:

1. Make a frequency table
2. Calculate arithmetic mean, median, and standard deviation
3. Make a cross tabulation
4. Calculate the standard score and the deviation score
5. Calculate the probability of the standard normal distribution
6. Calculate the point on the horizontal axis of the chi-square distribution
7. Calculate the correlation coefficient
8. Perform tests of independence

You can download these Excel files and follow along (get them at *http://www.nostarch.com/mg_statistics.htm*). Readers who are not familiar with Excel should try "Calculating Arithmetic Mean, Median, and Standard Deviation" on page 195 first.

1. MAKING A FREQUENCY TABLE

This exercise uses the ramen restaurant prices on page 33.

Step 1

Select cell J3.

	A	B	C	D	E	F	G	H	I	J
1		Price (yen)			Price (yen)					
2	Ramen shop 1	700		Ramen shop 26	780		Equal or greater	Less than	Equal or less	Frequency
3	Ramen shop 2	850		Ramen shop 27	590		500	600	599	
4	Ramen shop 3	600		Ramen shop 28	650		600	700	699	
5	Ramen shop 4	650		Ramen shop 29	580		700	800	799	
6	Ramen shop 5	980		Ramen shop 30	750		800	900	899	
7	Ramen shop 6	750		Ramen shop 31	800		900	1000	999	
8	Ramen shop 7	500		Ramen shop 32	550					
9	Ramen shop 8	890		Ramen shop 33	750					
10	Ramen shop 9	880		Ramen shop 34	700					
11	Ramen shop 10	700		Ramen shop 35	600					
12	Ramen shop 11	890		Ramen shop 36	800					
13	Ramen shop 12	720		Ramen shop 37	800					
14	Ramen shop 13	680		Ramen shop 38	880					
15	Ramen shop 14	650		Ramen shop 39	790					
16	Ramen shop 15	790		Ramen shop 40	790					
17	Ramen shop 16	670		Ramen shop 41	780					
18	Ramen shop 17	680		Ramen shop 42	600					
19	Ramen shop 18	900		Ramen shop 43	670					
20	Ramen shop 19	880		Ramen shop 44	680					
21	Ramen shop 20	720		Ramen shop 45	650					
22	Ramen shop 21	850		Ramen shop 46	890					
23	Ramen shop 22	700		Ramen shop 47	930					
24	Ramen shop 23	780		Ramen shop 48	650					
25	Ramen shop 24	850		Ramen shop 49	777					
26	Ramen shop 25	750		Ramen shop 50	700					

Step 2

Select **Insert ▶ Function**.

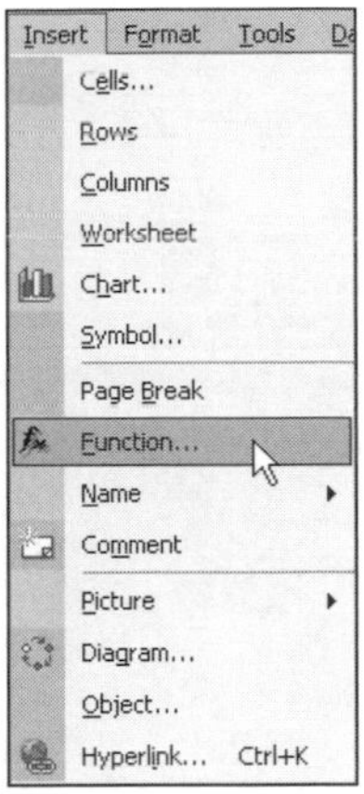

Insert
Format
Tools
Cells...
Rows
Columns
Worksheet
Chart...
Symbol...
Page Break
Function...
Name
Comment
Picture
Diagram...
Object...
Hyperlink... Ctrl+K

Step 3

Select **Statistical** from the category dropdown menu, and then select **FREQUENCY** as the name of the function.

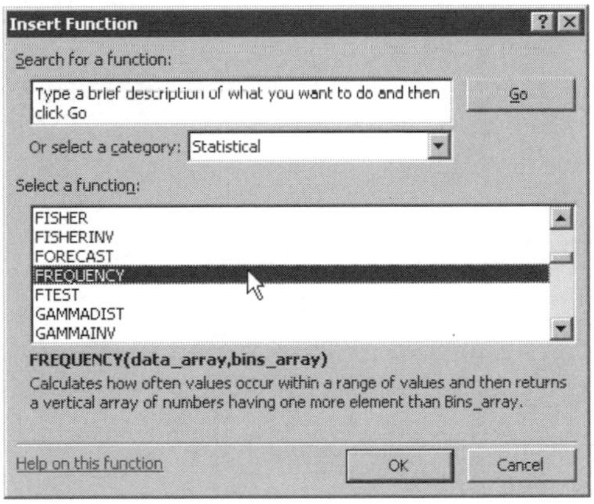

Insert Function
Search for a function:
Type a brief description of what you want to do and then click Go
Go
Or select a category: Statistical
Select a function:
FISHER
FISHERINV
FORECAST
FREQUENCY
FTEST
GAMMADIST
GAMMAINV
FREQUENCY(data_array,bins_array)
Calculates how often values occur within a range of values and then returns a vertical array of numbers having one more element than Bins_array.
Help on this function
OK
Cancel

Step 4

Select the area shown in the figure below, and click **OK**.

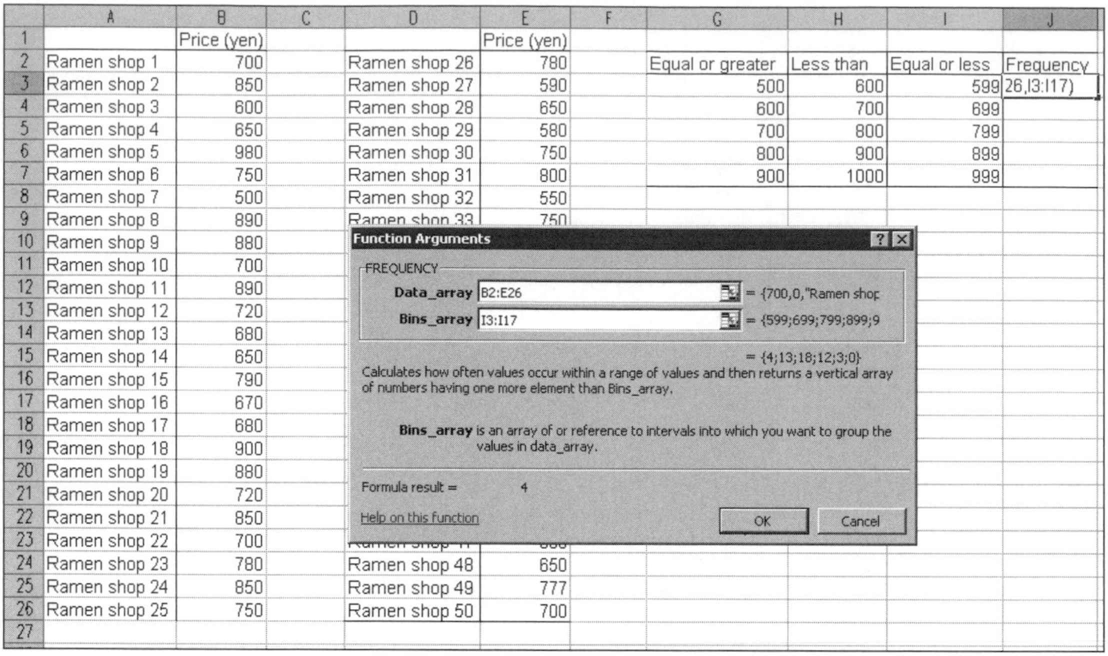
Function Arguments
FREQUENCY
Data_array B2:E26 = {700,0,"Ramen shop
Bins_array I3:I17 = {599;699;799;899;9
= {4;13;18;12;3;0}
Calculates how often values occur within a range of values and then returns a vertical array of numbers having one more element than Bins_array.
Bins_array is an array of or reference to intervals into which you want to group the values in data_array.
Formula result = 4
Help on this function
OK
Cancel

	A	B	C	D	E	F	G	H	I	J
1		Price (yen)			Price (yen)					
2	Ramen shop 1	700		Ramen shop 26	780		Equal or greater	Less than	Equal or less	Frequency
3	Ramen shop 2	850		Ramen shop 27	590		500	600	599	26,I3:I17)
4	Ramen shop 3	600		Ramen shop 28	650		600	700	699	
5	Ramen shop 4	650		Ramen shop 29	580		700	800	799	
6	Ramen shop 5	980		Ramen shop 30	750		800	900	899	
7	Ramen shop 6	750		Ramen shop 31	800		900	1000	999	
8	Ramen shop 7	500		Ramen shop 32	550					
9	Ramen shop 8	890		Ramen shop 33	750					
10	Ramen shop 9	880								
11	Ramen shop 10	700								
12	Ramen shop 11	890								
13	Ramen shop 12	720								
14	Ramen shop 13	680								
15	Ramen shop 14	650								
16	Ramen shop 15	790								
17	Ramen shop 16	670								
18	Ramen shop 17	680								
19	Ramen shop 18	900								
20	Ramen shop 19	880								
21	Ramen shop 20	720								
22	Ramen shop 21	850								
23	Ramen shop 22	700								
24	Ramen shop 23	780		Ramen shop 48	650					
25	Ramen shop 24	850		Ramen shop 49	777					
26	Ramen shop 25	750		Ramen shop 50	700					
27										

Step 5

Start with cell J3, and select the area from cell J3 to J7 as shown below.

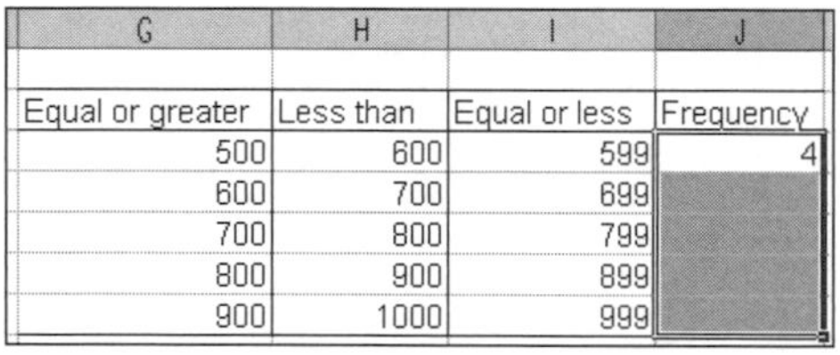

G	H	I	J
Equal or greater	Less than	Equal or less	Frequency
500	600	599	4
600	700	699	
700	800	799	
800	900	899	
900	1000	999	

Step 6

Click this part in the formula bar.

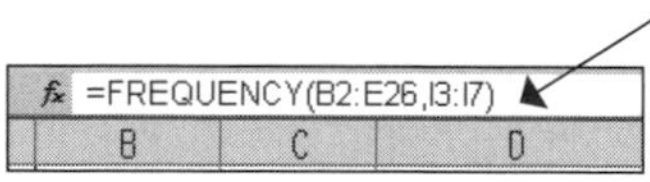
=FREQUENCY(B2:E26,I3:I7)
B
C
D

Step 7

Press ENTER while holding down the SHIFT key and CTRL key at the same time.

Step 8

Now you have the frequency of each class!

G	H	I	J
Equal or greater	Less than	Equal or less	Frequency
500	600	599	4
600	700	699	13
700	800	799	18
800	900	899	12
900	1000	999	3

2. CALCULATING ARITHMETIC MEAN, MEDIAN, AND STANDARD DEVIATION

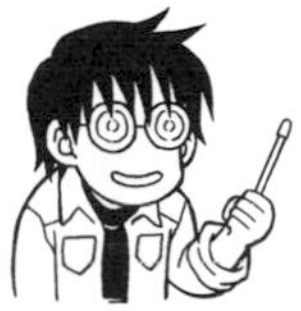

This data comes from Rui's classmates' bowling scores on page 41.

Step 1

Select cell B10.

	A	B
1		Team A
2	Rui-Rui	86
3	Jun	73
4	Yumi	124
5	Shizuka	111
6	Touko	90
7	Kaede	38
8		
9		
10	Average	
11	Median	
12	Standard Deviation	

Step 2

Select **Insert ▸ Function**.

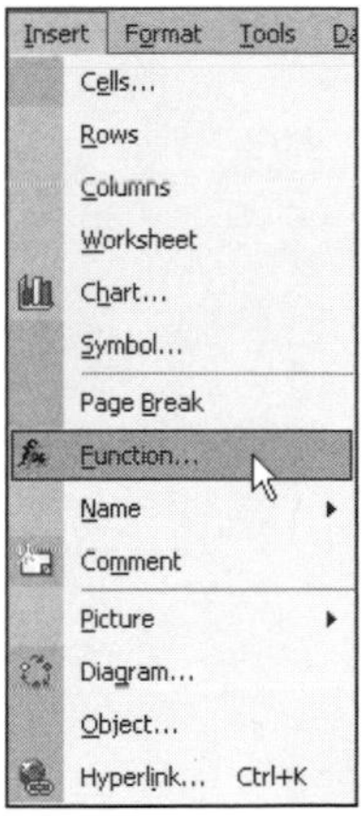
Insert
Format
Tools
Cells...
Rows
Columns
Worksheet
Chart...
Symbol...
Page Break
Function...
Name
Comment
Picture
Diagram...
Object...
Hyperlink... Ctrl+K

Step 3

Select **Statistical** in the category dropdown, and then select **AVERAGE**.

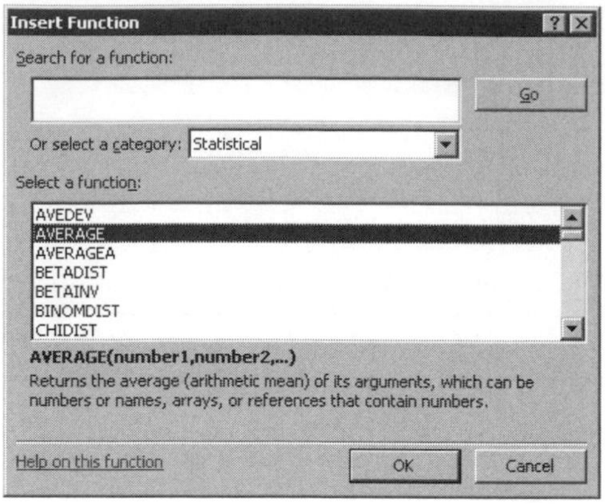
Insert Function
Search for a function:
Go
Or select a category: Statistical
Select a function:
AVEDEV
AVERAGE
AVERAGEA
BETADIST
BETAINV
BINOMDIST
CHIDIST
AVERAGE(number1,number2,...)
Returns the average (arithmetic mean) of its arguments, which can be numbers or names, arrays, or references that contain numbers.
Help on this function
OK
Cancel

Step 4

Type the range shown in the figure below, and click **OK**.

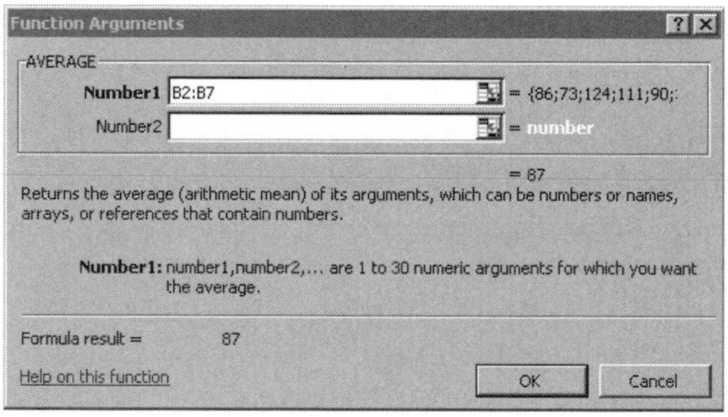
Function Arguments
AVERAGE
Number1 B2:B7
= {86;73;124;111;90;
Number2
= number
= 87
Returns the average (arithmetic mean) of its arguments, which can be numbers or names, arrays, or references that contain numbers.
Number1: number1,number2,... are 1 to 30 numeric arguments for which you want the average.
Formula result = 87
Help on this function
OK
Cancel

Step 5

Now you have the average score for the team.

	A	B
1		Team A
2	Rui-Rui	86
3	Jun	73
4	Yumi	124
5	Shizuka	111
6	Touko	90
7	Kaede	38
8		
9		
10	Average	87
11	Median	
12	Standard Deviation	

You can calculate the median and standard deviation by following steps 1 through 5 and using the functions MEDIAN and STDEVP in step 2.

3. MAKING A CROSS TABULATION

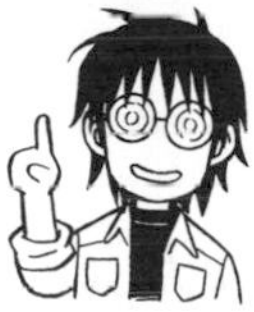

The data for this table is Rui's classmates' responses to the new uniform design, found on page 61.

Step 1

Select cell F20, then select **Insert ▸ Function**.

	A	B	C	D	E	F	G	H
1		Response			Response			Response
2	1	like		16	neither		31	neither
3	2	neither		17	like		32	neither
4	3	like		18	like		33	like
5	4	neither		19	like		34	dislike
6	5	dislike		20	like		35	like
7	6	like		21	like		36	like
8	7	like		22	like		37	like
9	8	like		23	dislike		38	like
10	9	like		24	neither		39	neither
11	10	like		25	like		40	like
12	11	like		26	like			
13	12	like		27	dislike			
14	13	neither		28	like			
15	14	like		29	like			
16	15	like		30	like			
17								
18								
19						Frequency		
20					like			
21					neither			
22					dislike			

Step 2

Select **Statistical** in the category dropdown, and then select **COUNTIF** as the name of the function.

Step 3

Select the area shown in the figure below, type *like* in the Criteria text box, and then click **OK**.

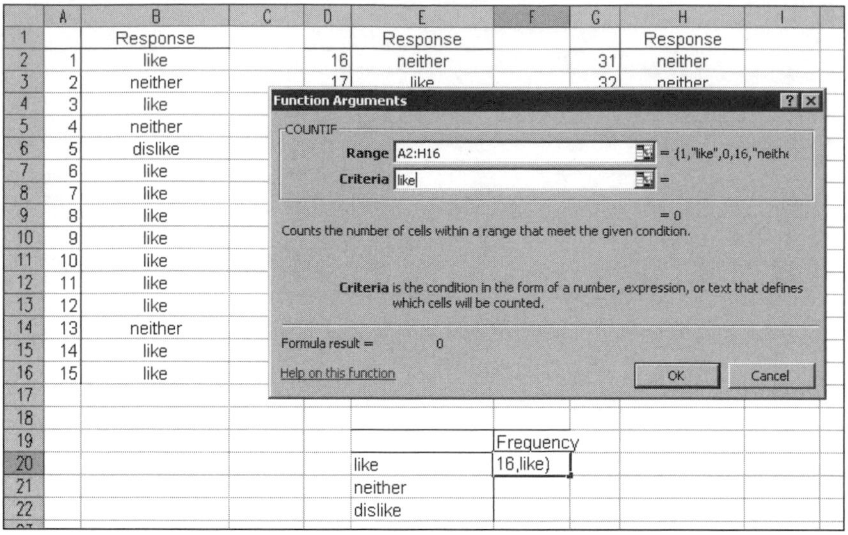

	A	B	C	D	E	F	G	H	I
1		Response			Response			Response	
2	1	like		16	neither		31	neither	
3	2	neither		17	like		32	neither	
4	3	like							
5	4	neither							
6	5	dislike							
7	6	like							
8	7	like							
9	8	like							
10	9	like							
11	10	like							
12	11	like							
13	12	like							
14	13	neither							
15	14	like							
16	15	like							
17									
18									
19						Frequency			
20					like	16,like)			
21					neither				
22					dislike				

Function Arguments   ? ✕

COUNTIF

Range A2:H16 = {1,"like",0,16,"neithe

Criteria like =

= 0

Counts the number of cells within a range that meet the given condition.

Criteria is the condition in the form of a number, expression, or text that defines which cells will be counted.

Formula result =   0

Help on this function   [OK]   [Cancel]

Step 4

Now you have the total number of Rui's classmates who like the new uniform.

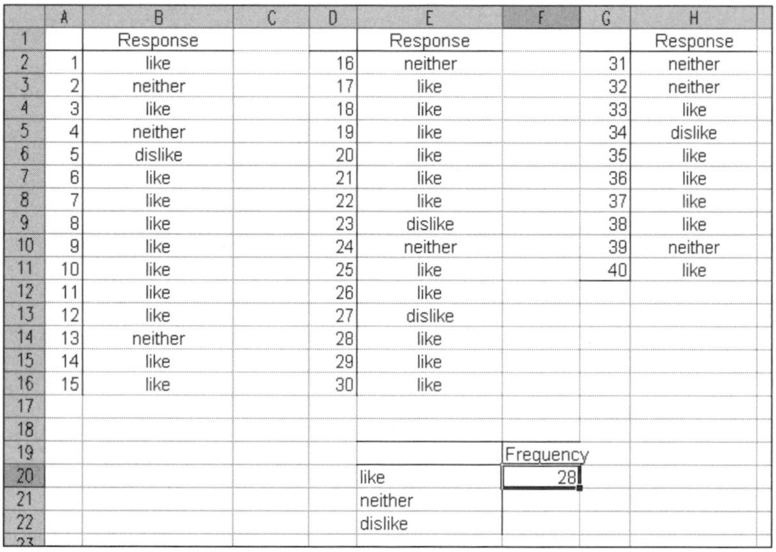

	A	B	C	D	E	F	G	H	I
1		Response			Response			Response	
2	1	like		16	neither		31	neither	
3	2	neither		17	like		32	neither	
4	3	like		18	like		33	like	
5	4	neither		19	like		34	dislike	
6	5	dislike		20	like		35	like	
7	6	like		21	like		36	like	
8	7	like		22	like		37	like	
9	8	like		23	dislike		38	like	
10	9	like		24	neither		39	neither	
11	10	like		25	like		40	like	
12	11	like		26	like				
13	12	like		27	dislike				
14	13	neither		28	like				
15	14	like		29	like				
16	15	like		30	like				
17									
18									
19						Frequency			
20					like	28			
21					neither				
22					dislike				

Step 5

You can obtain the frequency of *neither* and *dislike* by following steps 1 through 4 and typing those words instead of *like* in step 3.

4. CALCULATING THE STANDARD SCORE AND THE DEVIATION SCORE

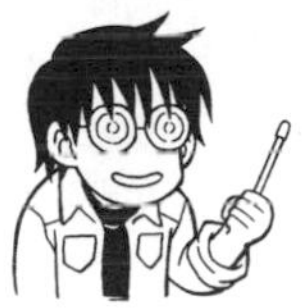

This exercise uses the test data from page 72.

Steps 1 through 8 show the process for obtaining the standard score.

Steps 9 through 11 show the process for obtaining the deviation score. There is an Excel function for calculating standard score, but there is no function for calculating deviation score. However, the deviation score can be calculated fairly easily if we use the result of the standard score calculation.

Step 1

Select cell E2.

	A	B	C	D	E	F
1		History			Standard Score	Deviation Score
2	Rui	73		Rui		
3	Yumi	61		Yumi		
4	A	14		A		
5	B	41		B		
6	C	49		C		
7	D	87		D		
8	E	69		E		
9	F	65		F		
10	G	36		G		
11	H	7		H		
12	I	53		I		
13	J	100		J		
14	K	57		K		
15	L	45		L		
16	M	56		M		
17	N	34		N		
18	O	37		O		
19	P	70		P		
20	Average	53				
21	Standard Deviation	22.7				

Step 2

Select **Insert ▸ Function**. Then select **Statistical**, and then select **STANDARDIZE** as the name of the function.

Step 3

Select cell B2.

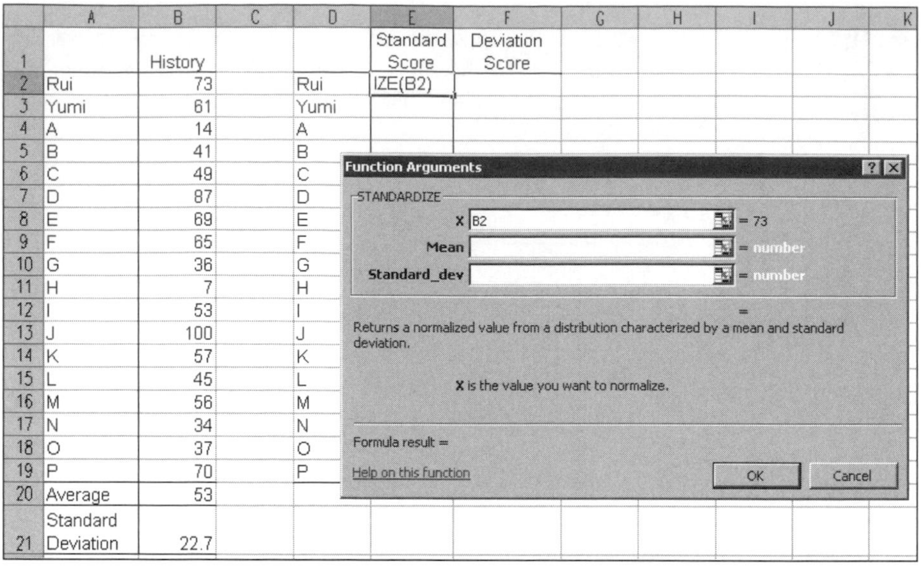

	A	B	C	D	E	F	G	H	I	J	K
1		History			Standard Score	Deviation Score					
2	Rui	73		Rui	IZE(B2)						
3	Yumi	61		Yumi							
4	A	14		A							
5	B	41		B							
6	C	49		C							
7	D	87		D							
8	E	69		E							
9	F	65		F							
10	G	36		G							
11	H	7		H							
12	I	53		I							
13	J	100		J							
14	K	57		K							
15	L	45		L							
16	M	56		M							
17	N	34		N							
18	O	37		O							
19	P	70		P							
20	Average	53									
21	Standard Deviation	22.7									

Function Arguments ? X

STANDARDIZE

X | B2 | = 73
Mean | | = number
Standard_dev | | = number

=

Returns a normalized value from a distribution characterized by a mean and standard deviation.

X is the value you want to normalize.

Formula result =

Help on this function   OK   Cancel

Step 4

Select B20 for Mean, press F4 once, and confirm that B20 has changed to B20.

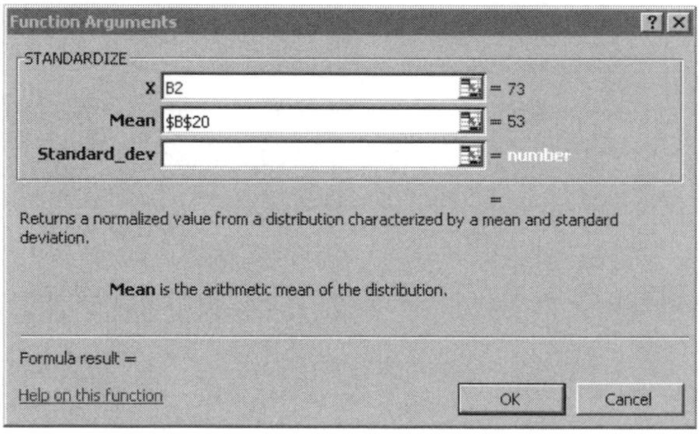

Function Arguments ? X

STANDARDIZE

X | B2 | = 73
Mean | B20 | = 53
Standard_dev | | = number

=

Returns a normalized value from a distribution characterized by a mean and standard deviation.

Mean is the arithmetic mean of the distribution.

Formula result =

Help on this function   OK   Cancel

Step 5

Select cell B21 for Standard_dev, press F4 once, and after confirming that B21 has changed to B21, click **OK**.

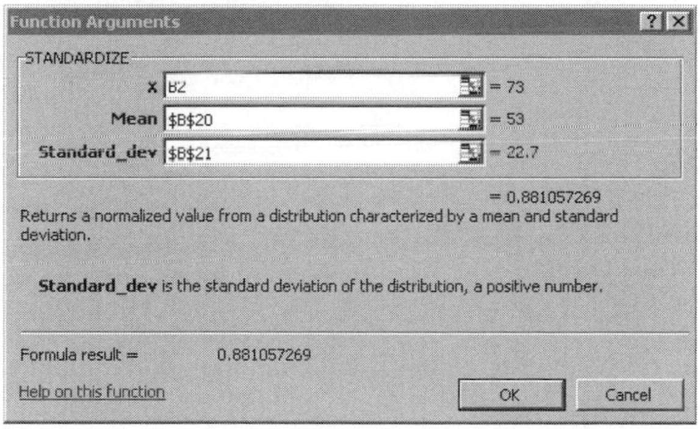
Function Arguments
STANDARDIZE
X B2 = 73
Mean B20 = 53
Standard_dev B21 = 22.7
= 0.881057269
Returns a normalized value from a distribution characterized by a mean and standard deviation.
Standard_dev is the standard deviation of the distribution, a positive number.
Formula result = 0.881057269
Help on this function
OK
Cancel

Step 6

Confirm that Rui's standard score has been calculated.

	A	B	C	D	E	F
1		History			Standard Score	Deviation Score
2	Rui	73		Rui	0.88	
3	Yumi	61		Yumi		
4	A	14		A		
5	B	41		B		
6	C	49		C		
7	D	87		D		
8	E	69		E		
9	F	65		F		
10	G	36		G		
11	H	7		H		
12	I	53		I		
13	J	100		J		
14	K	57		K		
15	L	45		L		
16	M	56		M		
17	N	34		N		
18	O	37		O		
19	P	70		P		
20	Average	53				
21	Standard Deviation	22.7				

Step 7

Put the point of the arrow near the bottom-right side of cell E2, confirm that the arrow has changed to a black cross, drag down to cell E19 by holding down the left button of the mouse, and let go of the button when you finish dragging.

	A	B	C	D	E	F
1		History			Standard Score	Deviation Score
2	Rui	73		Rui	0.88	
3	Yumi	61		Yumi		
4	A	14		A		
5	B	41		B		
6	C	49		C		
7	D	87		D		
8	E	69		E		
9	F	65		F		
10	G	36		G		
11	H	7		H		
12	I	53		I		
13	J	100		J		
14	K	57		K		
15	L	45		L		
16	M	56		M		
17	N	34		N		
18	O	37		O		
19	P	70		P		
20	Average	53				
21	Standard Deviation	22.7				

Step 8

Now you should have everyone's standard score!

	A	B	C	D	E	F
1		History			Standard Score	Deviation Score
2	Rui	73		Rui	0.88	
3	Yumi	61		Yumi	0.35	
4	A	14		A	-1.72	
5	B	41		B	-0.53	
6	C	49		C	-0.18	
7	D	87		D	1.50	
8	E	69		E	0.70	
9	F	65		F	0.53	
10	G	36		G	-0.75	
11	H	7		H	-2.03	
12	I	53		I	0.00	
13	J	100		J	2.07	
14	K	57		K	0.18	
15	L	45		L	-0.35	
16	M	56		M	0.13	
17	N	34		N	-0.84	
18	O	37		O	-0.70	
19	P	70		P	0.75	
20	Average	53				
21	Standard Deviation	22.7				

Step 9

Select cell F2 and type *=E2*10+50*, then press ENTER.

	A	B	C	D	E	F
1		History			Standard Score	Deviation Score
2	Rui	73		Rui	0.88	=E2*10+50
3	Yumi	61		Yumi	0.35	
4	A	14		A	-1.72	
5	B	41		B	-0.53	
6	C	49		C	-0.18	
7	D	87		D	1.50	
8	E	69		E	0.70	
9	F	65		F	0.53	
10	G	36		G	-0.75	
11	H	7		H	-2.03	
12	I	53		I	0.00	
13	J	100		J	2.07	
14	K	57		K	0.18	
15	L	45		L	-0.35	
16	M	56		M	0.13	
17	N	34		N	-0.84	
18	O	37		O	-0.70	
19	P	70		P	0.75	
20	Average	53				
21	Standard Deviation	22.7				

Step 10

Drag down to cell F19, as you did in step 7.

Step 11

Now you have the class's deviation score.

	A	B	C	D	E	F
1		History			Standard Score	Deviation Score
2	Rui	73		Rui	0.88	58.81
3	Yumi	61		Yumi	0.35	53.52
4	A	14		A	-1.72	32.82
5	B	41		B	-0.53	44.71
6	C	49		C	-0.18	48.24
7	D	87		D	1.50	64.98
8	E	69		E	0.70	57.05
9	F	65		F	0.53	55.29
10	G	36		G	-0.75	42.51
11	H	7		H	-2.03	29.74
12	I	53		I	0.00	50.00
13	J	100		J	2.07	70.70
14	K	57		K	0.18	51.76
15	L	45		L	-0.35	46.48
16	M	56		M	0.13	51.32
17	N	34		N	-0.84	41.63
18	O	37		O	-0.70	42.95
19	P	70		P	0.75	57.49
20	Average	53				
21	Standard Deviation	22.7				

5. CALCULATING THE PROBABILITY OF THE STANDARD NORMAL DISTRIBUTION

For this example, we'll use the data from page 93.

Step 1

Select cell B2.

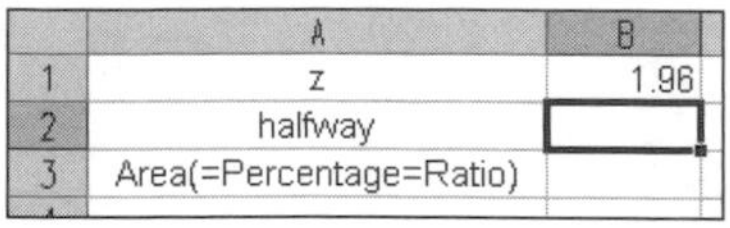

A
B
1
z
1.96
2
halfway
3
Area(=Percentage=Ratio)

Step 2

Select **Insert ▸ Function**, then select **Statistical**, and then select **NORMSDIST**.

Step 3

Select cell B1, and click **OK**.

A
B
C
D
E
F
G
H
I
z
1.96
halfway
ST(B1)
Area(=Percentage=Ratio)
Function Arguments
NORMSDIST
Z B1
= 1.96
= 0.975002175
Returns the standard normal cumulative distribution (has a mean of zero and a standard deviation of one).
Z is the value for which you want the distribution.
Formula result = 0.975002175
Help on this function
OK
Cancel

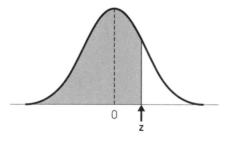

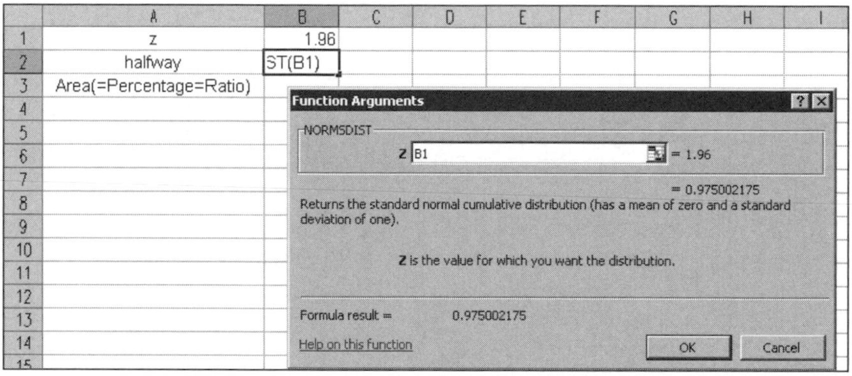

In fact, NORMSDIST is a function to calculate the probability shown in the figure below.

0
z

Step 4

Type *=B2-0.5* in cell B3.

	A	B
1	z	1.96
2	halfway	0.975
3	Area(=Percentage=Ratio)	=B2-0.5

Step 5

Now you have the area.

	A	B
1	z	1.96
2	halfway	0.975
3	Area(=Percentage=Ratio)	0.475

6. CALCULATING THE POINT ON THE HORIZONTAL AXIS OF THE CHI-SQUARE DISTRIBUTION

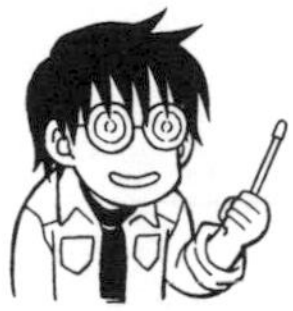

The data for this exercise comes from page 104.

Step 1

Select cell B3.

	A	B
1	P	0.05
2	Degrees of freedom	1
3	Chi-square	

Step 2

Select **Insert ▸ Function**, then select **Statistical**, and then select **CHIINV**.

Step 3

Select cells B1 and B2, and then click **OK**.

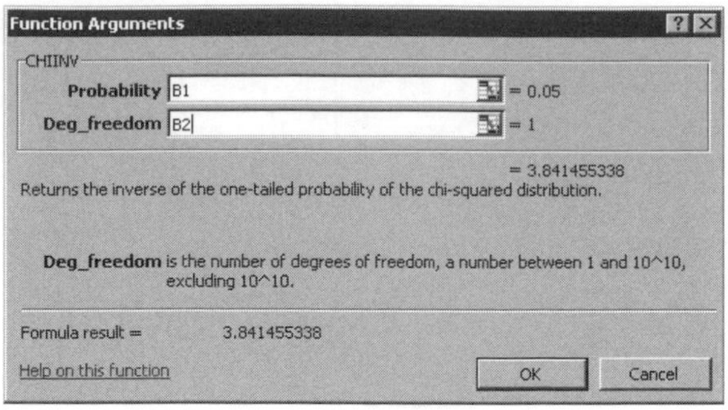

Function Arguments
CHIINV
Probability B1 = 0.05
Deg_freedom B2 = 1
= 3.841455338
Returns the inverse of the one-tailed probability of the chi-squared distribution.
Deg_freedom is the number of degrees of freedom, a number between 1 and 10^10, excluding 10^10.
Formula result = 3.841455338
Help on this function
OK
Cancel

Step 4

Now you're done.

	A	B
1	P	0.05
2	Degrees of freedom	1
3	Chi-square	3.84146

7. CALCULATING THE CORRELATION COEFFICIENT

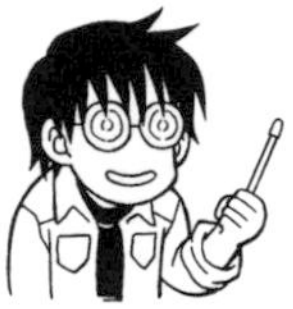

This data comes from the *P-Girls Magazine* survey found on page 116.

Step 1

Select cell B14.

	A	B	C
1		Amount spent on makeup (yen)	Amount spent on clothes (yen)
2	Ms. A	3000	7000
3	Ms. B	5000	8000
4	Ms. C	12000	25000
5	Ms. D	2000	5000
6	Ms. E	7000	12000
7	Ms. F	15000	30000
8	Ms. G	5000	10000
9	Ms. H	6000	15000
10	Ms. I	8000	20000
11	Ms. J	10000	18000
12			
13			
14	Correlation coefficient		

Step 2

Select **Insert ▸ Function**, then select **Statistical**, and then select **CORREL**.

Step 3

Select the area shown in the figure below, and then click **OK**.

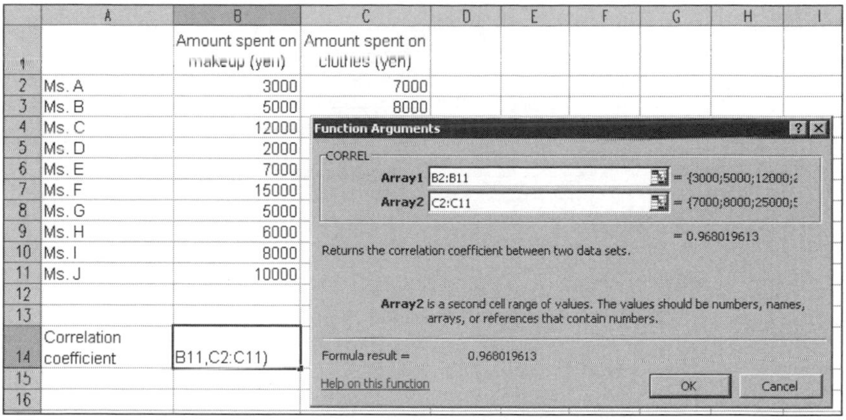

	A	B	C	D	E	F	G	H	I
1		Amount spent on makeup (yen)	Amount spent on clothes (yen)						
2	Ms. A	3000	7000						
3	Ms. B	5000	8000						
4	Ms. C	12000							
5	Ms. D	2000							
6	Ms. E	7000							
7	Ms. F	15000							
8	Ms. G	5000							
9	Ms. H	6000							
10	Ms. I	8000							
11	Ms. J	10000							
12									
13									
14	Correlation coefficient	B11,C2:C11)							
15									
16									

Function Arguments

CORREL
Array1 B2:B11 = {3000;5000;12000;2
Array2 C2:C11 = {7000;8000;25000;5

= 0.968019613

Returns the correlation coefficient between two data sets.

Array2 is a second cell range of values. The values should be numbers, names, arrays, or references that contain numbers.

Formula result = 0.968019613

Help on this function   OK   Cancel

Step 4

Now you have the correlation coefficient.

	A	B	C
1		Amount spent on makeup (yen)	Amount spent on clothes (yen)
2	Ms. A	3000	7000
3	Ms. B	5000	8000
4	Ms. C	12000	25000
5	Ms. D	2000	5000
6	Ms. E	7000	12000
7	Ms. F	15000	30000
8	Ms. G	5000	10000
9	Ms. H	6000	15000
10	Ms. I	8000	20000
11	Ms. J	10000	18000
12			
13			
14	Correlation coefficient	0.968019613	

NOTE *Unfortunately, there are no Excel functions for calculating the correlation ratio or the Cramer's coefficient.*

8. PERFORMING TESTS OF INDEPENDENCE

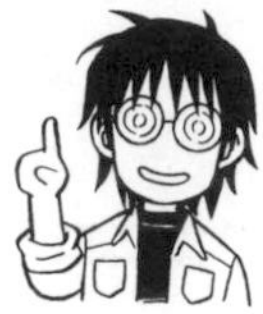

This data is from the dating survey on page 157.

Step 1

Select cell B8.

	A	B	C	D	E
1		Phone	E-mail	Face to face	Sum
2	Female	34	61	53	148
3	Male	38	40	74	152
4	Sum	72	101	127	300
5					
6					
7		Phone	E-mail	Face to face	
8	Female				
9	Male				
10					
11					
12	P-value				

Step 2

Type *=E2*B4/E4* in cell B8. Do not press ENTER yet.

	A	B	C	D	E
1		Phone	E-mail	Face to face	Sum
2	Female	34	61	53	148
3	Male	38	40	74	152
4	Sum	72	101	127	300
5					
6					
7		Phone	E-mail	Face to face	
8	Female	=E2*B4/E4			
9	Male				
10					
11					
12	P-value				

Step 3

Select *E2* in the equation you just typed, press F4 three times, and confirm that *E2* has changed to *$E2*. Do not press ENTER yet.

	A	B	C	D	E
1		Phone	E-mail	Face to face	Sum
2	Female	34	61	53	148
3	Male	38	40	74	152
4	Sum	72	101	127	300
5					
6					
7		Phone	E-mail	Face to face	
8	Female	=$E2*B4/E4			
9	Male				
10					
11					
12	P-value				

Step 4

Select *B4* in the equation in cell B8, press F4 twice, and confirm that *B4* has changed to *B$4*. Select *E4* in the equation in cell B8, press F4 once, and confirm that *E4* has changed to *E4*. Then press ENTER.

	A	B	C	D	E
1		Phone	E-mail	Face to face	Sum
2	Female	34	61	53	148
3	Male	38	40	74	152
4	Sum	72	101	127	300
5					
6					
7		Phone	E-mail	Face to face	
8	Female	=$E2*B$4/E4			
9	Male				
10					
11					
12	P-value				

Step 5

Select cell B8, put the point of the arrow near the bottom right side of cell B8, confirm that the arrow has changed to a black cross, drag down to cell D8 by holding down the left button of the mouse, and let go of the button when you finish dragging.

	A	B	C	D	E
1		Phone	E-mail	Face to face	Sum
2	Female	34	61	53	148
3	Male	38	40	74	152
4	Sum	72	101	127	300
5					
6					
7		Phone	E-mail	Face to face	
8	Female	35.52			
9	Male				
10					
11					
12	P-value				

Step 6

Select the area from cell B8 to D8, put the point of the arrow near the bottom right side of cell D8, confirm that the arrow has changed to a black cross, drag down to cell D9 by holding down the left button of the mouse, and let go of the button when you finish dragging.

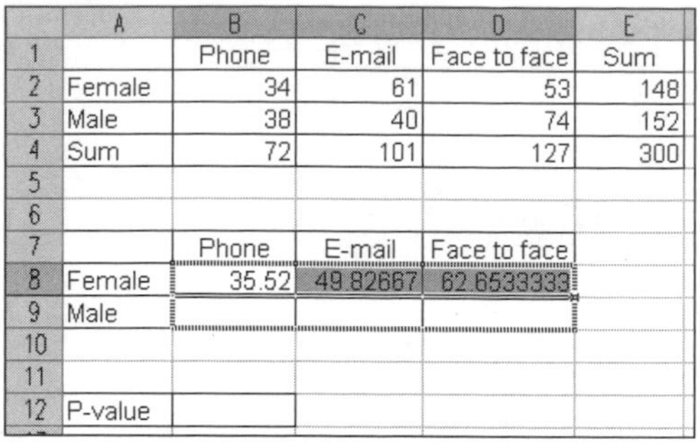

	A	B	C	D	E
1		Phone	E-mail	Face to face	Sum
2	Female	34	61	53	148
3	Male	38	40	74	152
4	Sum	72	101	127	300
5					
6					
7		Phone	E-mail	Face to face	
8	Female	35.52	49.82667	62.6533333	
9	Male				
10					
11					
12	P-value				

Step 7

Select cell B12, select **Insert ▸ Function**, then select **Statistical**, and then select **CHITEST**.

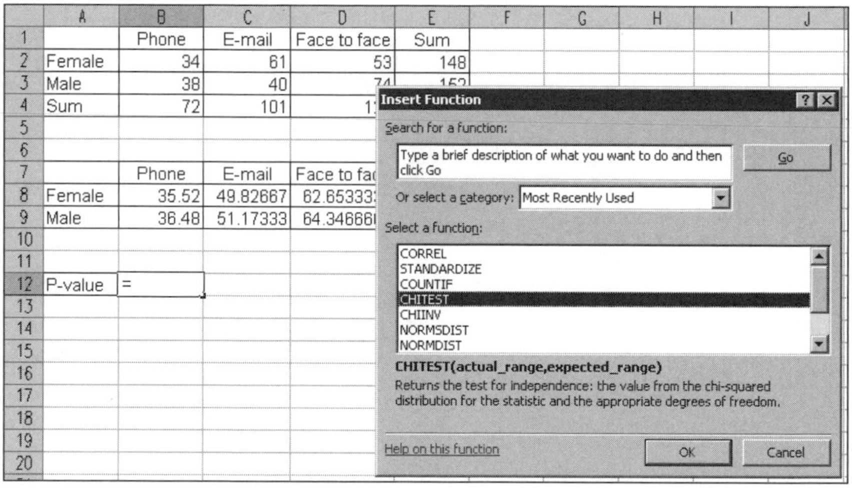

	A	B	C	D	E	F	G	H	I	J
1		Phone	E-mail	Face to face	Sum					
2	Female	34	61	53	148					
3	Male	38	40	74	152					
4	Sum	72	101	1						
5										
6										
7		Phone	E-mail	Face to fac						
8	Female	35.52	49.82667	62.65333						
9	Male	36.48	51.17333	64.34666						
10										
11										
12	P-value	=								
13										
14										
15										
16										
17										
18										
19										
20										

Insert Function ? ✕

Search for a function:

Type a brief description of what you want to do and then click Go   Go

Or select a category: Most Recently Used ▾

Select a function:

CORREL
STANDARDIZE
COUNTIF
CHITEST
CHIINV
NORMSDIST
NORMDIST

CHITEST(actual_range,expected_range)
Returns the test for independence: the value from the chi-squared distribution for the statistic and the appropriate degrees of freedom.

Help on this function   OK   Cancel

Step 8

Select the area shown in the figure below, and then click **OK**.

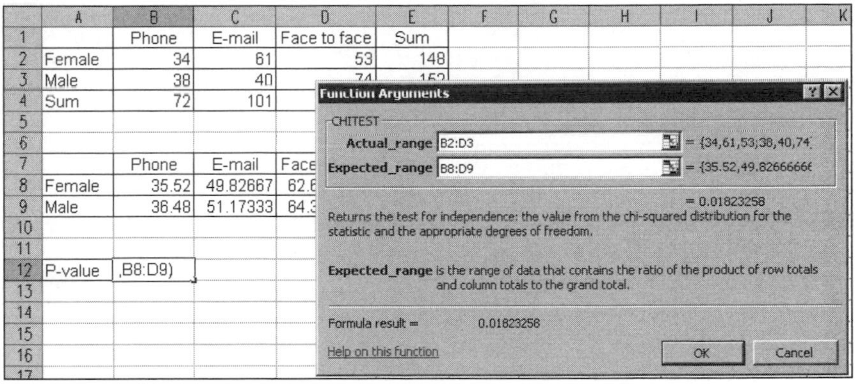

	A	B	C	D	E	F	G	H	I	J	K
1		Phone	E-mail	Face to face	Sum						
2	Female	34	61	53	148						
3	Male	38	40	74	152						
4	Sum	72	101								
5											
6											
7		Phone	E-mail	Face							
8	Female	35.52	49.82667	62.6							
9	Male	36.48	51.17333	64.3							
10											
11											
12	P-value	,B8:D9)									
13											
14											
15											
16											

Function Arguments

CHITEST

Actual_range B2:D3 = {34,61,53;38,40,74;

Expected_range B8:D9 = {35.52,49.8266666

= 0.01823258

Returns the test for independence: the value from the chi-squared distribution for the statistic and the appropriate degrees of freedom.

Expected_range is the range of data that contains the ratio of the product of row totals and column totals to the grand total.

Formula result =   0.01823258

Help on this function   OK   Cancel

Step 9

Now you're done. You can confirm that the calculated value is equal to the P-value on page 177.

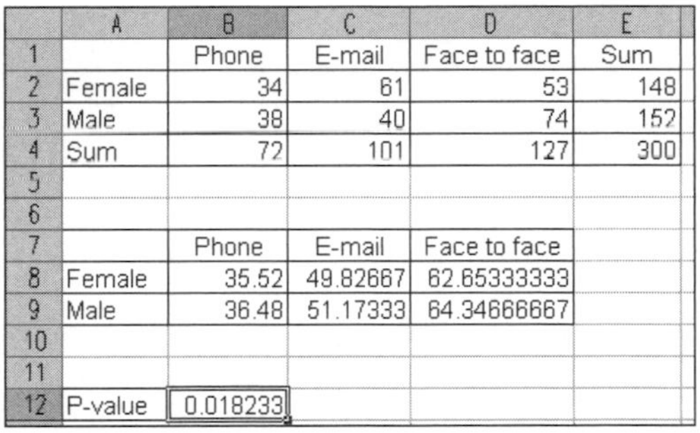

	A	B	C	D	E
1		Phone	E-mail	Face to face	Sum
2	Female	34	61	53	148
3	Male	38	40	74	152
4	Sum	72	101	127	300
5					
6					
7		Phone	E-mail	Face to face	
8	Female	35.52	49.82667	62.65333333	
9	Male	36.48	51.17333	64.34666667	
10					
11					
12	P-value	0.018233			

INDEX

A

actual measurement frequencies, 130, 131
alpha value (α), 159, 163
alternative hypothesis
  accuracy of, 166
  considerations, 174
  Cramer's coefficient, 186
  examples of, 161, 171–173
  overview, 170–174
  P-value and, 175–179
  test of difference between population ratios, 173
arithmetic mean, 43, 73, 74
average (mean). See mean
AVERAGE function, 196
average savings, 46–47

C

calculations. See Excel calculations
categorical data, 14–29
  correlation ratio, 121
  creating tables, 60–64
  cylinder charts, 114
  defined, 19
  examples of, 20, 23–26
  indexes, 117
  overview, 14–19
  as result of survey, 60–64
  scatter charts, 114
charts
  converting to graphs, 33–39
  correlation ratio, 126
  cylinder, 114
  degree of relation and, 115
  expenditure, 116–120
  scatter. See scatter charts
CHIDIST function, 107
CHIINV function, 107, 205–206
chi-square distribution, 99–105
  calculating, 130–133
  degrees of freedom, 99–108
  described, 99
  examples of, 99–105, 152
  points on horizontal axis, 205–206
chi-square symbol, 103
chi-square test of independence, 151–169

CHITEST function, 210–211
class midpoint, 36–39, 54, 56
classes
  calculating with Sturges' Rule, 55, 56, 58
  intraclass variance, 117, 123, 124, 126
  range of, 39, 54–57, 84
coefficient
  correlation, 116–120, 206–207
  Cramer's. See Cramer's coefficient
CORREL function, 207
correlation, 115, 119
correlation coefficient, 116–120, 206–207
correlation ratio, 117, 121–127, 207
COUNTIF function, 197–198
Cramer's coefficient, 127–138
  accuracy of, 147
  alternative hypothesis, 186
  calculating, 130–135, 141
  examples of, 127–136
  Excel and, 207
  indexes, 117, 129
  informal standard, 136
  making informed decision about, 147–148
  null hypothesis, 168, 186
  ratio of preference, 155
  variances in population, 145–150, 157, 186
Cramer's V. See Cramer's coefficient
critical region, 159, 165–167, 187
cross tabulation, 62–64, 128, 130, 135, 151, 153, 197–198
curve, grading on. See standard scores
cylinder charts, 114

D

data
  categorical. See categorical data
  collection of, 186
  immeasurable. See categorical data
  numerical. See numerical data
  "scattering of," 49, 58, 69, 70, 80
  unsuitable for correlation coefficient, 120
data point, 80

data types, 13–29, 117
degree of relation, 115, 116–120
degrees of freedom (df), 99–108
descriptive statistics, 57–58
deviation, standard, 48–53, 70–79
deviation scores, 74–80, 199–203
df (degrees of freedom), 99–108
distributions
  chi-square. See chi-square distribution
  Excel and, 107–109
  F, 106–107
  normal, 86–91
  standard normal, 89–98, 204–205
  t, 106

E

estimation theory, 57–58
Euler's number, 86
Excel calculations, 191–211
  chi-square distribution, 205–206
  correlation coefficient, 206–207
  cross tabulation, 197–198
  deviation scores, 74–80, 199–203
  distributions and, 107–109
  frequency tables, 192–195
  mean, 195–196
  median, 195–196
  standard deviation, 195–196
  standard normal distribution, 204–205
  standard scores, 199–202
  tests of independence, 208–211
Excel files, downloading, 192
Excel functions
  AVERAGE, 196
  CHIDIST, 107
  CHIINV, 107, 205–206
  CHITEST, 210–211
  CORREL, 207
  COUNTIF, 197–198
  FDIST, 107
  FINV, 107
  FREQUENCY, 193–194
  NORMDIST, 107
  NORMINV, 107
  NORMSDIST, 107, 204
  NORMSINV, 107

Excel functions, *continued*
    STANDARDIZE, 199–201
    TDIST, 107
expected frequencies, 130, 131
expenditure chart, 116–120

F

F distribution, 106–107
FDIST function, 107
FINV function, 107
freedom, degrees of, 99–108
frequency
    actual, 130, 131
    described, 36
    distribution tables, 32–39
    expected, 130, 131
    relative, 36–37, 39
FREQUENCY function, 193–194
frequency tables
    creating with Excel, 192–195
    range of class of, 54–56

G

geometric mean, 43
grading on a curve. *See* standard
        scores
graphs
    converting price charts to, 33–39
    converting surveys to, 62–64
    shape of, 100–101
    slope of, 101

H

harmonic mean, 43
histograms
    advantages of, 83
    examples of, 39, 83, 84, 154
    overview of, 38–39
    probability density function, 83–84
    range of class and, 84, 85
    variables, 39
homogeneity, test of, 184–186
horizontal axis, 39, 102, 107, 109, 125
    calculating points on, 107
hypothesis tests, 143–189. *See also*
        tests of independence
    alternative hypothesis. *See* alterna-
        tive hypothesis
    chi-square test of independence,
        151–169
    conclusions, 187

defined, 149
examples of, 149, 168–174
null hypothesis. *See* null hypothesis
overview of, 144–150
population considerations, 149, 186
procedure for, 150, 175–179
P-value, 163, 175–179, 189
tests of correlation, 149, 171
tests of correlation ratio, 149, 171
tests of difference between popula-
        tion means, 149, 171, 173
tests of difference between popula-
        tion ratios, 149, 171, 173
tests of homogeneity, 184–186
tests of independence, 149, 171
types of, 149, 171

I

immeasurable data. *See* categorical
        data
independent coefficient. *See* Cramer's
        coefficient
indexes
    correlation coefficient, 120
    Cramer's coefficient, 117, 129
    numerical data, 117
intraclass variance, 117, 123, 124, 126

L

linear relationships, 120

M

mean (average)
    arithmetic, 43, 73, 74
    calculating with Excel, 195–196
    defined, 43
    examples, 40–44
    geometric, 43
    harmonic, 43
    normal distribution and, 87–89
    standard normal distribution and,
        89–90
median
    calculating with Excel, 195–196
    defined, 45
    examples of, 45–47
    uses for, 44
Microsoft Excel. *See* Excel calculations,
        Excel files, *and* Excel functions
midpoint, class, 36–39, 54, 56
multiple-choice answers, 28

N

Napier's constant, 86
negative correlation, 119
non-linear relationships, 120
normal distribution, 86–91
normalization, 71–72
NORMDIST function, 107
NORMINV function, 107
NORMSDIST function, 107, 204
NORMSINV function, 107
null hypothesis
    considerations, 174
    Cramer's coefficient, 168, 186
    difficulty of proving, 174
    examples of, 167–174
    failing to reject, 150, 167, 178,
        179, 187
    overview, 170–174
    P-value and, 175–179
    rejecting, 158, 159, 178
    for tests of correlation, 172
    for tests of correlation ratio, 172
    for tests of difference between
        population ratios, 173
    for tests of independence, 172
numerical data, 14–29
    correlation ratio, 121
    defined, 19
    descriptive statistics, 57–58
    estimation theory, 57–58
    examples of, 21–23, 26
    frequency tables, 32–39, 54–56, 58
    histograms, 38–39, 54, 58
    indexes, 117
    mean (average), 40–43
    median, 44–47
    overview, 31–58
    scatter charts, 114
    standard deviation, 48–53, 70–79

P

Pearson's chi-square test statistic, 132,
        152–155, 158
percentage, 5, 37, 62, 64
population
    Cramer's coefficient, 145–150,
        157, 186
    defined, 6
    hypothesis tests and, 149, 186
    vs. sample, 52
    standard deviation, 52

status of, 4, 7, 57
variances in, 145–150, 157, 186
population ratios, 149, 171, 173
positive correlation, 119
price charts, 33–39
probability, 81–109
associated, 104
chi-square distribution, 99–105, 205–206
defined, 82
degrees of freedom (df), 99–108
distributions and Excel, 107–109
F distribution, 106–107
normal distribution, 86–89
standard normal distribution, 89–98, 204–205
t distribution, 106
test results, 83–84
probability density function, 82–85, 99, 107, 109
P-value
alternative hypothesis and, 175–179
hypothesis tests, 163, 175–179, 189
null hypothesis and, 175–179
tests of independence, 175

Q

qualitative data. *See* categorical data
quantitative data. *See* numerical data
questionnaires, 15–19

R

range, class, 39, 54–57, 84
relationships
correlation ratio, 117, 121–127
degree of, 115, 116–120
linear, 120
non-linear, 120
variables, 112–115
relative frequency, 36–37, 39

S

samples, 6, 7, 52, 57
scatter charts
correlation ratio, 122, 126
examples of, 114, 116
monthly expenditures, 116–120
scattering, of data, 49, 58, 69, 70, 80
scores
deviation, 74–80
evaluating, 71
standard, 65–80, 73, 199–202

significance level (α), 159, 163
slope, graph, 101
standard deviation
calculating with Excel, 195–196
normal distribution and, 87–91
numerical data, 48–53, 70–79
population, 52
standard normal distribution and, 89–90
standard normal distribution, 89–98, 204–205
standard scores, 65–80, 73, 199–202
standardization, 71–72, 80
STANDARDIZE function, 199–201
statistical hypothesis testing. *See* hypothesis tests
statistical significance, 187
statistics
defined, 4
descriptive, 57–58
estimation theory, 4–7
STEP test, 23–25
Sturges' Rule, 55, 56, 58
surveys, 4–7
categorical data, 60–64
converting to graphs, 62–64
limitations of, 4–7
tests of independence, 137, · 208–211

T

t distribution, 106
tables
categorical data, 60–64
chi-square distribution, 102–105, 205–206
cross tabulation, 128, 130, 135, 151, 153
frequency. *See* frequency tables
normal distribution and, 107
standard normal distribution, 92–93, 104, 108
TDIST function, 107
test results
normal distribution, 86–89
probability density function, 83–84
standard normal distribution, 89–98
tests of correlation, 149, 171, 172
tests of correlation ratio, 149, 171, 172
tests of difference between population means, 149, 171, 173

tests of difference between population ratios, 149, 171, 173
tests of homogeneity, 184 186
tests of independence, 208–211. *See also* hypothesis tests
chi-square, 151–169
examples of, 149, 171, 184–186
P-value, 175
vs. tests of homogeneity, 186
uses for, 137, 149
TINV function, 107

V

values
median, 44–47
P-value, 163, 175–179, 189
variables, 111–142
correlation coefficient, 116–120
correlation ratio, 121–127
Cramer's coefficient, 127–138, 141, 142
degree of relation, 115, 116–120
histograms, 39
relationships, 112–115
vertical axis, 39

W

weather forecasts, 82

Z

zero correlation, 119
z-score. *See* standard scores

NOTES

NOTES

NOTES

NOTES

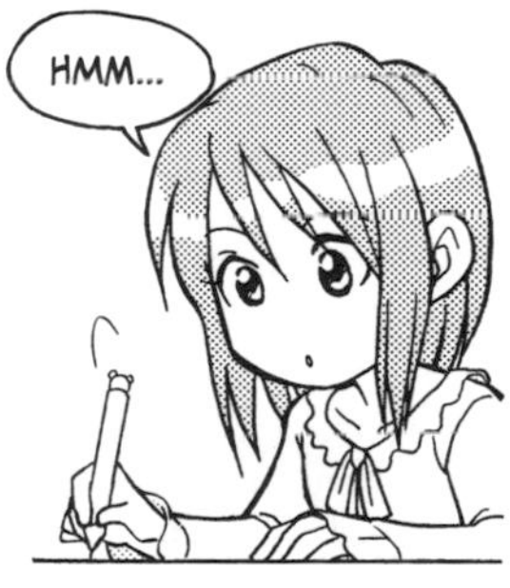
HMM...

NOTES

NOTES

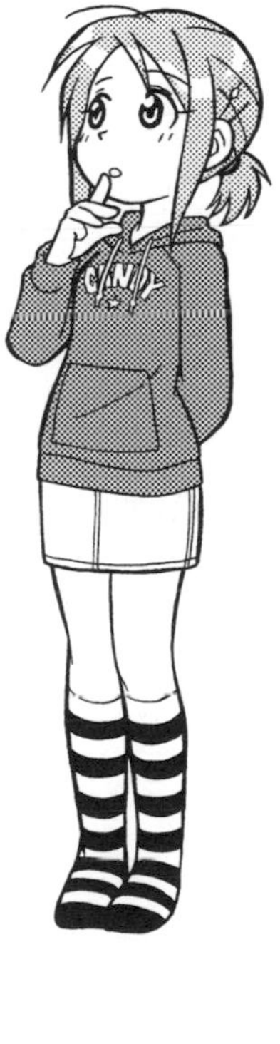

CANDY

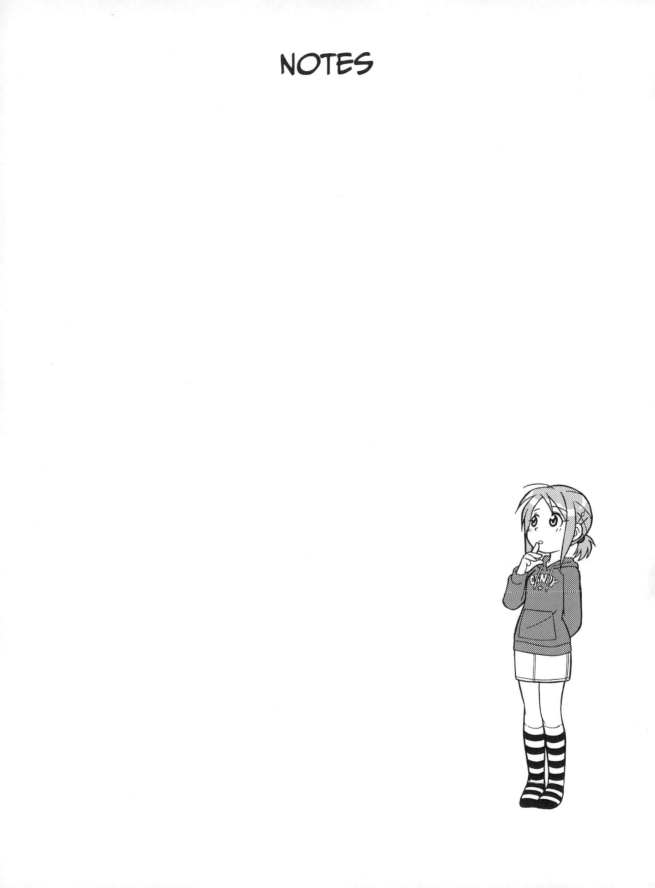

ABOUT THE AUTHOR

Shin Takahashi graduated from the Graduate School of Design of Kyushu University. Takahashi has worked as a lecturer and a data analyst and is currently active as a technical writer. He also wrote *Factor Analysis* and *Regression Analysis* in the Manga Guide series. Visit his website at *http://www.geocities.jp/sinta9695/*.

PRODUCTION TEAM FOR THE JAPANESE EDITION

Production: TREND-PRO Co., Ltd.

Founded in 1988, TREND-PRO produces newspaper and magazine advertisements incorporating manga for a wide range of clients from government agencies to major corporations and associations. Recently, TREND-PRO participated in advertisement and publishing projects using digital content. Some examples of past creations are publicly available at the company's website, *http://www.books-plus.jp/*.

Ikeden Bldg., 3F, 2-12-5 Shinbashi, Minato-ku, Tokyo, Japan

Telephone: 03-3519-6769; Fax: 03-3519-6110

Scenario Writer: re_akino

Artist: Iroha Inoue

MORE MANGA GUIDES

The *Manga Guide* series is a co-publication of No Starch Press and Ohmsha, Ltd. of Tokyo, Japan, one of Japan's oldest and most respected scientific and technical book publishers. Each title in the best-selling *Manga Guide* series is the product of the combined work of a manga illustrator, scenario writer, and expert scientist or mathematician. Once each title is translated into English, we rewrite and edit the translation as necessary and have an expert review each volume for technical accuracy. The result is the English version you hold in your hands.

Find more *Manga Guides* at your favorite bookstore, and learn more about the series at *http://www.edumanga.me/*.

THE MANGA GUIDE TO
DATABASES

THE MANGA GUIDE TO
ELECTRICITY

THE MANGA GUIDE TO
PHYSICS

THE MANGA GUIDE TO
MOLECULAR BIOLOGY

THE MANGA GUIDE TO
CALCULUS

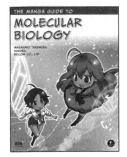

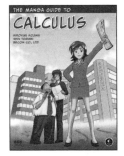

UPDATES

Visit *http://www.nostarch.com/mg_statistics.htm* for updates and errata, and to download the Excel files from the appendix.

The Manga Guide to Statistics is set in CCMeanwhile and Chevin. The book was printed and bound at Malloy Incorporated in Ann Arbor, Michigan. The paper is Glatfelter Spring Forge 60# Smooth Eggshell, which is certified by the Sustainable Forestry Initiative (SFI).

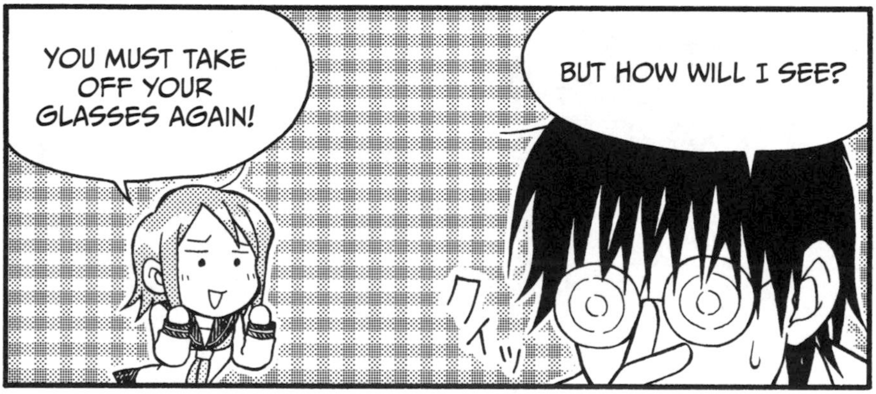
YOU MUST TAKE OFF YOUR GLASSES AGAIN!
BUT HOW WILL I SEE?
クイッ